# Lecture Notes in Mathematics

Edited by A. Dold and B. Eckmann

520

Roger H. Farrell

# Techniques of Multivariate Calculation

Springer-Verlag
Berlin · Heidelberg · New York 1976

**Author**
Roger H. Farrell
Department of Mathematics
Cornell University
Ithaca, New York 14850
USA

**Library of Congress Cataloging in Publication Data**

Farrell, Roger H 1929-
Techniques of multivariate calculation.

(Lecture notes in mathematics ; 520)
Bibliography: p.
Includes index.
1. Multivariate analysis. 2. Distribution (Probability theory) 3. Measure theory. I. Title. II. Series: Lecture notes in mathematics (Berlin) ; 520.
QA3.L28 no. 520 [QA278] 510'.8s [519.5'3]
76-14839

AMS Subject Classifications (1970): 62A05, 62E15, 62H10, 62J10

ISBN 3-540-07695-6 Springer-Verlag Berlin · Heidelberg · New York
ISBN 0-387-07695-6 Springer-Verlag New York · Heidelberg · Berlin



Printed in Germany.

Printing and binding: Beltz Offsetdruck, Hemsbach/Bergstr.

CONTENTS

LIST OF EXAMPLES

# Chapter 1. Introduction and brief survey.

## 1.1. The aspects of multivariate analysis.

Multivariate analysis originated with problems of statistical inference in the work of Pearson and Fisher, men with thorough grounding in applied statistics. The first important book on the subject, Anderson (1958) gives a balanced view of the subject by treating, in each case, first the question of inference, and then, the calculation of the multivariate density function of the resulting multivariate statistic. Of course, not all multivariate statistics have density functions but this book is limited to a discussion of statistics that have density functions relative to Lebesgue Measure.

Many of the hard mathematical problems are concerned with the problem of calculating the density functions. Anderson's book of 1958 was part of a general development of the 1950's of techniques for making these calculations. Random variable techniques found expression in Wijsman (1957, 1958), the use of Jacobians and change of variable found expression in Anderson's book, the use of differential forms on manifolds was developed by James (1954), and the use of invariance, matrix decompositions and maximal invariants found expression in Stein (1956c, 1959) and Karlin (1960). The use of Fourier transforms was widely known.

These techniques produced answers to many previously unanswered noncentral problems. See particularly James (1954, 1955a, 1960, 1961a, 1964), Constantine (1960, 1963, 1966), Constantine and James (1958), Karlin (1960), and Schwartz (1966a, 1967a). What was found was that the expression of the answer in many of these problems involved integrals of functions of a matrix argument, integrals that apparently cannot be evaluated in closed form in terms of the usual elementary functions. This was noted, especially by James (1955a, 1955b).

The development of the 1960's has centered about special functions and their use as an alternative to integrals as a means of representing the answers to these problems. One example of the use of special functions occurs in Mathai and Saxena (1969) where H-functions and Mellin transforms are used to study the distribution of a product of noncentral Chi-square random variables. But more in fashion today is the use of hypergeometric series which are sums of zonal polynomials multiplied by hypergeometric coefficients. This latter approach originated in Herz (1955) who defined the hypergeometric functions by successive use of Laplace and inverse Laplace transforms. Constantine (1963) showed that the functions defined by Herz were representable as weighted infinite sums of the zonal polynomials that were being developed by James (1960, 1961a, 1961b, 1964, 1968). Also, Herz (1955) and Constantine (1966) define Laguerre polynomials of a matrix argument and these polynomials are finding use in numerical analysis.

An alternative to the use of special functions is the use of approximate answers, i.e., asymptotic methods. A frequently cited early reference is Box (1949) who used inversion of Fourier transforms to obtain asymptotic expansions. Since the middle 1950's a vast literature of asymptotic expansions has appeared. A typical literature item gives several terms of an asymptotic expansion without provision for any error bound on the remainder. An exception to this pattern is Korin (1968) who obtains a complete asymptotic series which he then uses to check previous approximations by others. The statistical problem is that of making tests about the covariance matrix.

The subject is too vast for one book to make an inclusive treatment. It was decided to write a methods book that would attempt to illustrate methods that have been used in the literature but not treated in books on the subject. We assume therefore that the reader knows about Jacobians and changes of variable and has access to Anderson (1958) and perhaps Eaton (1972). We use Jacobians but do not go out of our way to discuss them in this book. The methods that we consider lead to a integral. If the integral requires special functions for evaluation, that is beyond the scope of this book. Likewise asymptotic series are not discussed. Inference is not discussed, nor are the statistical problems that underlay various statistics discussed.

As a methods book the book is very long. Yet it is sparsely written and assumes a great deal of its readers. Standard complex variable theory is needed in Chapter 2 and for the references to Chapter 12. Measure Theory equivalent to most of Halmos (1950) is assumed. If the reader has not seen a development of regular measures in metric spaces, locally compact Hausdorff spaces, and locally compact groups, then it is assumed the reader will take the sketch presented in this book and make good use of references like Halmos (1950), Loomis (1953) and Nachbin (1965). The theory of analytic manifolds cannot be done completely here and Dieudonné (1972) can be used for reference. On the algebraic side, a good understanding of quadratic forms, positive definite matrices, and canonical forms is essential. Often graduate students today do not learn this material, in which case the reader should find a suitable source and read. The algebra of this book is basically easy except for Chapter 12. In Chapter 12 we develop a theory of the mathematics of zonal polynomials and extensive reference will be made to Loomis (1953) and Weyl (1946) for material on algebras and the symmetric group, to Littlewood (1940, 1950) for material on group characters and symmetric functions, and Helgason (1962) on group representations and spherical functions. Nonetheless Chapter 12 is nearly self contained for we have tried to take the necessary material from these sources and make of coherent development (without group representations) of the Constantine-James theory of zonal polynomials. For the most part, the use of zonal polynomials is beyond the scope of this book, although a small amount of material is included in Chapter 13.

Chapter 13 consists almost totally of problems on the use of zonal polynomials together with some connective material and references to source. Many of the results of Chapter 2 are left as problems but are stated as problems in the overall context of the discussion rather than being segregated at the chapter's end. Every Chapter has some problems, except Chapter 12, since in this case Chapter 13 consists of the problem set for Chapter 12. In Chapters 3 through 11 the problems are collected together at the end of each Chapter. A number of items of theory needed at later stages of the book are stated as problems and in the sequel are referenced by their problem number. Thus the problems contain partly theory, partly calculations of intermediate results needed later, and partly of illustrative examples of the distributions of statistics used in the literature.

## 1.2. The literature.

Because of the mathematical nature of this book, many of the references in the bibliography are non-statistical. We have tried to include representative referencs to the various sides of the subject and a quick reading of the bibliography will often suffice to locate a starting point. But we have not tried to make a complete survey nor to write a complete summary. Other bibliographies include Gupta (1963) and Anderson, Gupta and Styon (1972). Other general sources on multivariate problems are Anderson (1958), Anderson, Gupta and Styon (1972), Bechhofer, Kiefer and Sobel (1968), Constantine (1960), Dempster (1969), Doob (1958), Eaton (1972), Gupta (1963), Karlin (1960), Kiefer (1966), Kullback (1959), Lehmann (1959), Miller (1964),

Olkin (1966), Scheffé (1959), Stein (1956a, 1956b, 1956c, 1959, 1966), Wijsman (1957, 1958, 1966), and Wilks (1962).

## 1.3. History.

The author has seen little historical writing other than Pearson (1968). It is to be hoped that the "histories" series published in Biometrika will provide this service.

## 1.4. Inference.

As noted earlier the subject of inference underlies all of multivariate analysis. The problems of inference may be classified in several ways which we mention here. Estimation and tests of hypotheses; invariance; maximum likelihood and Bayes; admissible, minimax, locally minimax, unbiased, asymptotically best; the analysis of variance, regression, linear hypotheses and the general linear hypothesis; design and combinatorial questions; ranking problems; and so on.

In much of the literature determination of a suitable statistic and its distribution (density function) is as far as the author carries a given development. Anderson (1958) follows this pattern: problem, likelihood ratio or maximum likelihood, distribution, without consideration of the "good" properties of the procedure.

The optimality literature is thin due to the difficulty of this side of the subject. See Kiefer (1966). We remark here on some special topics. The literature on unbiasedness and monotone properties of the power function seems to consist of Anderson (1955), Anderson and Das Gupta (1964), Cohen and Strawderman (1971), Gleser (1966), Lehmann (1959), Mudholkar (1965, 1966a, 1966b), Sugiura and Nagao (1968).

The relationship between being minimax or most stringent and being invariant is discussed in Lehmann (1959). From an example due to Stein published in Lehmann's book it follows that the interesting

Hunt-Stein theory fails to apply in many interesting statistical examples in which the transformation group is "too big." This includes the full linear group and nonparametric examples in which random variables are transformed by groups of monotone functions. An outstanding problem in the subject of most stringent tests is to show the UMP tests of $a = 0$ vs $a \neq 0$ using the Hotelling $T^2$ statistics are most stringent tests on orbits of the parameter space. Giri, Kiefer and Stein (1963) reduced this problem to that of solving an integral equation and by use of hypergeometric series solved the integral equation in the simplest case thus showing the Hotelling $T^2$ test to be minimax. Salaevskii (1971) has completely solved the integral equation in question for two dimensional random variables (c.f. Math. Reviews 42 5380) and the English translation is the version referenced. It has been rumored that the complete problem was solved by Salaevskii in an enormous calculation but we know of no published source. This is one of the simplest cases of an invariant multivariate statistic and the minimax question in harder cases is untouched.

In avoiding the minimax problem per se one approach that yields solutions is to find tests asymptotically minimax (most stringent) for small parameter values. For examples of this see Giri and Kiefer (1964), Giri (1968), and Schwartz (1967b).

The first complete class theorem for a multivariate problem was proven by Birnbaum (1955). Some years later Birnbaum's result was generalized by Farrell (1968). The sufficiency part of Birnbaum's proof was greatly simplified by Stein (1956b) who restated the form of the result and then used the restated form to prove admissibility of Hotelling's $T^2$-test. Stein's idea has since been used extensively by Schwartz (1966a, 1967a) and Farrell (1968). Roughly at the same time Kiefer and Schwartz (1965) showed many classical multivariate tests are Bayes tests and this was explored further by Schwartz

(1966a, 1966b). Stein's method of obtaining the density function of a maximal invariant, Stein (1956c), was developed by Schwartz (1966a) in order to get at complete class theorems for the general linear hypothesis.

A result following from the results of Kiefer and Schwartz, op. cit., is that many invariant multivariate tests are admissible tests. That the corresponding result for multivariate estimation might be false and is in fact false was first shown by Stein (1956a). There is now a considerable literature by, among others, J. Berger, L. Brown, A. Cohen, C. Stein, and J. Zidek, establishing the inadmissibility of many "classical" multivariate estimators. Corresponding admissibility results have been obtained recently by J. Berger, L. Brown, and S. Portnoy, among others. For tests of hypotheses the admissibility proofs have depended in an essential way on the actual distribution of the multivariate statistic in question whereas the results about estimators seem to have an almost non-parametric character about them in that the proofs seem to depend only on the "moment structure" of the problem, to borrow a term from Berger.

Ranking problems have been studied by Bechhofer, Kiefer and Sobel (1968) and new work on this important problem continues to appear.

Design problems have been studied extensively by Kiefer. See particularly Kiefer (1966). Several papers by Kiefer, to appear, will greatly extend and unify earlier results by placing them in a larger context.

To the extent that these problems in inference require knowledge of the multivariate density functions the chapters of this book may bear on these problems. In addition, some of the algebra developed plays a role in the algebra of multivariate design.

determined from the context. There is a tendency here to use H, U and V for orthogonal matrices, P, Q for idempotents, and W, X, Y, and Z for the values of random variables. The ordinary matrix sum of X and Y is X + Y and the product is XY. The transpose of X is $X^t$ and the small case t is not used otherwise in this book. Random matrices are always indicated by underlining, as $\underline{X}$ for example. Many matrices considered are in fact functions, so that random matrices are functions on some probability space, and in the discussion of manifolds the matrix entries are functions of local coordinates on the manifolds. A frequently used notation is to form the ij-entry function of a matrix as $(Z)_{ij}$, this being particularly useful when writing a differential form. dY would mean, compute the differential of each element of Y and form the matrix of these differentials, and $(dY)_{ij}$ is the ij-entry of this matrix. Because of the frequent use of integrals and differentials, the letter d is used only in the traditional meanings of differential or of derivative or of derivative of the measure. Traditionally in random matrices the row vectors represent independently distributed random vectors, and where relevant this tradition is kept in this book. Non square matrices are usually n×h or n×k representing n observations on h-dimensional or k-dimensional row vectors. All vectors in this book are column vectors and for this reason we write $a^t = (a_1, \ldots, a_n)$. Except in Chapters 12 and 13, square matrices are not of interest. In the exceptional chapters matrices are usually n×n.

Much of the book is concerned with multilinear forms. In the abstract discussions E becomes an n-dimensional vector space over the reals $\mathbb{R}$ or the complex numbers $\mathbb{C}$, or in Chapter 6 where the coefficients are functions, the coefficients for E are from a ring with unit thereby making E a free n-module. All discussion is relative to a fixed basis $e_1, \ldots, e_n$ of E and the canonical basis

$u_1,\ldots,u_n$ of the dual space to $E$, the space of 1-forms. In Chapter 6 $M(E^q, \mathcal{C})$ refers to the space of multilinear q-forms on $E^q$ with coefficients in the ring $\mathcal{C}$, and this notation is carried over to the development of multilinear algebra in Chapter 12. But in Chapter 12 the ring $\mathcal{C}$ is the ring of complex numbers whereas in Chapters 6 through 9 the ring is always a function ring such as a ring of globally defined $C_2$ functions. In Chapter 12 the main emphasis is on endomorphisms of $M(E^q, \mathbb{C})$ and the main new notation used is $\bigotimes_{i=1}^{m} X_i$ for the tensor product of matrices $X_1,\ldots,X_m$ and $\bigotimes_{i=1}^{m} X$ for the tensor product of $X$ with itself $m$ times.

Greek letters, $\mu$, $\nu$, and $\lambda$ are reserved for measures. Most often integrals are with respect to a differential form, in which case the integration notation of Section 6.5 is used, or with respect to a Haar measure on a group like the orthogonal matrices $\underline{O}(n)$ in which case $dU$ is written, or with respect to an invariant measure induced on a homogeneous space like the set of positive definite matrices $\underline{S}(n)$ in which case $dS$ is used. Because of this use of the Greek letters, if $\underline{X}$ is $n \times h$ with independently and identically distributed rows, each normal $(a, \Sigma)$, $\Sigma$ is the covariance matrix in this context and is the notation for a summation in other contexts. Likewise $\pi$ is the number in some contexts and the notation for a product in other places. In general we will write $E\underline{X} = M$, and will use the mean vector $a$ only in the rank one case.

Except for Chapter 12 and 13, matrices are with real entries. The special sets of matrices discussed are $GL(n)$, the general linear group of $n \times n$ matrices, $\underline{S}(n)$ and $\underline{O}(n)$ as above, $\underline{T}(n)$ the set of lower triangular $n \times n$ matrices with positive diagonal elements, and $\underline{D}(n)$ the set of diagonal $n \times n$ matrices with positive diagonal elements. These notations used throughout and are the other context in which underlining is used. Throughout, overlining such as $\overline{a}$ or $\overline{Y}$ means the complex conjugate of the vector $a$ or matrix Y. This does not become important until Chapter 12. The use of primes, X', does not mean transpose, but instead means X' is a matrix that is to be distinguished from X, etc.

In Chapter 7 the Grassman and Stiefel manifolds are defined and these together with $\underline{S}(n)$ are the homogeneous spaces discussed. Notations are introduced in Chapter 7 for the Grassman and Stiefel manifolds. $\underline{T}(n)$ is somewhat special in that it is both a locally compact group and a homogeneous space of $GL(n)/\underline{O}(n)$.

## Chapter 2. Transforms

### 2.0. Introduction

In this book we do not need to use the Levy continuity theorem, c.f. Feller (1966). For the uses illustrated below inversion theorems for the various transforms are sufficient. Inversion of Fourier transforms has been rendered almost trivial by the elegant calculation that appears in Feller, op. cit. Multidimensional uniqueness theorems for the (complex) Laplace transform are proven easily by induction on the dimension and these proofs are stated as problems in the sequel. For more detail than is presented in this chapter suitable references are Feller (1966), Widder (1941) and Wiener (1933).

Aside from the Laplace transform and its special case, the Fourier transform we discuss briefly the Mellin transform. Several authors of statistical literature have used inversions of Mellin transforms to determine multivariate density functions. Generally this last approach is used in cases where the moments have nice expressions but the Laplace transform appears to be intractable. Nonetheless the Mellin transform of a random variable $\underline{X}$ is $E(\underline{X})^{t-1}$ and so long as $\underline{X}$ is nonnegative an obvious change of variable reduces this to a Laplace transform. Inversion theorems for the Mellin transform may be found in Widder (1941).

Section 2.1 develops the necessary uniqueness theory. Section 2.2 gives a development of the distribution theory of multivariate normally distributed random variables. The author obtained this presentation from lectures of Doob (1958). In Section 2.3 transforms are used to derive the non-central Chi-square density function. In this section the noncentral F- and t-density functions are derived. Most of the presentation of Section 2.3 is stated as a series of problems. In Section 2.4 we have included a brief discussion of other inversion theorems and Hermite polynomials. Section 2.5

relates inversion of Laplace and Mellin transforms to inversion of the Fourier transform. Section 2.6 references a few pieces of significant literature which illustrate use of inverse transforms in the statistical literature.

## 2.1. Definitions and uniqueness

If $\mu$ is a finite signed measure defined on the Borel subsets of $(-\infty,\infty)$ then the Fourier transform of $\mu$ is

$$(2.1.1) \qquad \int \exp(itx)\mu(dx), \qquad i=\sqrt{-1}, \qquad t \text{ real} \;,$$

and the Laplace transform of $\mu$ is

$$(2.1.2) \qquad \int \exp(tx)\mu(dx), \qquad t \text{ any complex number.}$$

In this book we consider only those values of $t$ for which the integrals are absolutely convergent. There is an extensive literature on conditional convergence, some of which may be located using Widder (1941). If the support of $\mu$ is $[0,\infty)$ then the Mellin transform of $\mu$ is

$$(2.1.3) \qquad \int_0^\infty x^{t-1}\mu(dx), \qquad t \text{ a complex number.}$$

In this book integration is Lebesgue integration in the sense of Halmos (1950). Thus as noted above we consider only values of $t$ for which the integrals are absolutely convergent. This is automatic for (2.1.1). In terms of the positive part $\mu_+$ and negative part $\mu_+ - \mu = \mu_-$, c.f. Halmos (1950), page 121, we require for (2.1.2) that $\int \exp(tx)\mu_+(dx) < \infty$ and $\int \exp(tx)\mu_-(dx) < \infty$. The Mellin transform will be computed only for positive measures $\mu$ and we require $\int x^{t-1}\mu(dx) < \infty$ .

Problem 2.1.1. If the Laplace transform of a signed measure $\mu$ is absolutely convergent at $t=a_1+ib_1$ and $t=a_2+ib_2$, $a_1 < a_2$, then the integral (2.1.2) is absolutely convergent for all (complex) $t$

in the strip $a_1 < \operatorname{Re} t < a_2$ and is an analytic function in this strip.

Hint: Show that the line integrals $\oint dt \int \exp(tx)\mu(dx) = 0$ for every smooth enough closed curve in the strip. Why can the order of integration be changed?

Lemma 2.1.2. If $\varepsilon > 0$ and the Laplace transform $\int \exp(tx)\mu(dx) = \phi(t)$ of a signed measure $\mu$ is absolutely convergent in the strip $|\operatorname{Re} t| < \varepsilon$ then the Fourier transform $\int_{-\infty}^{\infty} \exp(itx)\mu(dx)$ uniquely determines $\phi$.

Proof. Two analytic functions which agree on a countable set with limit point are identical on simply connected domains which contain the limit point. #

Problem 2.1.3. Show $(\sigma\sqrt{2\pi})^{-1}\int_{-\infty}^{\infty} \exp(itx)\exp(-x^2/2\sigma^2)\,dx = \exp(-\sigma^2 t^2/2)$.

Hint: Complete the square in the exponent. Show that the problem is equivalent to showing $\sqrt{2\pi} = \int_{-\infty}^{\infty} \exp((x-it)^2/2)\,dx$, $t \in (-\infty, \infty)$. This may be shown using contour integration. #

Problem 2.1.4. Let $\{a_n, n \geq 1\}$ be a real number sequence, and $\lim_{n\to\infty} a_n = 0$. Let $\{\underline{X}_n, n \geq 1\}$ be a real valued random variable sequence such that $P(|\underline{X}_n| > a_n) \leq a_n$, $n \geq 1$. If $\underline{Y}$ is a random variable and $y$ is a continuity point of the distribution function of $\underline{Y}$, then $\lim_{n\to\infty} P(\underline{Y} + \underline{X}_n \leq y) = P(\underline{Y} \leq y)$.

The integral formula (2.1.4) stated next is obtained by Feller (1966), pages 480-481. Application of Problem 2.1.4 to the situation $\sigma \to 0$ shows that the right side of (2.1.4) converges to $F(s)$ at all continuity points $s$ of $F$. The left side of (2.1.4) is clearly a linear functional of the Fourier transform of $F$.

Lemma 2.1.5. Let $F$ be a probability distribution function. Then

$$(2.1.4) \quad \int_{-\infty}^{s} dt \int_{-\infty}^{\infty} d\zeta \int_{-\infty}^{\infty} (\sqrt{2\pi})^{-1} \exp(-\zeta^2/2\sigma^2)\exp(-i\zeta t)\exp(i\zeta x)F(dx)$$

$$= \int_{\infty}^{s}\int_{-\infty}^{\infty} \sigma(\sqrt{2\pi})^{-\frac{1}{2}} \exp(-\sigma^2(x-t)^2/2)F(dx)\,dt .$$

Of course, the right side of (2.1.4) is the distribution function of the sum of two random variables.

Problem 2.1.6. Let $f: (-\infty, \infty) \to (0, \infty)$ be a Borel measurable function. Suppose for all Borel sets $A$, that

$$(2.1.5) \qquad \int_A f(x)\, \mu(dx) = \int_A f(x)\, \nu(dx) ,$$

where $\mu, \nu$ are $\sigma$-finite Borel measures. Then $\mu = \nu$.

Theorem 2.1.7. (Uniqueness of the Laplace Transform) Let $\mu, \nu$ be two positive $\sigma$-finite Borel measures. Assume the Laplace transforms of $\mu$ and $\nu$ are absolutely convergent on the strip $a < \operatorname{Re} t < b$ and that on this strip the Laplace transforms are equal. Then $\mu = \nu$.

Proof. Let $a < t_o < b$, $t_o$ a real number, and define finite positive measures $\mu_o$ and $\nu_o$ by, if $A$ is a Borel set, then

$$\mu_o(A) = \int_A \exp(t_o x)\, \mu(dx) \quad \text{and} \quad \nu_o(A) = \int_A \exp(t_o x)\, \nu(dx) .$$

By Problem 2.1.6 it is sufficient to show that $\mu_o = \nu_o$. To show this, there exists $\varepsilon > 0$ such that the Laplace transforms of $\mu_o$ and $\nu_o$ are absolutely convergent on the strip $|\operatorname{Re} t| < \varepsilon$. In particular $\mu_o$ and $\nu_o$ are finite positive measures with the same Fourier transforms (i.e. $\operatorname{Re} t = 0$). From Lemma 2.1.5 and the equation (2.1.4) it follows that the Fourier transform uniquely determines the distribution function $F$ and hence the Borel measure induced by $F$. Applied here it follows that $\mu_o$ and $\nu_o$ must be the same. As noted, Problem 2.1.6 then implies that $\mu = \nu$. #

Problem 2.1.8. If $A \subset \mathbb{R}_n$ is an open set, if $1 \leq m < n$ and $t \in \mathbb{R}_m$, then the t-section $A_t = \{y \mid (y, t) \in A\}$ is an open subset of $\mathbb{R}_{n-m}$.

Theorem 2.1.9. (n-dimensional uniqueness theorem) Let $\mu$ and $\nu$ be positive n-dimensional Borel measures. Let $A \subset \mathbb{R}_n$ be an open set such that if $t \in A$ then $\int \exp(t.x)\mu(dx) = \int \exp(t.x)\nu(dx) < \infty$. Then $\mu = \nu$. ($t.x$ is the dot product of $t$ and $x$.)

Proof. We assume $n > 1$ as the Theorem holds for the case $n = 1$. We make a mathematical induction on the dimension $n$.

Choose a real number $t$ such that for some $y \in A$, $y^t = (y_1, \ldots, y_n)$ and $t = y_n$. Then there exist numbers $t_1 < t < t_2$ such that for some $y'$ and $y''$, $y'_n = t_1$ and $y''_n = t_2$. Define measures $\mu_o(\ ,t)$ and $\nu_o(\ ,t)$ by

$$(2.1.6) \qquad \mu_o(B,t) = \int \chi_B(x_1,\ldots,x_{n-1})\exp(x_n t)\mu(dx), \quad \text{and}$$

$$\nu_o(B,t) = \int \chi_B(x_1,\ldots,x_{n-1})\exp(x_n t)\nu(dx) .$$

In (2.1.6) $\chi_B$ is the indicator function of the n-1 dimensional Borel subset $B$. The Laplace transforms of $\mu_o$ and $\nu_o$ clearly satisfy, if $(s_1,\ldots,s_{n-1},t) \in A$ then

$$\begin{aligned} (2.1.7) \qquad & \int \exp(x_1 s_1 + \ldots + x_{n-1}s_{n-1})\mu_o(dx,t) \\ = \quad & \int \exp(x_1 s_1 + \ldots + x_{n-1}s_{n-1} + x_n t)\mu(dx) \\ = \quad & \int \exp(x_1 s_1 + \ldots + x_{n-1}s_{n-1} + x_n t)\nu(dx) \\ = \quad & \int \exp(x_1 s_1 + \ldots + x_{n-1}s_{n-1})\nu_o(dx,t) . \end{aligned}$$

By the inductive hypothesis it follows that if $t_1 < t < t_2$ then $\mu_o(\ ,t) = \nu_o(\ ,t)$.

Define measures $\mu_1(\ ,B)$ and $\nu_1(\ ,B)$ by

$$(2.1.8) \qquad \mu_1(C,B) = \int \chi_B(x_1,\ldots,x_{n-1})\chi_C(x_n)\mu(dx), \quad \text{and}$$

$$\nu_1(C,B) = \int \chi_B(x_1,\ldots,x_{n-1})\chi_C(x_n)\nu(dx) .$$

If $B$ is a n-1 dimensional Borel set and if $t_1 < t < t_2$ then

$$(2.1.9) \qquad \int \exp(zt)\mu_1(dz,B) = \mu_o(B,t) = \nu_o(B,t)$$

$$= \int \exp(zt)\nu_1(dz,B) .$$

By the uniqueness theorem for dimension $n = 1$, it follows that $\mu_1(C,B) = \nu_1(C,B)$ for all Borel subsets $C \subset \mathbb{R}_1$ and $B \subset \mathbb{R}_{n-1}$. By Fubini's Theorem, $\mu = \nu$ now follows. #

We state a second form of an n-dimensional uniqueness theorem as Problem 2.1.10. The proof proceeds by a similar induction on the dimension.

<u>Problem 2.1.10</u>. Let $\lambda$ be a nonatomic positive Borel measure on the Borel subsets of $\mathbb{R}$. Let $A \subset \mathbb{R}_n$ be a Borel subset such that $(\lambda x \ldots x\lambda)(A) > 0$. Let $\mu$ and $\nu$ be positive n-dimensional Borel measures such that if $t \in A$ then $\int \exp(t.x)\mu(dx) = \int \exp(t.x)\nu(dx)$. Then $\mu = \nu$.

Analogous results hold for signed measures $\mu = \mu_+ - \mu_-$, and $\nu = \nu_+ - \nu_-$. For if $\mu$ and $\nu$ have equal absolutely convergent Laplace transforms then

(2.1.10) $$\int \exp(t.x)\mu_+(dx) + \int \exp(t.x)\nu_-(dx)$$
$$= \int \exp(t.x)\mu_-(dx) + \int \exp(t.x)\nu_+(dx) .$$

By Theorem 2.1.9 it follows that $\mu_+ + \nu_- = \mu_- + \nu_+$ and hence that $\mu = \nu$. •

Since by change of variable the Mellin transform becomes a Laplace transform, corresponding uniqueness theorems hold. We have not stated a n-dimensional uniqueness theorem for the Fourier transform. Such a result may be proven either by induction following the proof of Theorem 2.1.9 or by using the multivariate analogue of (2.1.4). However see Problem 2.4.7.

## 2.2. The multivariate normal density functions.

Functions $K \exp(-x^tAx/2)$, $K > 0$ a real number, $x \in \mathbb{R}_n$, $A$ a $n \times n$ symmetric matrix, are considered here. In this and the remaining chapters of this book the transpose of a matrix $A$ is the matrix $A^t$.

Problem 2.2.1. $\int_{-\infty}^{\infty}\cdots\int_{-\infty}^{\infty}\exp(-x^tAx/2)\,dx$ is an absolutely convergent n-dimensional integral if and only if the $n\times n$ symmetric matrix $A$ is positive definite, denoted by $A>0$. In case $A>0$ the value of the integral is $(\sqrt{2\pi})^n(\det A)^{-1/2}$.

Hint: Let $U$ be an $n\times n$ orthogonal matrix such that $UAU^t$ is a diagonal matrix. Make the change of variable $y=Ux$ having Jacobian $\pm 1$. Integrate over spheres centered at $0$ so the region of integration does not change. #

Problem 2.2.2. The multivariate normal density function

$$(\det A)^{1/2}(2\pi)^{-n/2}\exp(-x^tAx/2) \tag{2.2.1}$$

has Laplace transform

$$\exp(s^tA^{-1}s/2)\ . \tag{2.2.2}$$

Hint: Complete the square in the exponent. #

Problem 2.2.3. If the random n-vector $\underline{X}$ has multivariate normal density function (2.2.1) then $E\underline{X}=0$ and $\operatorname{Cov}\underline{X}=A^{-1}$.

Hint: Compute the first and second order partial derivatives of the Laplace transform (2.2.2). #

If $E\underline{X}=0$, $\operatorname{Cov}\underline{X}=A^{-1}$, and $\underline{X}$ has a multivariate normal density function then the random vector $\underline{Y}=\underline{X}+a$, $a\in\mathbb{R}_n$, has as its Laplace transform

$$\begin{aligned} E\exp(\underline{Y}\cdot s) &= E\exp(\underline{X}+a).s \\ &= \exp(s.a+s^tA^{-1}s/2)\ . \end{aligned} \tag{2.2.3}$$

Clearly

$$E\underline{Y}=a \quad\text{and}\quad \operatorname{Cov}\underline{Y}=A^{-1} \tag{2.2.4}$$

so that the multivariate density function of $\underline{Y}$ is

(2.2.5) $$(\det A)^{1/2}(2\pi)^{-n/2}\exp(y-a)^t A(y-a) \quad .$$

In the sequel we will say that a random n-vector $\underline{Y}$ which has multivariate density function (2.2.5) is normal $(a, A^{-1})$ .

Problem 2.2.4. If $\underline{X}^t = (\underline{X}_1, \ldots, \underline{X}_n)$ has a multivariate normal density function then $\underline{Y}$ defined by $\underline{Y}^t = (\underline{X}_1, \ldots, \underline{X}_{n-1})$ has a multivariate normal density function.

Hint: Compute $E\exp(\Sigma_{i=1}^n s_i \underline{X}_i)$ and then set $s_n = 0$. If $\operatorname{Cov}\underline{X} = A^{-1}$ with $A^{-1} = \begin{pmatrix} B^{-1} & b \\ b^t & c \end{pmatrix}$ then since $A^{-1} > 0$, so is $B^{-1} > 0$ . #

Problem 2.2.5. Suppose $\underline{X}^t = (\underline{X}_1, \ldots, \underline{X}_n)$ has a multivariate normal density function. Suppose that $E\,\underline{X}_n = 0$ and $E\,\underline{X}_i\underline{X}_n = 0$, $1 \le i \le n-1$. Then $\underline{X}_n$ is stochastically independent of $(\underline{X}_1, \ldots, \underline{X}_{n-1})$ .

Hint: Show that the relevant Laplace transform factors. #

Theorem 2.2.6. Let $\underline{X}$ be normal $(a, A^{-1})$. Let $b_i^t = (b_{i1}, \ldots, b_{in})$, $1 \le i \le k$, be $k$ linearly independent vectors. Then the random k-vector $(b_1^t\underline{X}, \ldots, b_k^t\underline{X})^t$ has a multivariate normal density function with mean vector $(a^t b_1, \ldots, a^t b_k)^t$ and covariance matrix with $ij$ entry $b_i^t A^{-1} b_j$ .

Proof. The Laplace transform is

(2.2.6) $$E\exp(\Sigma_{i=1}^k s_i(b_i^t\underline{X})) = E\exp((\Sigma_{i=1}^k s_i b_i)^t \underline{X})$$

$$= \exp((\Sigma_{i=1}^k s_i b_i)^t A^{-1}(\Sigma_{i=1}^k s_i b_i)/2)\exp((\Sigma_{i=1}^k s_i b_i)^t a) \quad .$$

Since $b_1, \ldots, b_k$ are linearly independent, the exponent of the covariance part of the transform vanishes if and only if $s_1 = \ldots = s_k = 0$. From (2.2.3) the desired conclusion now follows. #

Theorem 2.2.7. If the random vector $(\underline{Y}, \underline{X}_1, \ldots, \underline{X}_n)^t$ has a multivariate normal density function with zero means then there exist constants $c_1, \ldots, c_n$ such that $\underline{Y} - (c_1\underline{X}_1 + \ldots + c_n\underline{X}_n)$ is stochastically independent of $\underline{X}_1,\dots,\underline{X}_n$ .

Proof. In view of Problem 2.2.5 it is sufficient to find constants $c_1,\dots,c_n$ such that $E\,\underline{X}_i(\underline{Y} - c_1\underline{X}_1 - c_2\underline{X}_2 - \dots - c_n\underline{X}_n) = 0$, $i = 1,\dots,n$. By Problem 2.2.4 the random vector $(\underline{X}_1,\dots,\underline{X}_n)^t$ has a nonsingular covariance matrix $A^{-1}$ . Thus the system of equations for $c_1,\dots,c_n$ has as matrix of coefficients the nonsingular matrix $A^{-1}$ and the equations have a unique solution. By Theorem 2.2.6, the random (n+1)-vector $(\underline{Y} - \Sigma_{i=1}^n c_i\underline{X}_i,\ \underline{X}_1,\dots,\underline{X}_n)^t$ has a multivariate normal density function with covariance matrix of the form $\begin{pmatrix} a^{-1} & 0 \\ 0^t & A^{-1}\end{pmatrix}$ . By Problem 2.2.5 or directly from the Laplace transform which is $\exp(ra^{-1}r/2)\exp(s^tA^{-1}s/2)$, independence follows. #

Problem 2.2.8. Let $(\underline{Y},\underline{X}_1,\dots,\underline{X}_n)$ have a joint normal probability density function with zero means and let constants $c_1,\dots,c_n$ be such that $\underline{Y} - \Sigma_{i=1}^n c_i\underline{X}_i$ and $(X_1,\dots,\underline{X}_n)^t$ are stochastically independent. Then the conditional expectation is

$$E(\underline{Y}\,|\,\underline{X}_1,\dots,\underline{X}_n) = \Sigma_{i=1}^n c_i\underline{X}_i \quad . \tag{2.2.7}$$

Problem 2.2.9. If $\underline{X}_1,\dots,\underline{X}_n$ are mutually independent random variables each normally distributed, then the random n-vector $(\underline{X}_1,\dots,\underline{X}_n)$ has a multivariate normal density function.

Hint: Write the product of the Laplace transforms. #

Problem 2.2.10. If $\underline{X}_1,\dots,\underline{X}_n$ are independently distributed random $h\times 1$ vectors, and if $\underline{X}_i$ is normal $(a_i, A^{-1})$, $i = 1,\dots,n$, then the random $n\times h$ matrix $\underline{X}$ with i-th row $\underline{X}_i^t$, $i = 1,\dots,n$, has a multivariate normal density function

$$(2\pi)^{-nh/2}(\det A)^{n/2}\exp(\operatorname{tr} A(X - M)^t(X - M)), \tag{2.2.8}$$

where $M = E\underline{X}$ and "tr" means trace of the matrix.

## 2.3. Noncentral Chi-square, F-, and t-density functions.

Although these are not multivariate random variables these random variables and their density functions play a central role not only in the analysis of variance but in parts of distribution theory. If the random $n \times 1$ vector $\underline{X}$ is normal $(a, A^{-1})$ and $A^{-1}$ is the identity matrix then the distribution theory problem that one wants an answer to is to write the density function of $\underline{X}^t B \underline{X}$, where $B$ is a $n \times n$ symmetric positive definite or positive semi-definite matrix. In the case of idempotents, $B = B^2$, the density function is the density function of a noncentral Chi-square random variable. Any choice of $B$ non-idempotent leads to a problem without a neat answer about which there is a growing literature. See Good (1969), Graybill and Milliken (1969), Press (1966), Shah (1970), and Shanbhag (1970) for contemporary literature.

In the following the basic argument is the same as in Section 2.2. The Laplace transform of a gamma density function is readily computed and this class of density functions includes all the central Chi-square density functions. The noncentral Chi-square (random variable) is defined as a sum of squares of independently distributed normal random variables and the corresponding product of Laplace transforms is readily inverted to obtain the density function of a noncentral Chi-square to be a weighted infinite sum of central Chi-square density functions.

Definition 2.3.1. If $\underline{X}_1, \ldots, \underline{X}_n$ are independently distributed real valued random variables such that $\underline{X}_i$ is normal $(a_i, 1)$, then the density function of $\underline{Y} = X_1^2 + \ldots + X_n^2$ is the noncentral Chi-square density function with non-centrality parameter $a = (a_1^2 + \ldots + a_n^2)/2$ and $n$ degrees of freedom. (It is shown below that the density function of $\underline{Y}$ depends only on $a$ and not individually on $a_1, \ldots, a_n$.) In speaking of a central Chi-square with $n$ degrees of freedom, we speak of the case $a = 0$ and will write $\chi_n^2$.

Definition 2.3.2. The two parameter family of density functions

$$(2.3.1)\qquad f_{a,b}(x) = (\Gamma(a)b^a)^{-1}x^{a-1}\exp(-x/b), \quad x > 0, \qquad = 0 \quad x \le 0,$$

is called the family of gamma density functions, named after the gamma function.

Problem 2.3.3. The gamma density function (2.3.1) has Laplace transform

$$(2.3.2)\qquad (1-bs)^{-a}, \quad \text{convergent if} \quad bs < 1.$$

Hint: Combine exponents in $\int_0^\infty (\Gamma(a)b^a)^{-1}\exp(-x/b)\exp(sx)x^{a-1}dx$ and determine the normalization required to make the integral equal one. #

Problem 2.3.4. If $\underline{X}_1$ is normal $(0,1)$ and $\underline{Y} = \underline{X}_1^2$ then $\underline{Y}$ has a gamma density function with parameters $a = \frac{1}{2}$ and $b = 2$. On the otherhand, by change of variable one may calculate that

$$(2.3.3)\qquad P(\underline{Y} \le y) = \int_0^y (2\pi x)^{-\frac{1}{2}}\exp(-x/2)\,dx,$$

and therefore

$$(2.3.4)\qquad \Gamma(\tfrac{1}{2}) = \sqrt{\pi}.$$

Corollary 2.3.5. The Laplace transform of $\chi_1^2$ is

$$(2.3.5)\qquad (1-2t)^{-\frac{1}{2}}$$

and the Laplace transform of $\chi_n^2$ is

$$(2.3.6)\qquad (1-2t)^{-n/2}.$$

Hence the density function of the central Chi-square with $n$ degrees of freedom is

$$(2.3.7)\qquad (\Gamma(n/2)2^{n/2})^{-1}x^{(n/2)-1}e^{-x/2}, \quad x > 0, \qquad 0, \quad x \le 0.$$

Problem 2.3.6. Let $\underline{X}_1,\ldots,\underline{X}_n$ be mutually independent random variables such that if $1 \le i \le n$ then $\underline{X}_i$ is normal $(a_i,1)$. Let $a=(a_1^2+\ldots+a_n^2)/2$. Then the Laplace transform of $\underline{X}_1^2+\ldots+\underline{X}_n^2$ is

$$(2.3.8.) \qquad (\exp(-a))\Sigma_{j=0}^{\infty}(1-2t)^{-((n/2)+j)}a^j/(j!) \quad .$$

Hint: Write $E\exp(t(\underline{X}_1^2+\ldots+\underline{X}_n^2))$ as a n-fold integral and complete the square in the exponent. The answer is

$$(2.3.9) \qquad (1-2t)^{-n/2}\exp(-\Sigma_{i=1}^{n}a_i^2/2)\exp((\Sigma_{i=1}^{n}a_i^2)/(2(1-2t))).$$

Substitute $a$ in (2.3.9) and expand $\exp(a/(1-2t))$ in a power series in the variable $x=a/(1-2t)$ to obtain (2.3.8). #

Problem 2.3.7. The numbers $\exp(-a)a^j/j!$ are the Poisson probabilities. Thus the Laplace transform (2.3.8) is a mixture of transforms of Chi-square density functions. One may invert the transform at once and read off the noncentral Chi-square density function to be

$$(2.3.10) \qquad = \Sigma_{j=0}^{\infty}\frac{x^{(n+2j)/2-1}e^{-x/2}}{2^{(n+2j)/2}\Gamma(\frac{n+2j}{2})}\frac{e^{-a}a^j}{j!}, \quad x>0,$$

$$= 0, \quad x \le 0 .$$

Definition 2.3.8. The parameter $a=\Sigma_{i=1}^{n}a_i^2/2$ of the density function (2.3.10) is called the noncentrality parameter.

Problem 2.3.9. Let $\underline{X}$ and $\underline{Y}$ be positive random variables, independently distributed, such that $\underline{X}$ has density function $f$ and $\underline{Y}$ has density function $g$ (relative to Lebesgue measure). Then the density function of $\underline{X}/\underline{Y}$ is

$$(2.3.11) \qquad h(s) = \int_0^{\infty} yf(sy)g(y)dy, \qquad s>0 \quad .$$

Definition 2.3.10. Let $\underline{X}$ and $\underline{Y}$ be independently distributed random variables such that $\underline{X}$ is a noncentral $\chi_n^2$ with noncentrality parameter $a$, and $\underline{Y}$ is a central $\chi_m^2$. Then the (normalized) ratio $(m/n)\ (\underline{X}/\underline{Y})$ has the distribution function of a noncentral $F_{n,m}$ - statistic.

Problem 2.3.11. Let $\underline{X}$ and $\underline{Y}$ be as in Definition 2.3.10. The random variable $\underline{Z} = \underline{X}/\underline{Y}$ (i.e., omit the normalization) has the following density function:

$$(2.3.12) \qquad \Sigma_{j=0}^{\infty} \frac{e^{-a}a^j}{j!} \frac{\Gamma((m+n+2j)/2)}{\Gamma((n+2j)/2)\Gamma(m/2)} \frac{z^{(n+2j-2)/2}}{(1+z)^{(m+n)/2+j}}, \quad z > 0 .$$

In the sequel we derive the density function of a noncentral t-statistic. A number of different ways of expressing this density function are available. C.f. Kruskal (1954).

Problem 2.3.12. If $\underline{Y}$ is a central $\chi_m^2$ random variable then $\sqrt{\underline{Y}}$ has the density function

$$(2.3.13) \qquad 2y^{m-1}\exp(-y^2/2)/2^{m/2}\Gamma(m/2), \quad y > 0 .$$

Definition 2.3.13. Let $\underline{X}$ and $\underline{Y}$ be independently distributed random variables such that $\underline{X}$ is normal $(a,1)$ and $\underline{Y}$ is a central $\chi_m^2$ random variable. Then the random variable $\underline{Z} = \underline{X}/\sqrt{\underline{Y}/m}$ has the density function of a noncentral t-statistic noncentrality parameter $a$ and $m$ degrees of freedom

Problem 2.3.14. Show that one form of the noncentral t-density is

$$(2.3.14) \qquad 2(\sqrt{2\pi}2^{m/2}\Gamma(m/2))^{-1}e^{-a^2/2}\int_0^{\infty} y^m e^{-(1+z^2)y^2/2} e^{azy}dy ,$$

$$- \infty < t < \infty .$$

## 2.4. Inversion of transforms and Hermite polynomials.

Theorems on inversion of transforms are readily available and this topic is not treated in much detail here. Standard references for such material are Widder (1941), Wiener (1933) together with more modern books like Gelfond (1971) and texts on complex variable which are directed towards physics and electrical engineering.

The result of Problem 2.1.3 can be rephrased as saying

$$(2.4.1) \qquad \int_{-\infty}^{\infty}(2\pi)^{-1/2}\exp(isx)(2\pi)^{-1/2}\exp(-x^2/2)$$

$$= (2\pi)^{-1/2}\exp(-s^2/2) \quad .$$

Or in words, the (normalized) Fourier transform of a normal (0,1) density function is the normal (0,1) density function. Relative to the weight function $e^{-x^2}$ we seek the sequence of polynomials $\{h_n, n \geq 0\}$ such that if $n \geq 0$ then $h_n$ is a polynomial of degree $n$, and if $m,n \geq 0$ then

$$(2.4.2) \qquad (2\pi)^{-1}\int_{-\infty}^{\infty} h_m(x)h_n(x)\exp(-x^2)\,dx = 0 \; , \quad m \neq n,$$
$$= 1 \; , \quad m = n.$$

This makes the sequence $\{h_n(x)(2\pi)^{-1/2}\exp(-x^2/2), n \geq 0\}$ an orthonormal sequence of functions relative to Lebesgue measure. Basic properties established in Wiener (1933), Chapter I, are

Theorem 2.4.1. The sequence $\{h_n(x)(2\pi)^{-1/2}\exp(-x^2/2), n \geq 0\}$ is a complete orthonormal system in $L_2$ of the real line. Each of these functions is its own Fourier transform, except for a change of normalization by a constant having absolute value one.

Theorem 2.4.2. (Plancherel's Theorem). If $f$ belongs to $L_2$ of the real line define the $L_2$ function $g$ by
$g(y) = \lim_{a\to\infty} \int_{-a}^{a} (2\pi)^{-1/2}\exp(iyx)f(x)\,dx$ . This limit exists in $L_2$ norm, and, in $L_2$ norm, $f(x) = \lim_{a\to\infty}\int_{-a}^{a}(2\pi)^{-1/2}\exp(-iyx)g(y)\,dy$ . The function $\hat{f} = g$ is called the Fourier transform of $f$ . The map $f \to \hat{f}$ is an isometry of period four of $L_2$ of the real line.

We now consider in detail the Lévy inversion theorems for functions of bounded variation of a real variable. For this purpose we need several lemmas.

Lemma 2.4.3. Let a contour $C$ in the complex plane be given by $z = r\exp(is)$, $0 \leq s \leq \pi$ . Then

$$(2.4.3) \qquad \left|\int_C \frac{\exp iz}{z}\,dz\right| < \pi/r\ .$$

Proof. In parametric form the line integral is

$$(2.4.4) \qquad a = \left|\int_0^\pi (\exp ir(\exp is))ds\right| \le \int_0^\pi \exp(-r \sin s)ds$$
$$= 2\int_0^{\pi/2} \exp(-r \sin s)ds\ .$$

By convexity, if $0 \le s \le \pi/2$, then $\sin s \ge 2s/\pi$ so that

$$(2.4.5) \qquad a \le 2\int_0^{\pi/2} \exp(-2rs/\pi)ds \le (\pi/r)\int_0^r \exp(-s)ds < \pi/r.\ \#$$

Lemma 2.4.4. $\left|\pi - \int_{-r}^r x^{-1}\sin x\,dx\right| < \pi/r$ .

Proof. Taken over a closed contour consisting of the line segment $[-r,r]$ together with the semicircle $C$ (see Lemma 2.4.3), we have

$$(2.4.6) \qquad 0 = \oint \frac{\exp(iz)-1}{z}\,dz = \int_{-r}^r \frac{\exp(is)-1}{s}\,ds + \int_C \frac{\exp(iz)-1}{z}\,dz\ .$$

By transposition we obtain

$$(2.4.7) \qquad i\int_{-r}^r x^{-1}\sin x\,dx = \int_{-r}^r x^{-1}(\exp(ix)-1)dx$$

$$= \int_C z^{-1}dz - \int_C z^{-1}\exp(iz)dz\ .$$

Thus from (2.4.3) we obtain

$$(2.4.8) \qquad \left|\int_{-r}^r x^{-1}\sin x\,dx - \pi\right| \le \left|\int_C z^{-1}\exp(iz)dz\right| < \pi/r.\ \#$$

In the sequel we need to consider the function

$$(2.4.9) \qquad \lim_{r\to\infty} (2\pi)^{-1}\int_{-r}^r \frac{(\exp(is(b-x))-\exp(is(a-x)))ds}{is},\ a < b\ .$$

When expressed in terms of sines and cosines, the cosine terms, being even functions, integrate to zero. Thus (2.4.9) is equal to

(2.4.10) $$g(x) = (2\pi)^{-1}\left[\int_{-\infty}^{\infty}\frac{\sin s(b-x)\,ds}{s} - \int_{-\infty}^{\infty}\frac{\sin s(a-x)\,ds}{s}\right] .$$

Using Lemma 2.4.4 it follows at once that

(2.4.11) if $a < x < b$ then $g(x) = 1$ ,

if $x = a$ or $x = b$ then $g(x) = 1/2$ ,

if $x < a$ or $x > b$ then $g(x) = 0$ .

Theorem 2.4.5. (Lévy). Suppose $F$ is monotonic, bounded, with normalized Fourier transform $\hat{F}$ . Define $F_1(s) = 1/2(F(s+)+F(s-))$. Then

(2.4.12) $$F_1(b)-F_1(a) = \lim_{r\to\infty} (2\pi)^{-\frac{1}{2}}\int_{-r}^{r}\hat{F}(s)\left[\frac{e^{-isb}-e^{-isa}}{is}\right]ds .$$

Proof. The double integral implied by (2.4.12) is absolutely convergent for each fixed value of $r$ . Let $\mu_F$ be the Borel measure determined by $F$ . By Fubini's theorem,

(2.4.13) $$\lim_{r\to\infty} (2\pi)^{-1}\int_{-r}^{r}\left[\frac{e^{-isb}-e^{-isa}}{is}\right]\int_{-\infty}^{\infty}e^{isx}\mu_F(dx)\,ds$$

$$= \lim_{r\to\infty} (2\pi)^{-1}\int_{-\infty}^{\infty}\int_{-r}^{r}\frac{\sin s(b-x) - \sin s(a-x)}{s}ds\ \mu_F(dx) .$$

By Lemma 2.4.4 the inner integral is a bounded function of $x$ and $r$ so that by the bounded convergence theorem, passing to the limit under the first integral sign, with $g$ defined in (2.4.10),

(2.4.14) $$= \int_{-\infty}^{\infty} g(x)\mu_F(dx) = F_1(b) - F_1(a) . \#$$

In case $F$ is absolutely continuous with density function $f$ of bounded variation, then defining $f_1$ by analogy to $F_1$, we have from (2.4.12) that if $f$ is absolutely continuous with derivative $f'$ and if $\lim_{|x|\to\infty} f(x) = 0$ then an integration by parts yields

$$(2.4.15)\qquad f_1(b) - f_1(a) = \lim_{r\to\infty} (2\pi)^{-\frac{1}{2}}\int_{-r}^{r} \left(\frac{e^{-ixb}-e^{-ixa}}{-ix}\right) dx \int_{-\infty}^{\infty} e^{ixy} f'(y)\,dy$$

$$= \lim_{r\to\infty} (2\pi)^{-1/2}\int_{-r}^{r} dx \int_{-\infty}^{\infty} (e^{-ixb}-e^{-ixa}) e^{ixy} f(y)\,dy$$

$$= \lim_{r\to\infty} (2\pi)^{-1/2}\int_{-r}^{r} (e^{-ixb}-e^{-ixa}) \hat{f}(x)\,dx \quad .$$

We summarize this in a slightly different form.

Problem 2.4.6. Suppose $f$ is absolutely integrable, and $f$ has a continuous absolutely integrable first derivative $f'$. Let $\hat{f}$ be the unnormalized Fourier transform of $f$. Then

$$(2.4.16)\qquad \lim_{r\to\infty} (2\pi)^{-1}\int_{-r}^{r} e^{-isx}\hat{f}(s)\,ds = f(x), \quad -\infty < x < \infty \quad .$$

Hint: The hypotheses imply that $\lim_{|x|\to\infty} f(x) = 0$. An integration by parts is required. At this step the constant of integration may be choosen in a helpful fashion. #

Without trying to develop a general n-dimension theory we note the useful obvious generalization of Problem 2.4.6.

Problem 2.4.7. Suppose $h\colon \mathbb{R}_n \to \mathbb{R}$ is in $L_1$ of Lebesgue measure, has zero integral,

$$(2.4.17)\qquad f(x_1,\ldots,x_n) = \int_{-\infty}^{x_1} \cdots \int_{-\infty}^{x_n} h(y_1,\ldots,y_n)\,dy_1\ldots dy_n \;;$$

$$\hat{f}(s_1,\ldots,s_n) = (2\pi)^{-n/2}\int_{-\infty}^{\infty} \cdots \int_{-\infty}^{\infty} \exp i(s_1x_1+\ldots+s_nx_n)$$

$$f(x_1,\ldots,x_n)\,dx_1\ldots dx_n,$$ assumed to be absolutely convergent. Then

$$(2.4.18)\qquad f(x_1,\ldots,x_n) = \lim_{r_1\to\infty} \cdots \lim_{r_n\to\infty} (2\pi)^{-n/2}\int_{-r_1}^{r_1} \cdots \int_{r_n}^{r_n}$$

$$\exp -i(s_1x_1+\ldots+s_nx_n)\hat{f}(s_1,\ldots,s_n)\,ds_1\ldots ds_n .$$

## 2.5. Inversion of the Laplace and Mellin transforms.

We discuss here the univariate case. The multivariate generalizations follow the argument of Problem 2.4.7.

We suppose that if $s_1 < s < s_2$ then $\int_{-\infty}^{\infty} \exp(sx)\, f(x)\, dx$ is an absolutely convergent integral. We will require in addition that $f$ be everywhere differentiable with derivative $f'$ which is Riemann integrable and that $\int_{-\infty}^{\infty} \exp(sx)|f'(x)|\, ds < \infty$. It then follows that if $s_1 < s_0 < s_2$ then the function $\exp(s_0 x) f(x)$ is of bounded variation. By Problem 2.4.6

$$\text{(2.5.1)} \qquad \lim_{r\to\infty} (2\pi)^{-1} \int_{-r}^{r} \exp(-(is+s_0)y)\, ds \int_{-\infty}^{\infty} \exp((is+s_0)x) f(x)\, dx$$

$$= \exp(-s_0 y)[\exp(s_0 y) f(y)] = f(y) \quad .$$

Theorem 2.5.1. Given the hypotheses of the preceeding paragraph, then (2.5.1) holds.

The Mellin transform $\int_0^{\infty} x^{s-1} f(x)\, dx$ becomes a Laplace transform $\int_{-\infty}^{\infty} \exp(sy) f(\exp y)\, dy$ under the change of variable $x = \exp y$. It then follows that

$$\text{(2.5.2)} \qquad \lim_{r\to\infty} (2\pi)^{-1} \int_{-r}^{r} \exp((-is-s_0)y)\, ds \int_0^{\infty} x^{is+s_0-1} f(x)\, dx = f(e^y),$$

so that with the choice $w = \log y$, we obtain

$$\text{(2.5.3)} \qquad \lim_{r\to\infty} (2\pi)^{-1} \int_{-r}^{r} y^{-is-s_0} \int_0^{\infty} x^{is+s_0-1} f(x)\, dx\, ds = f(y) \; .$$

The limit (2.5.3) will exist if enough smoothness holds. We will not make a formal statement of smoothness assumptions.

The inversion theorems stated here are adequate for most multivariate calculations. Less restricted assumptions may be found in Widder (1941).

## 2.6. Examples in the literature.

Kullback (1934) proved inversion theorems for Fourier transforms and used this method to calculate the probability density functions of products of independently distributed Chi-square random variables. In his paper it is shown that the distribution of Wilks' generalized variance (central case) is the same as the distribution of a product of Chi-squares. The answers are expressed in terms of residues of products of gamma functions without an explicit calculation of these residues being given.

Herz (1955) extended ideas of Bochner and defined a doubly infinite sequence of hypergeometric functions of complex symmetric matrix arguments. These functions were defined by using the Laplace transform and inverse Laplace transform to generate new functions. The hypergeometric functions of Herz have been given infinite series representations by Constantine (1963) in which the individual terms of the series are zonal polynomials, as defined by James (1961). It has been implied but never stated by James (1964), (1968), that the zonal polynomials (of a matrix argument) are spherical functions in the sense of Helgason (1962). We prove this fact in Chapter 12.

Box (1949) used inversion of Fourier transforms as a method of obtaining asymptotic series as the sample size tends to $\infty$. This is one of several methods currently in use in the literature of asymptotic approximations to the distributions of random variables.

Meijer's functions (c.f. Erdélyi, et.al., (1953)) have been generalized by Braaksma (1964). Meijer functions and inversion of Mellin transforms have been used by Consul (1969). Mathai and Rathie (1971) have extended the work of Consul, op. cit., and Mellin transforms, in the study of the distribution of products, claiming H-functions are the most general type of special function.

## Chapter 3. Locally compact groups and Haar measure.

### 3.0. Introduction.

This chapter is intended to summarize some results needed later. We state an existence and uniqueness theorem for Haar measure but rather than copy a proof we leave this result unproven. See for example Halmos (1950) or Loomis (1953). The applications made in these notes are to matrix groups in their usual metric topology. Hence all topologies used in applications are Hausdorff topologies with a countable base for open sets. More generality will be found in Loomis, op. cit., or Nachbin (1965). The manifolds discussed later are analytic manifolds for which the invariant measures can be given explicit representations using differential forms. Since the existence of invariant measures will usually be shown by explicit construction the part of the theory important to this book is usually the uniqueness part.

At various places in this chapter matrix groups of interest are defined and used to illustrate points of the discussion. It was decided to use differential forms in these examples in spite of the fact that the discussion of differential forms comes later. The matrices discussed in this chapter have real numbers for entries. Later, in Chapter 12, use of complex numbers becomes important. The manifolds discussed include $GL(n)$, the full linear group of $n \times n$ matrices; $\underline{O}(n)$, the group of $n \times n$ orthogonal matrices; $\underline{S}(n)$, the homogeneous space of $n \times n$ positive definite symmetric matrices; $\underline{T}(n)$ the group of lower triangular matrices with positive diagonal; $\underline{D}(n) \subset \underline{T}(n)$ the diagonal matrices. In each case matrix multiplication is the group operation so that the group identity is the identity matrix, and the group inverse is the same as the matrix inverse. These notations will be used throughout this book.

In the computation of differential forms it is convenient to compute $dX$, meaning, compute the differential of each entry of the matrix X and form the matrix of corresponding differentials. Then $(dX)_{ij}$ is the ij entry of dX and $\bigwedge_{i<j} (dX)_{ij}$ a wedge product of the indicated differentials. As will appear in the examples which follow, this short notation leads at once to the differential forms for the Haar measures of the matrix groups.

Section 3.1 gives a summary of basic point set topology for locally compact groups. Section 3.2 discusses quotient spaces. Section 3.3 gives the uniqueness theorems for invariant measures on locally compact groups and quotient spaces. This material provides the basis for a discussion of the factorization of invariant measures and the factorization of manifolds discussed in Section 3.4, and again in Chapter 10. Section 3.5 discusses the modular function, needed for Chapter 10. Section 3.6 discusses in the abstract the construction of differential forms for Haar measures on matrix groups. Section 3.7 discusses briefly the problem of cross sections, which provides one way of doing the theory of Chapter 10. Last, Section 3.8 briefly discusses material related to the Hunt-Stein theory of minimax invariant statistical procedures. No problems were written for this chapter.

## 3.1. Basic point set topology.

The set of group elements $\mathfrak{G}$ is assumed to have a locally compact Hausdorff topology. In this topology the map of $\mathfrak{G} \times \mathfrak{G} \to \mathfrak{G}$ given by $(x,y) \to xy$ is to be jointly continuous and the map $x \to x^{-1}$ is to be continuous. For each fixed y, the maps $x \to xy$ and $x \to yx$ are to be continuous. Thus these latter three maps are homeomorphisms of $\mathfrak{G}$.

If $V \subset \mathfrak{G}$, V is an open set, and the unit e of $\mathfrak{G}$ is in V, then $V^{-1} = \{v | v^{-1} \in V\}$ is an open set and $e = e^{-1} \in V^{-1}$. Given y, as noted, the map $x \to yx$ is a homeomorphism. Thus $y(V \cap V^{-1})$ is an open set containing $y$ and is said to be a <u>symmetrical neighborhood of $y$</u>. Further, if $U$ is an open set, then $U\cdot U = \{z \mid \text{exist } x,y \in U, z = xy\}$ is an open set since $U\cdot U = \bigcup_{x\in U} (xU)$. Thus the map $(x,y) \to xy$ is an open mapping.

Let $U$ be an open set and $e \in U$. Then the inverse image of $U$ under the map $(x,y) \to xy$ is an open set of $\mathcal{G} \times \mathcal{G}$ and hence there exists a set $V \times V$, $e \in V$, $V$ open, such that $V \times V$ is contained in the inverse image. Thus $e \in V\cdot V \subset U$. Similarly there exists $V_1$ open, $e \in V_1$, with $V_1^{-1} \cdot V_1 \subset U$, and there exists $V_2$ open, $e \in V_2$, with $V_2 \cdot V_2^{-1} \subset U$.

If $U$ and $V$ are compact subsets of $\mathcal{G}$ then $U \times V$ is a compact subset of $\mathcal{G} \times \mathcal{G}$ and hence $U\cdot V$ is a compact subset of $\mathcal{G}$. It follows that there exist compact symmetric neighborhoods of $e$.

In general, if $V$ is an open set and $y \in V$ then $e \in y^{-1}V$, an open set. Thus every neighborhood of $y$ has the form $yU$, $U$ open, $e \in U$.

<u>Lemma 3.1.1</u>. If $W \subset \mathcal{G}$ then the topological closure $\overline{W}$ of $W$ is $\overline{W} = \cap_V W\cdot V = \cap_V V\cdot W$ taken over all neighborhoods of $e$.

<u>Proof</u>. If $e \in V$ then $W \subset W\cdot V$. Let $x \in \overline{W}$ and $V$ be an open set, $e \in V$. Then $(xV) \cap W$ is nonempty and contains $w_0$, say. Then for some $v_0 \in V$, $xv_0 = w_0$ so that $x = w_0v_0^{-1} \in W\cdot V^{-1}$. Thus $x \in W\cdot V^{-1}$ for all open neighborhoods $V$ of $e$. That is, $\overline{W} \in \cap_V W\cdot V$. Conversely, suppose $x \in W\cdot V$ for all open neighborhoods $V$ of $e$. If $U$ is a neighborhood of $x$ then $x^{-1}U = V$ is a neighborhood of $e$, so that $x \in W\cdot V^{-1}$, $x = w_0v_0^{-1}$, and $w_0 = xv_0 \in U$. Hence $U \cap W$ is not empty. Thus $x \in \overline{W}$. Use of the map $x \to x^{-1}$ gives the second identity $\overline{W} = \cap_V V\cdot W$. #

## 3.2. Quotient spaces.

We let $H \subset \mathcal{G}$ be a subgroup and the points of $\mathcal{G}/H$ be the left cosets of $H$. The projection map $\pi$ is defined by

$$(3.2.1) \qquad \pi(x) = x \cdot H .$$

Topologize $\mathcal{G}/H$ with the finest topology such that the projection map is continuous. Thus $V \subset \mathcal{G}/H$ is an open set if and only if $\pi^{-1}(V)$ is an open set of $\mathcal{G}$.

$$(3.2.2) \qquad \pi^{-1}(V) = \{x \mid \pi(x) \in V\} = \{x \mid x \cdot H \in V\} = \bigcup_{x \in \pi^{-1}(V)} x \cdot H .$$

$\pi$ is an open mapping since if $U$ is an open subset of $\mathcal{G}$ then

$$(3.2.3) \qquad \pi^{-1}(\pi(U)) = \bigcup_{x \in U} x \cdot H = U \cdot H ,$$

which is a union of open sets. If $H$ is a closed subgroup of $\mathcal{G}$ then each coset $x \cdot H$ is a closed set in $\mathcal{G}$, so that points of $\mathcal{G}/H$ are closed (complements of points are open).

As shown, $H$ a closed subgroup implies $\mathcal{G}/H$ is a $T_1$-space. We now show that if $H$ is a compact subgroup then $\mathcal{G}/H$ is a Hausdorff space. Let $H$ be compact and $x \cdot H$ and $y \cdot H$ be disjoint sets. Then $y \cdot H = \bigcap_V V \cdot (y \cdot H)$, by Lemma 3.1.1, and for some neighborhood $V$ of $e$, and $h \in H$, $x \cdot h \notin V \cdot y \cdot H$. We may choose a neighborhood $V_1$ of $e$ such that $\overline{V}_1$ is compact and $V_1 \subset \overline{V}_1 \subset V$. Then $\mathcal{G} - \overline{V}_1 \cdot y \cdot H$ is a neighborhood of $x \cdot h$ disjoint from $y \cdot H$. The compact set $x \cdot H$ may thus be covered by a finite number of open sets $\mathcal{G} - \overline{V} \cdot y \cdot H$ and each is the inverse image under $\pi$ of an open subset of $\mathcal{G}/H$. #

<u>Lemma 3.2.1</u>. The projection map (3.2.1) under the topology described for $\mathcal{G}/H$ is an open continuous map. If $H$ is a closed subgroup then $\mathcal{G}/H$ is a $T_1$-space and if $H$ is a compact subgroup then $\mathcal{G}/H$ is a Hausdorff space.

Lemma 3.2.2. If the group $\mathcal{G}$ is a locally compact Hausdorff space and if $W \subset \mathcal{G}/H$ is a compact subset then there exists a subset $W_1 \subset \mathcal{G}$ which is compact and such that $\pi(W_1) = W$.

Proof. Choose an open set $V$ such that $e \in V$ and $V$ has compact closure. Choose points $w_1, \ldots, w_k$ of $W$ and $g_1, \ldots, g_k$ of $\mathcal{G}$ such that $w_i = g_i \cdot H$, $1 \leq i \leq k$, and such that $\bigcup_{i=1}^{k} \pi(g_i \cdot \overline{V})$ covers $W$. Then $W_1 = \pi^{-1}(W) \cap \bigcup_{i=1}^{k} (g_i \cdot \overline{V})$ is a compact subset of $\mathcal{G}$ and $\pi(W_1) = W$. #

## 3.3. Haar measure.

As noted in the introductory section the basic results are stated without proof. We use as a source for our definitions Halmos (1950).

Theorem 3.3.1. (Existence). If $\mathcal{G}$ is a locally compact group which is a Hausdorff space then there exist regular Borel measures $\mu$, $\nu$ such that $\mu$ and $\nu$ are nonzero, and,

(3.3.1) If $g \in \mathcal{G}$ then $\mu(g \cdot W) = \mu(W)$ and $\nu(W \cdot g) = \nu(W)$ for all Borel subsets $W$.

Regularity, being nonzero, together with (3.3.1) imply,

(3.3.2) If $V \subset \mathcal{G}$ is an open Borel subset then $\mu(V) > 0$ and $\nu(V) > 0$;

(3.3.3) Compact Borel subsets have finite measure.

A regular Borel measure with the left invariance (3.3.1) is called a left invariant Haar measure. A regular Borel measure with the right invariance (3.3.1) is called a right invariant Haar measure.

Theorem 3.3.2. (Uniqueness) If $\mu_1$ and $\mu_2$ are left invariant Baire measures then there exist constants $c_1$ and $c_2$ such that $c_1\mu_1 = c_2\mu_2$. If $\nu_1$ and $\nu_2$ are right invariant Baire measures then there exist constants $c_1$ and $c_2$ such that $c_1\nu_1 = c_2\nu_2$. (Note that Baire measures are regular. For us a measure is non-negative.)

We now develop a theory for invariant measures induced on $\mathcal{G}/H$. For more generality see the last chapter of Halmos (1950). The projection map (c.f. (3.2.1)) indices a measure $\mu\pi^{-1}$ on the Baire sets of $\mathcal{G}/H$ from a measure $\mu$ on $\mathcal{G}$ by means of the definition

$$(3.3.4) \qquad \mu(\pi^{-1}(W)) = (\mu\pi^{-1})(W) .$$

For (3.3.4) to be meaningful when $\mu$ is a Baire measure we need to show that $\pi^{-1}(W)$ is a Baire subset. At this point we suppose the group is $\sigma$-compact and $H$ is compact so that both $\mathcal{G}$ and $\mathcal{G}/H$ are Baire sets. If $f\colon \mathcal{G}/H \to \mathbb{R}$ is continuous then the composition $f\circ\pi\colon \mathcal{G} \to \mathbb{R}$ is continuous and for Borel subsets of the real numbers $A$, $(f\circ\pi)^{-1}(A) = \pi^{-1}(f^{-1}(A))$ is thus a Baire set. Thus if $\mathcal{C}$ is the set of all subsets $W$ of $\mathcal{G}/H$ such that $\pi^{-1}(W)$ is a Baire subset of $\mathcal{G}$ then $\mathcal{C}$ contains all open and compact Baire sets and is a monotone class. Thus $\mathcal{C}$ contains all the Baire subsets of $\mathcal{G}/H$. We summarize this in

Lemma 3.3.3. Suppose $\mathcal{G}$ is a $\sigma$-compact locally compact Hausdorff space, and $H$ is a compact subgroup. The map $\pi^{-1}$ maps the Baire subsets of $\mathcal{G}/H$ into the Baire subsets of $\mathcal{G}$. The induced measure $\mu\pi^{-1}$ defined by (3.3.4) is a well defined Baire measure.

$\mathcal{G}$ acts as a transformation group on $\mathcal{G}/H$ by means of the action

$$(3.3.5) \qquad \bar{g}(x\cdot H) = (gx)\cdot H .$$

The action $\bar{g}$ is well defined since $(gx)\cdot H = (gy)\cdot H$ if and only if $(gx)^{-1}(gy) \in H$ if and only if $x^{-1}y \in H$ if and only if $x\cdot H = y\cdot H$. Therefore

(3.3.6) $$\bar{g}\circ\pi = \pi\circ g .$$

It is clear that the composition of mappings satisfies $\overline{h\circ g} = \bar{h}\circ\bar{g}$, and that $\bar{e}$ is the identity of the induced group. Therefore the mapping $\mathcal{G} \to \bar{\mathcal{G}}$ is a group homomorphism. In particular $\pi^{-1}\bar{g}^{-1} = g^{-1}\pi^{-1}$ and $\bar{g}^{-1} = \overline{(g^{-1})}$. Last, from (3.3.6) it follows that

(3.3.7) $$\mu(\pi^{-1}\bar{g}(W)) = \mu(g\pi^{-1}(W)) = \mu(\pi^{-1}(W)). \#$$

Lemma 3.3.4. $\mathcal{G}$ acts on a transformation group on $\mathcal{G}/H$ by means of the actions defined in (3.3.5). The group $\bar{\mathcal{G}}$ of actions on $\mathcal{G}/H$ is a homomorphic image of $\mathcal{G}$ under the mapping $g \to \bar{g}$ which satisfies (3.3.6). If $H$ is a compact subgroup the measure $\mu\pi^{-1}$ induced by the left invariant Haar measure $\mu$ for $\mathcal{G}$ is invariant under the actions of $\bar{\mathcal{G}}$ on $\mathcal{G}/H$. (See also Lemma 3.3.6.)

Remark 3.3.5. $\bar{g}$ acts as the identity element if and only if $(gx)\cdot H = x\cdot H$ for all $x \in \mathcal{G}$. This holds if and only if $x^{-1}gx \in H$ for all $x \in \mathcal{G}$. It is clear that $\cap_{x\in\mathcal{G}} xHx^{-1}$ is a normal subgroup of $\mathcal{G}$ and in most of the examples considered the only proper normal subgroups are $\{e\}$, and in some cases, $\underline{D}(n)$. If the only proper normal subgroup is $\{e\}$ then, of course, $\mathcal{G}$ and $\bar{\mathcal{G}}$ are isomorphic.#

Lemma 3.3.6. If $\mu$ is a regular Baire measure on $\mathcal{G}$ and $H$ is a compact subgroup of $\mathcal{G}$ then the induced measure $\bar{\mu}$ is a regular Baire measure on $\mathcal{G}/H$. We assume $\mathcal{G}$ is $\sigma$-compact.

Proof. We have already shown in Lemma 3.3.3 that $\bar{\mu}$ is defined on the Baire subsets of $\mathcal{G}/H$. Lemma 3.3.3 implies and it is shown in the proof of Lemma 3.3.7 that if $W$ is a compact Baire subset of $\mathcal{G}/H$ then $\bar{\mu}(W) < \infty$. By Halmos (1950), Section 52, Theorem G, the Baire measure $\bar{\mu}$ is inner and outer regular, hence is regular. #

Lemma 3.3.7. If $H$ is a compact subgroup of the $\sigma$-compact group $\mathcal{G}$ and $\mu$ is a Baire measure on $\mathcal{G}$ then the induced measure $\bar{\mu}$ is a Baire measure on $\mathcal{G}/H$. If $\mu$ is a Haar measure then $\bar{\mu}$ gives positive mass to open sets.

Proof. Much of the proof is already contained in Lemma 3.3.3 and Lemma 3.3.6. Let $W \subset \mathcal{G}/H$ and $W$ be compact. By Lemma 3.3.2 there exists a compact set $W_1$ with $\pi(W_1) = W$. Since $W_1$ is compact so is $W_1 \cdot H$. Then there exists a compact Baire set $W_2$ such that $W_1 \cdot H \subset W_2 \subset \mathcal{G}$. Then $\bar{\mu}(W) = \mu(\pi^{-1}(\pi(W_1))) \leq \mu(W_2) < \infty$. If $\mu$ gives positive mass to open sets and $W \subset \mathcal{G}/H$ is open then $\bar{\mu}(W) = \mu(\pi^{-1}(W)) > 0$. #

Theorem 3.3.8. (Uniqueness). Suppose $\mathcal{G}$ has a countable base for the open sets. If $H$ is a compact subgroup of $\mathcal{G}$ then invariant Baire measures on $\mathcal{G}/H$ are uniquely determined up to multiplicative constants.

Proof. Note at the onset that $\mathcal{G}/H$ is a locally compact space. For if $\bar{g}$ has a neighborhood $W$, then $\pi^{-1}(W)$ is an open set and contains a point $g_1$ with $\pi(g_1) = \bar{g}$. Choose a neighborhood $V$ of $g_1$ such that $\bar{V}$ is compact, and $\bar{V} \subset \pi^{-1}(W)$. Then $\bar{g} \in \pi(V) \subset \pi(\bar{V}) \subset W$ and $\pi(V)$ is an open set, by Lemma 3.2.1. Thus $\overline{\pi(V)}$ is compact since in a Hausdorff space a closed subset of a compact set is compact.

Thus both $\mathcal{G}$ and $\mathcal{G}/H$ are metrizable. The proof following depends heavily on the fact that in a locally compact metric space with a countable base for the open sets, the Baire sets and the Borel sets are the same.

Let $\bar{\mu}$ be a regular invariant nonnegative measure on the Borel subsets of $\mathcal{G}/H$ (invariant for the group $\bar{\mathcal{G}}$), so that $\bar{\mu}$ is finite valued on compact subsets. Let $\mu_H$ be a nonzero Haar measure on the group $H$. Since $H$ is a compact metric space, the Baire and Borel subsets of $H$ are the same. Since $H$ is compact, $\mu_H(H) < \infty$.

Let $U$ be a Borel subset of $\mathcal{G}$ and $\chi_U$ be the indicator function of $U$. We consider the integral $\int_H \chi_U(xh)\mu_H(dh)$. By Fubini's theorem, since the map $(x,h) \to xh$ is jointly continuous, the integrand is jointly measurable and the integral is a measurable function of $x$. Clearly by invariance of the measure $\mu_H$, if $yh_o = x$ and $h_o \in H$ then

$$\int_H \chi_U(xh)\mu_H(dh) = \int_H \chi_U(yh)\mu_H(dh) \ . \tag{3.3.8}$$

Thus (3.3.8) defines a function $f$ of variables $\overline{x} \in \mathcal{G}/H$ and $U$ given by

$$f(\overline{x},U) = \int_H \chi_U(xh)\mu_H(dh) \ . \tag{3.3.9}$$

We need to know that if $U$ is a Borel subset of $\mathcal{G}$ then $f(\ ,U)$ is a $\mathcal{G}/H$ Borel measurable function. We now prove this. If $g$ is a continuous real valued function on $\mathcal{G}$ with compact support $C$ then

$$\int_H g(xh)\mu_H(dh) = f(\overline{x},g) \tag{3.3.10}$$

vanishes outside the compact set $C\cdot H$. Therefore using the fact that $\mathcal{G}$ is a metric space and using the bounded convergence theorem it follows that (3.3.10) is a continuous function of $x$. Thus, if $A$ is a closed set of real numbers, $0 \notin A$, then the inverse images

$$\{x|\int g(xh)\mu_H(dh) \in A\}, \text{ and} \tag{3.3.11}$$

$$\pi(\{x|\int g(xh)\mu_H(dh) \in A\})$$

are compact, hence measurable sets. It is easy to see that the inverse image under $f(\ ,g)$ of $A$ is the second set of (3.3.11). Thus (3.3.10) is a Borel measurable function on $\mathcal{G}/H$. By successive applications of the monotone convergence theorem it now follows that if $U$ is the intersection of a compact and open set then (3.3.9) is a Borel measurable function on $\mathcal{B}/H$. The sets which are intersections of a compact and open set, or which are finite unions of these, form a set ring. And the set of $U$ such that (3.3.9) is measurable is clearly a monotone class. Hence (3.3.9) is measurable for the $\sigma$-ring closure, i.e., for all Borel subsets $U$.

As a function of the variable $U$, (3.3.9) is nonnegative, countably additive, and finite on compact sets. Define a set function $\nu$ by

$$(3.3.12) \qquad \nu(U) = \int f(\bar{x},U)\bar{\mu}(d\bar{x}).$$

Then

$$(3.3.13) \qquad \nu(gU) = \int f(\bar{x},gU)\bar{\mu}(d\bar{x}) = \int f(\bar{g}^{-1}\bar{x},U)\bar{\mu}(d\bar{x})$$

$$= \int f(\bar{x},U)\bar{\mu}(d\bar{x}) = \nu(U) \ .$$

From (3.3.9), if $U$ is compact then $f(\ ,U)$ has compact support so that $\nu(U) < \infty$. Thus $\nu$ is a left invariant measure, finite on compact subsets, and hence is zero or is a left invariant Haar measure on the group $\mathcal{B}$. As a reference measure let $\mu$ be a nonzero left invariant Haar measure on $\mathcal{B}$ so that there exists a constant $c$ such that $c\nu = \mu$, provided $\nu \neq 0$.

For Borel subsets of the form $U = V\cdot H$ a direct computation in (3.3.9) shows that

$$(3.3.14) \qquad f(\bar{x},V\cdot H) = \int \chi_{V\cdot H}(xh)\mu_H(dh) = \mu_H(H)\chi_{\pi(V\cdot H)}(\bar{x}).$$

By choice of $\mu_H$ the number $\mu_H(H)$ is nonzero and finite. If $\bar{\mu}_1$ and $\bar{\mu}_2$ are two regular left invariant non zero measures on the Borel subsets of $\mathcal{B}/H$ then the preceding discussion together with integration of (3.3.14) shows there exist constants $c_1$ and $c_2$ such that

(3.3.15) $$c_1\bar{\mu}_1(\pi(V\cdot H))\mu_H(H) = c_1\nu_1(V\cdot H) = \mu(V\cdot H)$$

$$= c_2\nu_2(V\cdot H) = c_2\bar{\mu}_2(\pi(V\cdot H)).$$

The arbitrary compact subset of $\mathfrak{G}/H$ has the form $\pi(V) = \pi(V\cdot H)$, as shown in Lemma 3.2.2. By regularity of the measures, $c_1\bar{\mu}_1 = c_2\bar{\mu}_2$.#

Example 3.3.9. Matrices $X \in GL(n)$ (real entries) have factorizations $X = AS$ with $A \in \underline{O}(n)$ and $S \in \underline{S}(n)$. One such factorization is given by $A = X(X^tX)^{-1/2}$ and $S = (X^tX)^{1/2}$. Clearly any factorization must satisfy $S^2 = X^tX$ which is nonsingular. (Problem) If $T$ is another positive definite matrix and $S^2 = T^2$ it follows that $S = T$. It follows that for the compact subgroup $\underline{O}(n)$ each right coset contains a unique positive definite matrix. Thus $\underline{S}(n) \simeq GL(n)/\underline{O}(n)$ and $\underline{S}(n)$ may be interpreted as a homogeneous or quotient space. A similar interpretation holds for $\underline{T}(n)$, and this example is discussed below.

The cosets of this example are right cosets, so $GL(n)$ is to be interpreted as acting on the right. Thus $X = AS$ and if $B \in GL(n)$ then $XB$ has the symmetric matrix component $(B^tX^tXB)^{1/2} = (B^tS^2B)^{1/2}$. Thus the action of $GL(n)$ is $S \to (B^tS^2B)^{1/2}$. This is different from the more usual group action $S \to B^tSB$. This second action arises as follows. The function $f(S) = S^2$ defined on $\underline{S}(n)$ is one to one. Let $g_B$ be the function $g_B(S) = (B^tS^2B)^{1/2}$. Then $(fg_Bf^{-1})(S) = B^tSB$. The transformations $fg_Bf^{-1}$ form a group isomorphic to the group of transformations $g_B$. Haar measures $\mu$ on $GL(n)$ induce invariant measures $\bar{\mu}$ on $\underline{S}(n)$ and invariant measures $\bar{\mu}(f^{-1}(\cdot)) = \bar{\nu}$ under the group actions $fg_Bf^{-1}$. Show that $g_B = g_C$ if and only if $B = \pm C$ so that $GL(n)/A$ is isomorphic to the group with elements $g_B$ and is isomorphic to the group with elements $fg_Bf^{-1}$, $A = \{I, -I\}$.

After the theory of differential forms is developed in a later chapter we will see that the invariant measures $\overline{\nu} = \overline{\mu}(f^{-1}(\cdot))$ on $\underline{S}(n)$ are given by differential forms

$$(3.3.16) \qquad \frac{c \bigwedge_{j \leq i} d\, s_{ij}}{(\det S)^{n+1}}, \quad S = (s_{ij}),$$

where $c$ is a constant. By Theorem 3.3.8 the invariant measures $\overline{\mu}$ are determined up to a multiplicative constant. Since to every measure $\overline{\nu}$ there is a uniquely determined $\overline{\mu} = \overline{\nu}(f(\cdot))$, the measures invariant under the actions $fg_B f^{-1}$ are uniquely determined up to multiplicative constant. Therefore every invariant measure $\overline{\nu}$ is representable by a differential form (3.3.16).

If $X \in GL(n)$ we write $X_T$ to be the matrix in $\underline{T}(n)$ such that $X_T \in \underline{O}(n)X$. This amounts to using a Gram-Schmidt orthogonalization process on $X$. If $X_T \in \underline{T}(n)$ and $A \in \underline{O}(n)$ we let $f(X_T, A) \in \underline{T}(n)$ represent the coset $\underline{O}(n)X_T A$. Then

$$(3.3.17) \qquad \underline{O}(n)XY = \underline{O}(n)X_T A Y_T = \underline{O}(n) f(X_T, A) Y_T \text{ and } f(X_T, A)Y_T \in \underline{T}(n)$$

represents the coset $\underline{O}(n)XY$. By invariance the induced measure satisfies

$$(3.3.18) \qquad \overline{\mu}(\{T \mid T \in U\}) = \overline{\mu}(\{T \mid f(T,A) \in U\}).$$

This property may be verified directly from the differential form for the invariant measure $\overline{\mu}$.

In the next section we need the following Lemma.

Lemma 3.3.10. Suppose $H$ is a compact subgroup of $\mathfrak{G}$, $\mu$ is a non-zero left invariant Haar measure for $\mathfrak{G}$, and $\overline{\mu}$ is the induced measure on $\mathfrak{G}/H$. Let $\overline{f}: \mathfrak{G}/H \to \mathbb{R}$ be measurable and $f: \mathfrak{G} \to \mathbb{R}$ be defined by $\overline{f}(\pi(x)) = f(x)$. Then

$$(3.3.19) \qquad \int f(x)\mu(dx) = \int \bar{f}(\bar{x})\bar{\mu}(d\bar{x}) = \int \bar{f}(\pi(x))\mu(dx).$$

Proof. If $\bar{A} \subset \mathcal{G}/H$ is a Baire set and $\bar{\mu}(\bar{A}) < \infty$ then

$$(3.3.20) \qquad \int \chi_{\bar{A}}(\bar{x})\bar{\mu}(d\bar{x}) = \bar{\mu}(\bar{A}) = \mu(\pi^{-1}(\bar{A})) = \int \chi_{\pi^{-1}(\bar{A})}(x)\mu(dx).$$

Hence, for any simple functions $f$ and $\bar{f}$, it follows that (3.3.19) holds. By monotone convergence it then follows at once that (3.3.19) holds for any integrable $\bar{f}$ and corresponding $f$. #

Remark 3.3.11. If $\mathfrak{M}$ is a manifold on which $\mathcal{G}$ acts and if $\mathfrak{M}$ admits an invariant nonzero (positive) invariant measure $\mu$, then the induced measure $\bar{\mu}$ on the orbit space $\mathfrak{M}/\mathcal{G}$ satisfies (3.3.19).

## 3.4. Factorization of measures.

The results needed in these notes are for locally compact spaces with countable base for the open sets. These spaces are metrizable and every compact subset of a metric space is both a Baire set and a Borel set. In this section we suppose $H_1$ and $H_2$ are locally compact metric spaces with countable base and $H = H_1 \times H_2$, the Cartesian product in the product topology. Let the group $\mathcal{G}$ act as transformations on $H_1$ and define $\bar{\mathcal{G}}$ on $H$ by

$$(3.4.1) \qquad \text{if } g \in \mathcal{G}, \text{ then } \bar{g}(h_1,h_2) = (gh_1,h_2).$$

Then $\mathcal{G}$ and $\bar{\mathcal{G}}$ are "canonically" isomorphic.

In the sequel we will say left invariant Baire measures on $H_1$ satisfy the uniqueness property if such measures form a one-dimensional space.

Theorem 3.4.1. Suppose $H_1$ and $H_2$ are locally compact Hausdorff spaces with countable bases and $H = H_1 \times H_2$. Let $\mathcal{G}$ and $\bar{\mathcal{G}}$ be as above. Let $\mu \neq 0$ be a regular Borel measure for $H$ and assume $\mu$ is left invariant under the actions of $\bar{\mathcal{B}}$. Suppose left invariant regular Borel measures for $H_1$ have the uniqueness property. Then there exist regular Borel measures $\nu_1$ and $\nu_2$ defined on the Borel subsets of $H_1$ and $H_2$ respectively such that $\nu_1$ is left invariant under the actions of $\mathcal{B}$ and such that if $A \subset H_1$ and $B \subset H_2$ are Borel subsets then

$$(3.4.2) \qquad \mu(A \times B) = \nu_1(A)\nu_2(B) .$$

<u>Proof</u>. Since $\mu \neq 0$ there must exist a compact set $B \subset H_2$ and an open set $A_0$ with compact closure in $H_1$ for which $0 < \mu(A_0 \times B) < \infty$. For by Fubini's Theorem, if $\mu$ vanishes for all such compact sets then $\mu$ vanishes identically. Define $\mu_B$ by $\mu_B(A) = \mu(A \times B)$. Then $\mu_B$ is a nonnegative countably additive measure on the Borel subsets of $H_1$ such that $\mu_B(A) < \infty$ for every compact subset $A$ of $H_1$. Since $H_1$ has a countable base, the Borel and Baire sets coincide, so it follows that $\mu_B$ is a regular Borel measure. Further, if $g \in \mathcal{B}$ then

$$(3.4.3) \qquad \mu_B(gA) = \mu((gA) \times B) = \mu(\bar{g}(A \times B)) = \mu(A \times B),$$

so that $\mu_B$ is a regular left invariant Borel measure not identically zero. Pick a nonzero regular left invariant measure $\nu_1$ for $H_1$ as a reference measure. Then we may use the uniqueness hypothesis to define a function $f$ on the Borel subsets of $H_2$ by the definition

$$(3.4.4) \qquad f(B)\nu_1(A) = \mu_B(A) = \mu(A \times B).$$

The equation (3.4.4) clearly defines a nonnegative countably additive function $f$ on the Borel subsets of $H_2$ and $f(B) < \infty$ for compact sets $B$. Therefore $\nu_2 = f$ is a regular measure on the Borel subsets of $H_2$. #

Example 3.4.2. We continue Example 3.3.9. Justification of the use of differential forms will be made later.

$X \in GL(n)$ has a decomposition $X = AT$ with $A \in \underline{O}(n)$ and $T \in \underline{T}(n)$, by use of a Gram Schmidt orthogonalization process. Consequently a nonzero Haar measure $\mu$ on $GL(n)$ (using both right and left invariance) factors into $\mu = \nu_1\nu_2$ with $\nu_1$ a left invariant Haar measure on $\underline{O}(n)$ and $\nu_2$ a right invariant Haar measure on $\underline{T}(n)$. As is well known (and discussed in Section 3.5) $\underline{O}(n)$ is a unimodular group and the right and left invariant Haar measures are the same. On the other hand, the group $\underline{T}(n)$ is not unimodular and the measure $\nu_2$ is not left invariant.

In terms of differential forms this becomes the following.

$$(3.4.5) \qquad X = UT \quad \text{and} \quad dX = (dU)T + U(dT);$$

$$U^t dX\, T^{-1} = U^t dU + (dT)T^{-1}.$$

By computation of a wedge product for the individual matrix entries

$$(3.4.6) \qquad \wedge_{i=1}^n \wedge_{j=1}^n (U^t dXT^{-1})_{ij} = \epsilon_1 \bigwedge_{i<j} (u_i^t du_j) \bigwedge_{j\le i} (dT\, T^{-1})_{ij},$$

where $u_i$ is the i-th column of $U$ and $\epsilon_1$ is a real number in absolute value 1. The left side of (3.4.6) is

$$(3.4.7) \qquad (\det U)^n (\det T^{-1})^n \bigwedge_{i=1}^n \bigwedge_{j=1}^n d\, x_{ij},\ X = (x_{ij}),$$

and

$$(3.4.8) \qquad \det X = (\det T)(\det U^{-1}).$$

Therefore

$$(3.4.9) \qquad \frac{\bigwedge_{i=1}^n \bigwedge_{j=1}^n dx_{ij}}{(\det X)^n} = \epsilon_1 \bigwedge_{i<j} (u_i^t\, du_j) \bigwedge_{j\le i} (dT\, T^{-1})_{ij}.$$

The differential forms in (3.4.9) are differential forms for invariant measures on $GL(n)$, $\underline{O}(n)$ and $\underline{T}(n)$ respectively. We do not compute the correct sign as it is not needed in the integrations. Subsequent integrations will use unsigned measures so the absolute value of the differential forms will be used. #

Example 3.4.3. $H$ will be the set of $n \times k$ matrices with real entries such that the $k \times k$ minor consisting of the first $k$ rows, called $X_1$, is nonsingular. We set $H_1 = GL(k)$ and $H_2$ the set of $(n-k) \times k$ matrices, so that the decomposition becomes

$$(3.4.10) \qquad X = YG,\ G \in GL(k) \text{ and } Y^t = (I_k, Z^t),$$

with $I_k$ the $k \times k$ identity matrix and $Z$ a $(n-k) \times k$ matrix. The group $GL(k)$ acts on the right and the regular right invariant measure $\mu$ factors to $\mu = \nu_1\nu_2$ where $\nu_2$ is a Haar measure on $GL(k)$.

In terms of differential forms we have

$$(3.4.11) \qquad X = YG,\ dX = (dY)G + Y(dG)$$

$$(\det G^{-1})^n \bigwedge_{i=1}^{n} \bigwedge_{j=1}^{k} d\,x_{ij} = \bigwedge_{i=1}^{n} \bigwedge_{j=1}^{k} (dY + Y\,dG\,G^{-1})_{ij}$$

$$= \epsilon_2 \bigwedge_{i=1}^{k} \bigwedge_{j=1}^{k} (dG\,G^{-1})_{ij} \bigwedge_{i=k+1}^{n} \bigwedge_{j=1}^{k} dz_{ij}\ .$$

In order to obtain an invariant measure on the left side of (3.4.11) the required normalization is

$$(3.4.12) \qquad (\det G^{-1})^n (\det(I_k + Z^tZ))^{-n/2} = (\det X^tX)^{-n/2}, \text{ so that}$$

$$(3.4.13) \qquad (\det X^tX)^{-n/2} \bigwedge_{i=1}^{n} \bigwedge_{j=1}^{n} d\,x_{ij}$$

$$= \epsilon_2 \frac{\bigwedge_{i=k+1}^{n} \bigwedge_{j=1}^{k} dz_{ij}}{(\det(I_k + Z^tZ))^{n/2}} \bigwedge_{i=1}^{k} \bigwedge_{j=1}^{k} (dG\,G^{-1})_{ij}\ .\ \#$$

## 3.5. Modular functions.

If $\mu$ is a left invariant and regular measure for a locally compact group $\mathcal{G}$ and $g \in \mathcal{G}$ then the measure $\nu$ defined by

$$(3.5.1) \qquad \nu(A) = \mu(Ag)$$

is a nonzero left invariant regular measure defined on the Borel subsets of $\mathcal{G}$. By the uniqueness theorem, Theorem 3.3.2, there exists a nonzero constant $m(g)$ such that if $A$ is a Borel subset of $\mathcal{G}$ then

$$(3.5.2) \qquad \nu(A) = m(g)\mu(A) = \mu(Ag) .$$

The function $m$ is called the modular function. This section is used to state and prove a few needed properties of modular functions. A Haar measure and the associated group are said to be unimodular if the modular function is identically one. Otherwise we say the measure and or the group are not unimodular.

Lemma 3.5.1. Let $f$ be a continuous function with compact support $C$, $f: \mathcal{G} \to \mathbb{R}$, where $\mathcal{G}$ is a locally compact group. Given $\epsilon > 0$ there exists a compact neighborhood $U$ of $e$, the identity of $\mathcal{G}$, such that

$$(3.5.3) \qquad \sup_{x \in \mathcal{G}} \sup_{g \in U} |f(gx) - f(x)| < \epsilon .$$

Proof. Suppose the contrary is true. To every compact neighborhood $U$ of $e$, $\{(g,x) \mid |f(gx) - f(x)| \geq \epsilon$ and $g \in \overline{U}\}$ is a compact set contained in the compact set $\overline{U} \times (\overline{U}^{-1} \cdot C)$. These sets clearly have the finite intersection property. Hence there is a point $(g,x)$ in the intersection of all these sets and this point must have $g = e$. Thus $f(x) - f(x) \neq 0$. This contradiction shows that the required compact neighborhood of $e$ must exist. #

<u>Remark 3.5.2</u>. This Lemma may be used to generalize the proofs of several earlier results in which the existence of a countable base was assumed.

<u>Lemma 3.5.3</u>. The modular function $m$ is a continuous group homomorphism of $\mathcal{G}$ to $(0,\infty)$. If $f: \mathcal{G} \to \mathbb{R}$ is a $\mu$-integrable function then

$$\int f(hg^{-1})\mu(dh) = m(g) \int f(h)\mu(dh) . \tag{3.5.4}$$

If $\mathcal{G}$ is compact then $m(g_1) = m(g_2) = 1$ for all $g_1, g_2 \in \mathcal{G}$ .

<u>Proof</u>. Choose a Baire set $A$ such that $0 < \mu(A) < \infty$. Then $m(g_1g_2)\mu(A) = \mu(A(g_1g_2)) = \mu((Ag_1)g_2) = m(g_1)m(g_2)\mu(A)$. This implies $m(g_1)m(g_2) = m(g_1g_2)$.

We now prove continuity of the function $m$. Let $f$ be a continuous function with compact support, $f: \mathcal{G} \to \mathbb{R}$. By Lemma 3.5.1 it follows that since $\mu$ restricted to compact sets is a totally finite measure,

$$(\lim_{g \to e} m(g)) \int f(h)\mu(dh) = \int f(h)\mu(dh) . \tag{3.5.5}$$

We have used here (3.5.4) which is proven in the next paragraph. As is well known, continuity of a group homomorphism at $e$ implies continuity at all points.

To establish (3.5.4) we use a standard approximation argument. If $\chi_A$ is the indicator function of the Borel subset $A$ then

$$m(g)\mu(A) = \int \chi_{Ag}(h)\mu(dh) = \int \chi_A(hg^{-1})\mu(dh). \tag{3.5.6}$$

By linearity it follows that (3.5.4) holds for all simple functions and by monotone convergence it follows that (3.5.4) holds for all $\mu$-integrable functions $f$.

If $\mathcal{G}$ is a compact group then the function $f(g) = 1$ for all $g \in \mathcal{G}$ is integrable and $0 < \mu(\mathcal{G}) < \infty$. By (3.5.4) it follows that $m(g) = 1$, $g \in \mathcal{G}$. #

Lemma 3.5.4. If the group $\mathcal{G}$ is Abelian then the modular function is identically equal one.

Proof. $m(g)\mu(A) = \mu(Ag) = \mu(gA) = \mu(A)$. #

Example 3.5.5. The invariant measures on $GL(n)$ given by the differential forms $\bigwedge_{i=1}^{n} \bigwedge_{j=1}^{n} (G^{-1}dG)_{ij}$ and $\bigwedge_{i=1}^{n} \bigwedge_{j=1}^{n} ((dG)G^{-1})_{ij}$ are the same since both differential forms compute to be $\bigwedge_{i=1}^{n} \bigwedge_{j=1}^{n} (dG)_{ij}/(\det G)^n$. Thus the Haar measures on $GL(n)$ are unimodular. By Lemma 3.5.3, the Haar measures on $\underline{O}(n)$ are unimodular. If $\underline{D}(n)$ is the group of $n \times n$ diagonal matrices with positive diagonal entries, then $\underline{D}(n)$ is Abelian and by Lemma 3.5.4 the Haar measures are unimodular. The differential forms are, if $L \in \underline{D}(n)$ and $L = \begin{pmatrix} \lambda_1 & \cdots & 0 \\ \vdots & \ddots & \vdots \\ 0 & & \lambda_n \end{pmatrix}$, then

$$\bigwedge_{i=1}^{n} (L^{-1}dL)_{ii} = \bigwedge_{i=1}^{n} (d\lambda_i/\lambda_i) = \bigwedge_{i=1}^{n} ((dL)L^{-1})_{ii},$$

verifying unimodularity in this case. $\underline{T}(n)$ is not unimodular.

Problem 3.5.6. Compute the modular function of a left invariant Haar measure on $\underline{T}(n)$.

## 3.6. A remark on matrix groups.

Some of the matrix groups of interest to us are named in Example 3.5.5. In all these examples the group operation is matrix multiplication. Given $G, H \in \mathcal{G}$, then $(GH)_{ij} = \Sigma_{k=1}^{n} G_{ik}H_{kj}$ so under the topology of coordinatewise convergence the map $(G,H) \to GH$ is clearly jointly continuous. The entries of $G^{-1}$ are rational functions of the entries of $G$ so the map $G \to G^{-1}$ is continuous in the above topology. Since the coordinatewise convergence topology is induced by embedding $\mathcal{G}$ in an Euclidean space the topology is a locally compact topology. The topology is clearly a metric topology.

In the case of each group we specify a set $I$ which determines a maximal set of local coordinates. Then if $G = (g_{ij})$ the differential form $\bigwedge_{(i,j)\in I} d\, g_{ij}$ has maximal degree and $\bigwedge_{(i,j)\in I} (G^{-1}dG)_{ij}$ gives the differential form of a left invariant (regular) measure, i.e., a Haar measure. See Example 3.5.5.

## 3.7. Cross-sections.

Suppose $\mathcal{G}$ acts as a transformation group on a manifold $H$. Then $H$ may be factored via $\mathcal{G}$ into the space $H/\mathcal{G}$ of orbits. The problem of cross-sections is to choose from each orbit $\mathcal{G}h$ a single element in some "nice" way. We let $H_1$ be the subset of chosen elements. $H_1$ is to be topologized in such a way that $\mathcal{G} \times H_1$ is measure isomorphic, or homeomorphic, or diffeomorphic to $H$, depending on the problem.

The construction of cross-sections and their use in factoring measures in the sense of Section 3.4 has been discussed by Wijsman (1966) and Koehn (1970). We do not pursue the abstract treatment of cross-sections since in the examples considered in this book there are obvious choices of cross-sections $H_1$.

Example 3.7.1. The action of $GL(k)$ on the set $H_0$ of $n \times k$ matrices given by multiplication on the right was discussed in Example 3.4.3. We let $N$ be the subset of $H_0$ consisting of those $n \times k$ matrices having a linearly dependent set of first $k$ rows. Then $N$ is clearly an invariant subset under the action of $\mathcal{G}$, hence so is $H = H_0 - N$. If $X \in H$ we may use the factorization

$$X^t = G^t(I_k, Z^t) , \qquad (3.7.1)$$

with the notation the same as in Example 3.4.3. The matrix $(I_k, Z^t)^t$ is then the choice of a matrix in the orbit $X$. We call this set $H_1$ and give it the usual Euclidean topology. Then $H_1$ is a Borel subset of $H$ and has a locally compact topology. Note that $H_1$ has nk-dimensional Lebesgue measure zero. #

As is illustrated in Example 3.7.1, in general an invariant null set will be constructed having zero measure relative to a given measure which is left (or right) invariant. Thus $\mu$ is also an invariant measure on $H_0 - N = H$. The residual set $H$ will factor in an obvious way to

$$H = H_0 - N \cong \mathfrak{H} \times H_1 . \tag{3.7.2}$$

By the results of Section 3.4 the measure $\mu$ then factors to $\mu = \nu_1\nu_2$ with $\nu_1$ a left invariant measure for $\mathfrak{H}$, and in cases where it is meaningful, $\nu_2$ a right invariant measure on $H_1$. These factors are regular measures.

Example 3.7.2. The process of factoring, described above, should be compared with inducing a measure on the quotient space $\mathfrak{H}/H$ in the sense discussed in Sections 3.2 and 3.3. Consider $\mathfrak{H} = \mathbb{R} \times \mathbb{R}$ and $H = \{0\} \times \mathbb{R}$, $\mu$ = Lebesgue measure on $\mathfrak{H}$. The induced measure is infinite on all open subsets of $\mathfrak{H}/H$ while the factored measure is one-dimensional Lebesgue measure. #

## 3.8. Solvability.

This section has been included because of its relevance to use of the Hunt-Stein theory as presented in Lehmann (1959). As a notation in this section $A \oplus B$ means the symmetric difference of sets defined by

$$A \oplus B = (A-B) \cup (B-A). \tag{3.8.1}$$

The hypotheses of the Hunt-Stein theory required the existence of a sequence $\{U_n, n \geq 1\}$ of Borel subsets of the group $\mathcal{G}$ satisfying if $n \geq 1$ then $U_n \subset U_{n+1}$, $0 < \mu(U_n) < \infty$, and $\lim_{n\to\infty} U_n = \mathcal{G}$, where $\mu$ is a right invariant Haar measure on the group $\mathcal{G}$. In addition the following condition must hold.

(3.8.2) If $g \in \mathcal{G}$ then $\lim_{n\to\infty} \mu(U_n g \oplus U_n)/\mu(U_n) = 0.$

Example 3.8.1. $\mathcal{G} = \mathbb{R}$, $U_n = [-n,n]$, $\mu$ = Lebesgue measure. Then if $g \geq 0$, $(U_n + g) \oplus U_n = [-n,-n+g) \cup (n,n+g]$ and this set has Lebesgue measure $2g$. Since $\lim_{n\to\infty} 2g/2n = 0$, condition (3.8.2) holds. #

If the group $\mathcal{G}$ acts on $\mathfrak{X}$, the point set of a measure space, such that the map $(g,x) \to g(x)$ is jointly measurable, then it is shown in Lehmann, op. cit., that for a bounded measurable function $\varphi$

(3.8.3) $$\lim_{n\to\infty} \frac{\int_{U_n} \varphi(gx)\mu(dg)}{\mu(U_n)} = \psi(x)$$

exists as a weak limit in $L_\infty$ of $\mathfrak{X}$ and $\psi$ is a $\mathcal{G}$ invariant function.

This process may be carried out in stages for groups $\mathcal{G}$ with the following structure. Suppose $\mathcal{G}$ has subgroups $\mathcal{G}_1$ and $\mathcal{G}_2$ such that $\mathcal{G}_1 \cap \mathcal{G}_2 = \{e\}$ and $\mathcal{G}_1$ is a normal subgroup. If $\varphi$ is a $\mathcal{G}_1$ invariant function and $g \in \mathcal{G}_2$ then $\varphi_1$ defined by $\varphi_1(x) = \varphi(gx)$ is again a $\mathcal{G}_1$ invariant function. For $\varphi_1(g_1x) = \varphi(gg_1x) = \varphi(g_1'gx) = \varphi_1(x)$. In this argument we use the assumption that $\mathcal{G}_1$ is a normal subgroup. It then follows that if $\mu$ is a Borel measure for $\mathcal{G}_2$ such that $\int_{\mathcal{G}_2} \varphi(gx)\mu(dg)$ is meaningful, one expects this to also be a $\mathcal{G}_1$ invariant function. Thus if (3.8.2) holds for $\mathcal{G}_2$ also, a limit (3.8.3) may be taken to generate a $\mathcal{G}_2$ invariant function that is also $\mathcal{G}_1$, hence because of normality, $\mathcal{G}$ invariant. The argument for this last step requires further specification of the measures for $\mathcal{X}$. See Lehmann, op.cit.

Example 3.8.2. Let the group $\mathcal{G} = \underline{T}(n)$ and let $\underline{T}_1(n)$ be the subgroup of those matrices with diagonal elements equal 1. Then $\underline{T}_1(n)$ is a normal subgroup and

$$(3.8.4) \qquad \underline{T}(n) = \underline{T}_1(n) \cdot \underline{D}(n).$$

A decomposition of $\underline{T}_1(n)$ is given by a chain

$$(3.8.5) \qquad \underline{T}_1(n) = \mathcal{G}_n \supset \mathcal{G}_{n-1} \supset \dots \supset \mathcal{G}_1$$

together with subgroups $\mathcal{H}_i$, $2 \leq i \leq n$ such that $\mathcal{H}_i$ is isomorphic to the additive group of $\mathbb{R}_{i-1}$. Then

$$(3.8.6) \qquad \mathcal{G}_i = \mathcal{H}_1 \dots \mathcal{H}_i, \; 1 \leq i \leq n ,$$

and $\mathcal{H}_i$ is a normal subgroup of $\mathcal{G}_i$, $2 \leq i \leq n$. This is easily seen from the identity

$$(3.8.7) \qquad \begin{pmatrix} I & 0 \\ a^t & 1 \end{pmatrix}\begin{pmatrix} T & 0 \\ 0 & 1 \end{pmatrix}\begin{pmatrix} I & 0 \\ b^t & 1 \end{pmatrix}\begin{pmatrix} T^{-1} & 0 \\ 0 & 1 \end{pmatrix}\begin{pmatrix} I & 0 \\ -a^t & 1 \end{pmatrix}$$

$$= \begin{pmatrix} I & 0 \\ b^t T^{-1} & 1 \end{pmatrix} . \; \#$$

Example 3.8.3. Let $\mathcal{G}$ be the set of $2 \times 2$ matrices $\begin{pmatrix} a & 0 \\ b & 1 \end{pmatrix}$ with subgroups $\mathcal{G}_1$ the matrices $\begin{pmatrix} 1 & 0 \\ b & 1 \end{pmatrix}$ and $\mathcal{G}_2$ the matrices $\begin{pmatrix} a & 0 \\ 0 & 1 \end{pmatrix}$. Then $\mathcal{G}_1$ is a normal subgroup and (3.8.2) may be verified for $\mathcal{G}_1$ as in Example 3.8.1. For (3.8.2) in the case of $\mathcal{G}_2$ we let $U_n^{(2)}$ be those matrices with $1/n \leq a \leq n$ and use the Haar measure for the multiplicative group on $(0,\infty)$. One may compute that

$$(3.8.8) \qquad \begin{pmatrix} a & 0 \\ 0 & 1 \end{pmatrix}\begin{pmatrix} 1 & 0 \\ b & 1 \end{pmatrix}\begin{pmatrix} a^{-1} & 0 \\ 0 & 1 \end{pmatrix} = \begin{pmatrix} 1 & 0 \\ a^{-1}b & 1 \end{pmatrix} .$$

Then if $g_2 = \begin{pmatrix} a & 0 \\ 0 & 1 \end{pmatrix}$ and $a \neq 1$, then

$$(3.8.9) \qquad \lim_{n\to\infty} \mu_1(g_2^{-1}U_n^{(1)}g_2 \cap U_n^{(1)})/\mu_1(U_n^{(1)}) \neq 1.$$

Here $U_n^{(1)} = [-n,n]$. #

Example 3.8.4. Another decomposition of $\underline{T}(n)$ results if the multiplicative part is taken out in $n$ steps rather than simultaneously. In this decomposition the alternation of normal subgroups is more complex than in Example 3.8.2. We have

$$(3.8.10) \qquad \underline{T}(n) = \mathcal{B}_{2n-1} \supset \mathcal{B}_{2n-2} \supset \ldots \supset \mathcal{B}_1$$

together with subgroups $\mathcal{H}_i$, $1 \leq i \leq 2n-1$, such that if $i = 2j$ then $\mathcal{H}_i$ is isomorphic to $\mathbb{R}_j$ while if $i = 2j-1$ then $\mathcal{H}_i$ is isomorphic to the multiplicative group on $(0,\infty)$. Then

$$(3.8.11) \qquad \mathcal{B}_i = \mathcal{H}_1 \cdot \ldots \cdot \mathcal{H}_i, \ 1 \leq i \leq 2n-1.$$

In this decomposition, at the odd numbered steps, when a multiplicative subgroup is factored, the normal subgroup is $\mathcal{B}_{2j}$, while at the even numbered steps when an additive subgroup is factored, the normal subgroup is the subgroup $\mathcal{H}_{2j-1}$.

# Chapter 4. Wishart's Paper

## 4.0. Introduction

Some of the earlier writers, notably J. Wishart and R.A. Fisher, made extensive use of geometric reasoning in order to compute the volume elements that arise from changes of variable. We have selected the paper by Wishart (1928) to illustrate this type of argument. The Wishart density function so obtained plays a central role in the discussions which follow in later chapters.

With minor changes of notation Wishart writes as follows. $\underline{X}$ is a random $n \times h$ matrix having joint density function $f(X^tX)$ relative to Lebesgue measure $dx_{11} \cdots dx_{nh}$ on nh-dimensional space. One wishes to find the density function of $\underline{X}^t\underline{X}$ expressed in terms of Lebesgue measure on the $h(h+1)/2$ - dimensional space of the variables $(\xi_{ij}) = X^tX$. Since $f$ is already a function of the new variables the change of variable followed by integration out of the extra variables introduces a multiplying factor (i.e. volume element) and the desired density function has the form

$$g(X^tX)f(X^tX) \prod_{j \le i} d\xi_{ij} \quad . \tag{4.0.1}$$

The function $g$ is independent of the function $f$ so that by computing $g$ in one case we solve the problem of computing (4.0.1) in all similar problems. This fact was noted and used by James (1955a). We will return to a discussion of James (1955a) in Chapter 5. Section 4.2 states a number of problems.

## 4.1. Wishart's argument.

Wishart thought of the problem as follows. $X_1, \ldots, X_h$ are the columns of $X$, are vectors in $\mathbb{R}_n$. We suppose $X_1, \ldots, X_h$ are linearly independent and leave the proof of this as Problem 4.2.4 at the end of this chapter. Let $S_h$ be the linear span of

$X_1,\ldots,X_h$ and let $Y_1,\ldots,Y_h$ be an orthonormal basis of $S_h$. Then we write, using a Gram-Schmidt orthogonalization,

$$X_i = a_{i1}Y_1 + \ldots + a_{ii}Y_i \,, \quad 1 \leq i \leq h \,, \tag{4.1.1}$$

and setting $a_i^t = (a_{i1},\ldots,a_{ii})$ we compute

$$X_i^t X_j = a_i^t a_j \,, \quad 1 \leq i,j \leq h \,. \tag{4.1.2}$$

As is well known the h-dimensional parallelopiped with sides $X_1,\ldots,X_h$ has volume $V_h$ given by

$$(\det X^t X)^{1/2} = (\det(a_i^t a_j))^{1/2} = V_h \,. \tag{4.1.3}$$

More generally if $1 \leq i \leq h$ we will write $V_i$ for the volume of the i-dimensional parallelopiped determined by $X_1,\ldots,X_i$. Then the length of the perpendicular projection of $X_h$ on the span of $X_1,\ldots,X_{h-1}$ is $V_h/V_{h-1}$.

In h-dimensional space introduce new coordinates for $X_h$ defined by considering $X_1,\ldots,X_{h-1}$ as fixed and setting

$$\xi_{hi} = a_h^t a_i = \Sigma_{j=1}^h a_{hj} a_{ij} \,. \tag{4.1.4}$$

The Jacobian of this transformation is easily computed to be

$$\frac{\partial(\xi_{h1},\ldots,\xi_{hh})}{\partial(a_{h1},\ldots,a_{hh})} = \begin{vmatrix} a_{11} & 0 & \ldots & 0 \\ a_{21} & a_{22} & \ldots & 0 \\ & & & \\ 2a_{h1} & 2a_{h2} & \ldots & 2a_{hh} \end{vmatrix} = 2V_h \,. \tag{4.1.5}$$

We now discuss the problem of integrating out extra variables. At the first step we consider $X_1,\ldots,X_{h-1}$ as fixed and introduce new variables $\xi_h^t = (\xi_{h1},\ldots,\xi_{hh})$. In addition let us write

$$\xi_i^t = (\xi_{i1},\ldots,\xi_{ih}) \,, \quad 1 \leq i \leq h \,. \tag{4.1.6}$$

If we fix $\xi_1,\ldots,\xi_h$ then we fix the lengths of $X_1,\ldots,X_h$ and the angles between them. In particular if $X_h'$ and $X_h$ are two possible choices of the h-th vector then

$$(4.1.7) \qquad X_i^t(X_h' - X_h) = 0 , \qquad 1 \leq i \leq h-1 .$$

Therefore $X_h' - X_h$ is orthogonal to the span of $X_1,\dots,X_{h-1}$. Further, any such $X_h'$ when adjoined to $X_1,\dots,X_{h-1}$ gives rise to a parallelopiped of volume $V_h$ since this volume is given by (4.1.3). If $S_{h-1}$ is the linear span of $X_1,\dots,X_{h-1}$ and $S_{h-1}^{\perp}$ is its orthogonal complement in n-dimensional space, if $S_{h-1,\lambda}^{\perp}$ are those vectors in $S_{h-1}^{\perp}$ of length $\lambda$, where we take $\lambda = V_h/V_{h-1}$, then for fixed $X_1,\dots,X_{h-1}$, the set of $X_h'$ having coordinates $\xi_{h1},\dots,\xi_{hh}$ are just the sum of $a_{h1}Y_1+\dots+a_{h\,h-1}Y_{h-1}$ depending on $X_1,\dots,X_{h-1}$ and the vectors in $S_{h-1,\lambda}^{\perp}$. Thus the required volume element is the surface area of this sphere. Note that $a_{hh} = V_h/V_{h-1}$.

The required surface area is given by

$$(4.1.8) \qquad (2\pi)^{(n-h+1)/2}(\Gamma((n-h+1)/2))^{-1}(V_h/V_{h-1})^{n-h} .$$

Derivation of this formula is given as Problem 4.2.3. Therefore the multiplier introduced by the change of variables (4.1.4) followed by integration is

$$(4.1.9) \qquad \frac{2\pi^{(n-h+1)/2}}{\Gamma((n-h+1)/2)}\left(\frac{V_h}{V_{h-1}}\right)^{n-h}\frac{1}{2V_h} = \frac{\pi^{(n-h+1)/2}V_h^{n-h-1}}{\Gamma((n-h+1)/2)V_{h-1}^{n-h}} .$$

Having made the change of variable for $X_h$ we proceed inductively for $X_{h-1},\dots,X_1$ and obtain as an end result

$$(4.1.10) \qquad \frac{\pi^{n/2}\cdots\pi^{(n-h+1)/2}}{\Gamma(\frac{n}{2})\cdots\Gamma(\frac{n-h+1}{2})}\,\frac{V_h^{n-(h+1)}}{V_{h-1}^{n-h}}\cdot\frac{V_{h-1}^{n-h}}{V_{h-2}^{n-(h-1)}}\cdots\frac{V_1^{n-2}}{V_0^{n-1}}\prod_{j\leq i} d\xi_{ij} .$$

<u>Theorem 4.1.1</u> (Wishart). Let the random $n\times h$ matrix $\underline{X}$ have a density function $f(X^tX)$ relative to Lebesgue measure. Then the

joint density function of the $h(h+1)/2$ random variables $\underline{X}^t\underline{X}$ is

$$(4.1.11)\qquad \frac{\pi^{hn/2 - h(n-1)/4}}{\Gamma(\frac{n}{2})\cdots\Gamma(\frac{n-h+1}{2})} |X^tX|^{(n-h-1)/2} f(X^tX) \prod_{j\leq i} d(X_i^tX_j) .$$

4.2. Related problems.

In the sequel we present as a sequence of problems part of the above calculation. Write $V_n(r)$ as the volume of the set $\{(x_1,\ldots,x_n) | x_1^2 + \ldots + x_n^2 \leq r^2\}$ and $A_{n-1}(r)$ as the surface area of the sphere $\{(x_1,\ldots,x_n) | x_1^2 + \ldots + x_n^2 = r^2\}$. Required in the first step, (4.1.8), was the quantity $A_{n-h-1}(V_h/V_{h-1})$ .

Problem 4.2.1. Show that

$$(4.2.1)\qquad V_n(r) = 2^n \int_{\substack{x_1\geq 0,\ldots,x_{n-1}\geq 0 \\ x_1^2+\ldots+x_{n-1}^2 \leq r^2}} \cdots\cdots \int (r^2-(x_1^2 + \ldots + x_{n-1}^2))^{1/2} dx_1\ldots dx_{n-1}$$

$$= 2^n r^n \prod_{i=1}^{n-1} \int_0^1 (1-y^2)^{i/2} dy \quad .$$

Hint. Make successive changes of variable

$$(4.2.2)\qquad x_j = (r^2-(x_1^2 + \ldots + x_{j-1}^2))^{1/2}, \quad 2\leq j\leq n \quad . \#$$

Problem 4.2.2. Show that

$$(4.2.3)\qquad \int_0^1 (1-y^2)^{i/2}\, dy = \frac{1}{2}\int_0^1 t^{i/2}(1-t)^{-1/2} dt$$

$$= \frac{1}{2}\Gamma((i/2)+1)\Gamma(\tfrac{1}{2}) / \Gamma((i/2)+(3/2)) \ .$$

Hint. Make the change of variable $t = 1 - y^2$ . The result is a Beta integral. #

Problem 4.2.3. Show that

$$(4.2.4)\qquad V_n(r) = r^n\pi^{n/2} / \Gamma(1 + (n/2)) \ .$$

Show the area function satisfies

$$(4.2.5) \qquad A_{n-1}(r) = (d/dr)V_n(r) = 2\, r^{n-1}\pi^{n/2}/\Gamma(n/2)\ .$$

A side issue to Wishart's calculation is the question why $A_{n-1}(r)$ is the right quantity to use. We do not deal with this question. #

The calculation as presented assumes $\underline{X}_1,\ldots,\underline{X}_n$ are linearly independent except on a set of Lebesgue measure zero. We develop several such results here as problems.

Problem 4.2.4. Let $\mu_1,\ldots,\mu_n$ be regular $\sigma$-finite Borel measures on $\mathbb{R}_n$ each of which gives zero mass to proper hyperplanes of $\mathbb{R}_n$. Then

$$(4.2.6) \qquad 0 = (\mu_1 x \ldots x \mu_n)(\{(x_1,\ldots,x_n) \mid x_1,\ldots,x_n \in \mathbb{R}_n,\ \det(x_1,\ldots,x_n) = 0\}).$$

Hint. Fix $x_1,\ldots,x_{n-1}$ and expand the determinant on the last column. The equation (4.2.6) then defines a hyperplane in $\mathbb{R}_n$ which has $\mu_n$ measure zero. Use Fubini's Theorem. #

Problem 4.2.5. Continuation of Problem 4.2.4. Let $h \leq n$. If $x_1,\ldots,x_h$ are dependent with positive $\mu_1 x \ldots x \mu_h$ measure then $\det(x_1,\ldots,x_n) = 0$ with positive $\mu_1 x \ldots x \mu_n$ measure.

Problem 4.2.6. Continuation of Problem 4.2.4. The event that $\sum_{i=1}^{q} x_i x_i^t$ have a double eigenvalue $\lambda$ such that $\lambda \neq 0$ has $\mu_1 x \ldots x \mu_q$ measure zero.

Hint. By induction on $q$. The case $q = 1$ is the empty event. If $\lambda$ is a (random) double eigenvalue then $\sum_{i=1}^{q-1} x_i x_i^t - \lambda I_n$ is a singular $n \times n$ matrix. Choose $U \in O(n)$ depending measurably on $x_1,\ldots,x_{q-1}$ so that

$$(4.2.7) \qquad U\left(\sum_{i=1}^{q-1} x_i x_i^t\right) U^t = D$$

is an $n \times n$ diagonal matrix. Then in the (random) $(i,i)$-position on the diagonal of $D$ occurs $\lambda$. One of the following two cases hold. Show that each has zero measure.

Case 1. $\sum_{i=1}^{q-1} x_i x_i^t$ has $\lambda$ as a double eigenvalue.

Case 2. If Case 1 fails to hold then for some eigenvector $a$ for $\lambda$, $a^t U x_q \neq 0$. From (4.2.7) it then follows that the $i$ th entry of $Ux_q = 0$, so that the event in question is a subset of $\bigcup_{i=1}^{n} \{x \mid (Ux)_i = 0\}$. By hypothesis on $\mu_n$, this set has zero measure. Use Fubini's Theorem. #

Problem 4.2.7. Let $p \geq 1$, $q \geq 0$ and $p+q \geq n$. Suppose $x_1,\ldots,x_p \in \mathbb{R}_n$ and $y_1,\ldots,y_q \in \mathbb{R}_n$ and $\mu_1,\ldots,\mu_{p+q}$ are $\sigma$-finite regular Borel measures on the Borel subsets of $\mathbb{R}_n$, each giving zero measure to proper hyperplanes of $\mathbb{R}_n$. The equation

$$(4.2.8) \qquad 0 = \det(\lambda_0 \Sigma_{i=1}^{p} x_i x_i^t + \Sigma_{i=1}^{q} y_i y_i^t)$$

is said to have a double root (at least) $\lambda_0$ if there exist two linearly independent vectors $a_1$ and $a_2$ such that

$$(4.2.9) \qquad 0 = (\lambda_0 \sum_{i=1}^{p} x_i x_i^t + \Sigma_{i=1}^{q} y_i y_i^t) a_i, \quad i = 1,2 .$$

Show the event that equation (4.2.8) have at least a double root $\lambda_0 \neq 0$ has $\mu_1 \times \cdots \times \mu_{p+q}$ measure zero.

Hint. Any double root $\lambda_0$ of (4.2.8) must be at least a single root $\lambda_0$ of

$$(4.2.10) \qquad \lambda_0 \sum_{i=1}^{p} x_i x_i^t + \sum_{i=1}^{q-1} y_i y_i^t .$$

Choose $G \in GL(n)$ such that

(4.2.11) $$G(\sum_{i=1}^{p} x_i x_i^t)G^t = D\ , \quad \text{a diagonal matrix, and}$$

$$G(\sum_{i=1}^{q-1} y_i y_i^t)G^t = I_n - D\ .$$

Let $z_q = Gy_q$

The case $\lambda_0 = 1$ is Problem 4.2.4. We suppose $\lambda_0 \neq 1$. For a root $\lambda_0$ of (4.2.10) suppose the $(i,i)$ element of $D$ is $1/(1-\lambda_0)$.

Case 1. Let $a \in \mathbb{R}_n$ be a solution of (4.2.9) such that $z_q^t a \neq 0$. Since the $(i,i)$ position of $\lambda_0 D + I_n - D$ is zero, if $z_q^t = (z_{q1}, \ldots, z_{qn})$ it follows that $z_{qi} = 0$. This says that the i-th row of $G$ into $y_q$ is zero.

Case 2. If $a$ is any solution of (4.2.9) then $a^t z_q = 0$. Then there exist nonzero vectors $a$ such that $a^t z_q = 0$ and $(\lambda_0 D + I_n - D)a = 0$.

That is,

(4.2.12) $$\lambda_0 \sum_{i=1}^{p} x_i x_i^t + \sum_{i=1}^{q-1} y_i y_i^t$$

has a double root not zero. Make an induction on $q$ with the first step for the induction being $p + q = n$. See the next case.

Case 3. In the case that $p + q = n$ we may choose $G \in GL(n)$ such that

(4.2.13) $$GX_i = e_i \quad \text{and} \quad Gy_j = e_{p+j},\ 1 \le i \le p \ \text{and}\ 1 \le j \le q\ .$$

Then

(4.2.14) $$G(\lambda_0 \sum_{i=1}^{p} x_i x_i^t + \sum_{i=1}^{q} y_i y_i^t)G^t = \begin{pmatrix} \lambda_0 & & & 0 \\ & \lambda_0 & & \\ & & 1 & \\ 0 & & & 1 \end{pmatrix} .$$

If $\lambda_0 \neq 0$ then this matrix is nonsingular. This establishes the first step of the induction. #

Problem 4.2.8. Let the real valued random variables $\underline{X}_1,\dots,\underline{X}_n$ have a joint density function relative to Lebesgue measure of the form

$$f(x_1^2+\dots+x_n^2), \qquad f \text{ continuous.} \tag{4.2.15}$$

Let $\underline{Y}=(\underline{X}_1^2+\dots+\underline{X}_n^2)^{1/2}$. Show that $\underline{Y}$ has a density function of the form

$$2y^{n-1}\pi^{n/2}f(y^2)\,/\Gamma(n/2)\;. \tag{4.2.16}$$

Use this to derive the density function of a $\chi_n^2$ random variable.

Hint. Show that $P(y<\underline{Y}\leq y+s)\sim f(y^2)(V_n(y+s)-V_n(y))$. Divide by $s$ and take a limit as $s\to 0$. Use (4.2.5). #

Problem 4.2.9. Substitute in (4.1.11) the joint normal density function of Chapter 2, means zero, covariance matrix $A^{-1}$. That is, write the Wishart density function.

Problem 4.2.10. If in Problem 4.2.9 $E\underline{X}=M\neq 0$, does Theorem 4.1.1 still apply? See Chapter 5.

## Chapter 5. The noncentral Wishart density function

### 5.0. Introduction.

Herz (1955) and James (1955a) at roughly the same time (they were fellow students at Princeton University) obtained expressions for the noncentral Wishart density function thereby generalizing previous partial results of Anderson (1946). In this short chapter we review the idea of James.

The noncentral problem is the problem of finding the density function of $\underline{X}^t\underline{X}$ when $\underline{X}$ is a random $n \times h$ matrix with independently and identically distributed rows each normal $(M, \Sigma)$, $M \neq 0$. If $M \neq 0$ the joint density function of $\underline{X}$ is not a function of $\underline{X}^t\underline{X}$ and the discussion of Chapter 4 fails to apply. James, op. cit., had the following idea.

### 5.1. James' method.

Let $H_1, \ldots, H_k$ be $n \times n$ orthogonal matrices and $(\alpha_1, \ldots, \alpha_k)$ a probability vector. Then

$$(5.1.1) \qquad \sum_{i=1}^{k} \alpha_i \, P(H_i\underline{X} \in A)$$

is a probability measure on the Borel subsets of $\mathbb{R}_{nh}$. The measure (5.1.1) induces a measure on $\underline{S}(n)$, the space of $n \times n$ positive definite matrices, by the rule, if $B \subset \underline{S}(n)$ is a Borel subset then

$$(5.1.2) \qquad A = \{X \mid X^tX \in B\}.$$

Then if $\underline{Y}$ has the probability measure (5.1.1) the random variable $\underline{Y}^t\underline{Y}$ has the probability measure

$$(5.1.3) \qquad P(\underline{Y}^t\underline{Y} \in B) = P(\underline{Y} \in A) = \sum_{i=1}^{k} \alpha_i P(H_i\underline{X} \in A)$$

$$= \sum_{i=1}^{k} \alpha_i \, P((H_i\underline{X})^t(H_i\underline{X}) \in B) = (\alpha_1 + \ldots + \alpha_k) \, P(\underline{X}^t\underline{X} \in B)$$

$$= P(\underline{X}^t\underline{X} \in B).$$

Since (5.1.3) holds for all finite convex mixtures, we take a sequence of probability vectors and orthogonal matrices tending in the limit to Haar measure of unit mass on $\underline{O}(n)$ and obtain

$$(5.1.4) \qquad P(\underline{X}^t\underline{X} \in B) = \int_{\underline{O}(n)} P(H\underline{X} \in A)\, dH,$$

where dH represents the Haar measure of unit mass.

This sketch is not, of course, a proof. It is easy to check dH-integrability of $P(H\underline{X} \in A)$ and that the integral in (5.1.4) is a probability measure. Then the steps of (5.1.3) can be repeated under the integral to obtain (5.1.4). We now consider the technical points which arise. The measurability of $P(H\underline{X} \in A)$ as a function of H follows from the joint measurability of the map $(H,X) \to HX$. The integral (5.1.4) is a double integral with respect to a product measure so Fubini's Theorem applies. Measurability of $P(H\underline{X} \in A) = E\chi_A(H\underline{X})$ then follows. As a second consideration, in the sequel we are interested in the cases where $\underline{X}$ has a density function, say g. Suppose $\underline{Y}$ has density function

$$(5.1.5) \qquad \int_{\underline{O}(n)} g(HX)\, dH\,.$$

Measurability as a function of X follows from the above remarks. It then follows that $\underline{Y}^t\underline{Y}$ and $\underline{X}^t\underline{X}$ have the same density function which is obtained from (5.1.5) by introducing appropriate normalization a la Chapter 4. Some further details are given in Problems 5.3.9 and 5.3.10.

In this section we need to know that (5.1.5) is a function of $(X^tX)^{1/2}$. We now prove this in the case $h \leq n$ of taking n observations with n at least the dimension of observation space. We ignore the set of Lebesgue measure zero on which the columns of X are linearly dependent, Problem 4.2.4. Choose $X_1$ a $n \times (n-h)$ matrix such that the $n \times n$ matrix $Z = (X, X_1)$ is nonsingular. Further suppose that $X^t X_1 = 0$. Then $H = Z(Z^t Z)^{-1/2}$ is an orthogonal matrix and

$$(5.1.6) \qquad H^t Z = (H^t X,\ H^t X_1) = \begin{pmatrix} (X^t X)^{1/2} & \\ 0 & (X_1^t X_1)^{1/2} \end{pmatrix}.$$

We let $X_1 \to 0$ in such a way that the corresponding $H$'s converge to some $H_0$ ($\underline{O}(n)$ is a compact metric space) to obtain

$$(5.1.7) \qquad H_0 X = \begin{pmatrix} (X^t X)^{1/2} \\ 0 \end{pmatrix}.$$

By invariance of Haar measure ($O(n)$ is unimodular), (5.1.5) becomes

$$(5.1.8) \qquad \int_{\underline{O}(n)} g(HX)\, dH = \int_{\underline{O}(n)} g(HH_0 X)\, dH = \int_{\underline{O}(n)} g\left(H \begin{pmatrix} (X^t X)^{1/2} \\ 0 \end{pmatrix}\right) dH.$$

Let $g(X) = f((X-M)^t(X-M))$. The change of variable $Z = HX$ has Jacobian $\underline{+}1$ so $\underline{Z}$ has density function

$$(5.1.9) \qquad f(Z^t Z - Z^t HM - M^t H^t Z + M^t M),$$

and integration with respect to $H$ by Haar measure of unit mass on $\underline{O}(n)$ gives

$$(5.1.10) \qquad \int_{\underline{O}(n)} f(Z^t Z - Z^t HM - H^t M^t Z + M^t M)\, dH .$$

By our earlier discussion if a random variable $\underline{Z}$ has density function (5.1.10) then $\underline{X}^t\underline{X}$ and $\underline{Z}^t\underline{Z}$ have the same probability law. By the discussion of Chapter 4, the density function of the probability law is simply the function (5.1.10) multiplied by a normalization that represents integration out of the remaining variables. From Chapter 4 it follows that $\underline{X}^t\underline{X}$ has density function

$$(5.1.11)\quad \frac{\pi^{hn/2 - h(h-1)/4}}{\Gamma(n/2)\ldots\Gamma((n-h+1)/2)} |X^tX|^{(n-h-1)/2}$$

$$\times \int_{\underline{O}(n)} f(X^tX-X^tHM-M^tH^tX+M^tM)\, dH.$$

Specifically for the multivariate normal density function with $\Sigma = I_n$,

$$(5.1.12)\quad f((X-M)^t(X-M)) = (2\pi)^{-nh/2} \exp(-\mathrm{tr}\, M^tM/2)$$

$$\times \exp(-\mathrm{tr}\, X^tX/2)\exp(\mathrm{tr}\, X^tM).$$

The noncentral Wishart density is, then,

$$(5.1.13)\quad \frac{\pi^{-h(h-1)/4}}{2^{hn/2}\Gamma(n/2)\ldots\Gamma((n-h+1)/2)} |X^tX|^{(n-h-1)/2}\exp(-\mathrm{tr}(X^tX+M^tM)/2)$$

$$\times \int_{\underline{O}(n)} \exp(\mathrm{tr}\, X^tHM)\, dH.$$

The integral appearing in (5.1.13) can be phrased as

$$(5.1.14)\quad \int_{\underline{O}(n)} \exp(\mathrm{tr}(XM^tMX^t)^{1/2}H)\, dH$$

and this evaluation of (5.1.14) is equivalent to computing the Laplace transform of Haar measure. The answer in simple closed form is unknown. Transforms of functions of a matrix argument was part of the subject treated by Herz (1955). (5.1.14) was explicitly evaluated by Anderson (1946) when $XM^t$ is a rank two matrix. James (1955b) has given in series an evaluation of (5.1.14) in the case $XM^t$ is a rank three matrix. See Section 5.2 for a summary of this work. Subsequent papers by James (1964) and Constantine (1963) on zonal polynomials give a different type of series evaluation of (5.1.14) as a infinite sum of zonal polynomials.

## 5.2. James on series, rank 3.

Our source is James (1955b). We set

$$(5.2.1) \qquad \psi(Z) = \int_{\underline{O}(n)} \exp(\mathrm{tr}\, H^t Z)\, dH,$$

$Z \in GL(n)$. Since Haar measure on $\underline{O}(n)$ is unimodular, see Lemma 3.5.3, and since $\mathrm{tr}\, AB = \mathrm{tr}\, BA$, from (5.2.1) we obtain

$$(5.2.2) \qquad \int_{\underline{O}(n)} \exp(\mathrm{tr}\, H^t Z)\, dH = \int_{\underline{O}(n)} \exp(\mathrm{tr}\, H^t H_1 Z H_2)\, dH,$$

and

$$\psi(Z) = \psi(H_1\, Z\, H_2) \quad \text{for all} \quad H_1,\, H_2 \in \underline{O}(n).$$

We take $Z$ nonsingular and set $H_1 = H_2^t\, (Z^t Z)^{-1/2}\, Z^t$ and choose $H_2$ so that $H_2^t (Z^t Z)^{1/2}\, H_2 = D$ is a diagonal matrix. From (5.2.1) and (5.2.2) we obtain

$$(5.2.3) \qquad \psi(Z) = \psi(D),$$

that is, the value of $\psi(Z)$ depends only on the eigenvalues of $(Z^t Z)^{1/2}$. Since

$$(5.2.4) \qquad \psi(D) = \psi(P^t D P),\ P \in \underline{O}(n) \text{ a permutation matrix},$$

it follows that $\psi(Z)$ is a symmetric function of the eigenvalues of $(Z^t Z)^{1/2}$. Since $\exp(\mathrm{tr}\, H^t D)$ has a infinite series expansion, it follows that $\psi(D)$ is an entire function of the $n$ eigenvalues of D.

The method of James (1955b) was to derive differential equations for $\psi$ by taking special types of matrices $Z$. For example, if

$$\Delta = \frac{\partial^2}{\partial z_{11}^{\,2}} + \ldots + \frac{\partial^2}{\partial z_{n1}^{\,2}}, \qquad \text{then} \quad \psi = \Delta\psi.$$

From these differential equations recursion relations were developed. We don't pursue this further but give James' result.

Theorem 5.2.1. (James (1955b)). Let $r_1,\ldots,r_n$ be the eigenvalues of $(Z^tZ)^{1/2}$ and define

$$(5.2.5)\qquad \xi(r_1,\ldots,r_n) = \int_{\underline{O}(n)} \exp(\operatorname{tr} H^tZ)\,dH.$$

Then

$$(5.2.6)\qquad \xi(r_1,r_2,r_3,0,\ldots,0) =$$

$$\Sigma_{j_1j_2j_3} \frac{\Gamma(\frac{n}{2})\Gamma(\frac{n-1}{2})\Gamma(\frac{n-2}{2})(n+j_1+2j_2+4j_3-3)!}{2^{2j_1+4j_2+6j_3}j_1!j_2!j_3!\Gamma(\frac{h}{2}+j_1+2j_2+3j_3-3)}$$

$$\times \frac{r_1^{j_1}\ r_2^{j_2}\ r_3^{j_3}}{\Gamma(\frac{n-2}{2}+j_3)\Gamma(\frac{n-1}{2}+j_2+2j_3)(n+j_1+2j_2+3j_3-3)!}.$$

James noted and is quoted by Herz, op. cit., as saying the computation got out of hand if $r_4 \neq 0$ was allowed. Please check the original source before using (5.2.6).

## 5.3. Problems

Problems 5.3.1 to 5.3.4 are about noncentral Chi-square random variables. Problems 5.3.5 to 5.3.8 are about integrals of polynomials in the entries of orthogonal matrices with respect to Haar measure. Problems 5.3.9 and 5.3.10 are about the density functions of integrated probability measures.

Problem 5.3.1. See Problem 4.2.8. Suppose $\underline{X}_1,\ldots,\underline{X}_n$ have a joint density function $f(x_1^2+\ldots+x_n^2)$. Let $\underline{Y} = \underline{X}_1^2+\ldots+\underline{X}_n^2$. All random variables here are real valued. Then

$$(5.3.1)\qquad \lim_{s\to 0} s^{-1}P(y < \underline{Y} \leq y+s) = \frac{y^{(n/2)-1}\,\pi^{n/2}\,f(y)}{\Gamma(n/2)}.$$

Problem 5.3.2. Let X have density function $(2\pi)^{-n/2}\exp(-\sum_{i=1}^{n}(x_i-a_i)^2/2)$. Let $a = \sum_{i=1}^{n} a_i^2/2$. Let $\underline{Y} = \underline{X}^t\underline{X}$ and apply the ideas of this chapter. Show the density function of $\underline{Y}$ is

(5.3.2) $$\frac{y^{(n/2)-1}e^{-y/2}e^{-a}}{2^{n/2}\Gamma(n/2)}\int_{\underline{O}(n)}\exp(b^tHx)\,dH,$$

where $b^t = (a_1,\ldots,a_n)$.

Problem 5.3.3. Compare (5.3.2) with the noncentral Chi-square density (2.3.10). Infer that

(5.3.3) $$\int_{\underline{O}(n)}\exp(b^tHx)\,dH = \sum_{j=0}^{\infty}\frac{\Gamma(n/2)r^ja^j}{\Gamma((n+2j)/2)2^jj!},$$

with $r = x^tx$, and $a = b^tb/2 = \sum_{i=1}^{n} a_i^2/2$.

Problem 5.3.4. Use the fact that there exist $H_1$ and $H_2 \in \underline{O}(n)$ such that $b^tH_1 = (\|b\|,0,\ldots,0)$ and $x^tH_2 = (\|x\|,0,\ldots,0)$. Show that

(5.3.4) $$\int_{\underline{O}(n)}\exp(b^tHx)\,dH = \int_{\underline{O}(n)}\exp(\|x\|\,\|b\|\,H_{11})\,dH,$$

where $H_{11}$ is the (1,1)-element of $H$. Expand in power series and match coefficeints. In this way obtain,

(5.3.5) $$\text{if } j \geq 0,\ \int_{\underline{O}(n)} H_{11}^{2j}\,dH = \frac{\Gamma(n/2)\Gamma(2j+1)}{\Gamma((n+2j)/2)\Gamma(j+1)2^{2j}}.$$

Problems 5.3.5 to 5.3.8 are on evaluation of integrals $\int_{\underline{O}(n)} f(H)\,dH$ where $f$ is a polynomial. The identities following are based on the following observation.

(5.3.6) $$\text{If } P,Q \in \underline{O}(n) \text{ then } \int_{\underline{O}(n)} f(H)\,dH = \int_{\underline{O}(n)} f(PHQ)\,dH.$$

Observe that if $P \in \underline{O}(n)$ is a diagonal matrix with all diagonal entries = 1 except for the (1,1)-entry which is -1, then for the function $f(H) = H_{11}^m$

$$(5.3.7) \qquad \int_{\underline{O}(n)} H_{11}^m \, dH = (-1)^m \int_{\underline{O}(n)} H_{11}^m \, dH.$$

The identity (5.3.7) implies that if $m$ is odd then $\int_{\underline{O}(n)} H_{11}^m = 0$.

More generally, if $P$ is diagonal and $Q$ induces the permutation $\sigma$ of $1,\ldots,n$ on the rows (multiplication from the left by $Q$) then

$$(5.3.8) \qquad \int_{\underline{O}(n)} H_{11}^{m_1} \cdots H_{nn}^{m_n} \, dH = \int_{\underline{O}(n)} H_{\sigma(1)1}^{m_1} \cdots H_{\sigma(n)n}^{m_n}$$

$$= \int_{\underline{O}(n)} H_{\sigma(1)\sigma(1)} \cdots H_{\sigma(n)\sigma(n)} \, dH,$$

and similarly for the sign changes induced by $P$.

<u>Problem 5.3.5</u>. Let $m = m_1 + \ldots + m_n$. Then

$$(5.3.9) \quad \int_{\underline{O}(n)} H_{11}^{2m_1} \cdots H_{1n}^{2m_n} = \frac{m!}{(2m)!} \frac{(2m_1)!}{m_1!} \cdots \frac{(2m_n)!}{m_n!} \int_{\underline{O}(n)} H_{11}^{2m} \, dH .$$

<u>Hint</u>. Let $P \in \underline{O}(n)$ have a first row $u_1,\ldots,u_n$ and let the function $f$ be $f(H) = H_{11}^m$. Then $(u_1^2 + \ldots + u_n^2)^m \int_{\underline{O}(n)} H_{11}^{2m} \, dH =$ $\int_{\underline{O}(n)} (u_1 H_{11} + \ldots + u_n H_{1n})^{2m} \, dH$. Expand both sides using the multinomial theorem and match coefficients. Note that both sides are homogeneous of degree $2m$ and hence the identity holds for <u>all</u> real numbers $u_1,\ldots,u_n$. #

<u>Problem 5.3.6</u>. Let $q = q_2 + \ldots + q_n$. Then

$$(5.3.10) \quad \int_{\underline{O}(n)} H_{11}^{2m} H_{22}^{2q_2} \cdots H_{2n}^{2q_n} = \frac{q!}{(2q)!} \frac{(2q_2)!}{q_2!} \cdots \frac{(2q_n)!}{q_n!}$$

$$\times \int_{\underline{O}(n)} H_{11}^{2m} H_{22}^{2q} \, dH.$$

Problem 5.3.7.

$$(5.3.11) \qquad \int_{\underline{O}(n)} H_{11}^{2m-2}\, dH = (1 + (n-1)/(2m-1)) \int_{\underline{O}(n)} H_{11}^{2m}\, dH.$$

Hint. Integrate $H_{11}^{2m-2}(H_{11}^2 + \ldots + H_{n1}^2) = H_{11}^{2m-2}$. Use (5.3.9) to obtain an identity. #

Problem 5.3.8. Show that (5.3.5) and (5.3.11) agree.

Problem 5.3.9 and Problem 5.3.10 develop details of the theory for density functions refered to in Section 5.1.

Problem 5.3.9. If the $n \times h$ random variable $\underline{X}$ has density function $g_1$ relative to Lebesgue measure on $\mathbb{R}_{nh}$ then the random variable $\underline{Y} = H\underline{X}$, $H \in GL(n)$, has density function

$$(5.3.12) \qquad g_2(X) = g_1(H^{-1}X)/|\det H|^h,$$ where $|\det H|$ means the absolute value of the determinant.

Hint. $P(\underline{Y} \in A) = \int \chi_A(HX)\, g_1(H^{-1}HX)\, dX$. Make the change of variable $Y = HX$ and compute the Jacobian of the transformation. #

Problem 5.3.10. If the $n \times h$ random matrix $\underline{X}$ has density function $g$ then the probability measure

$$(5.3.13) \qquad \int_{\underline{O}(n)} P(H\underline{X} \in A)\, dH$$

has density function

$$(5.3.14) \qquad \int_{\underline{O}(n)} g(H^tX)\, dH\, .$$

It is assumed that density functions are integrated by Lebesgue measure.

Hint. Write (5.3.14) as a double integral using (5.3.13). Apply Fubini's Theorem. #

# Chapter 6. Manifolds and exterior differential forms.

## 6.0. Introduction

Our discussion of differential forms is spread over the next four chapters. The fundamentals of the algebraic theory and theory of integration of differential forms form the substance of Chapter 6. This material is a necessary expansion of the introductory material in James (1954). Chapter 7 is concerned with explicit derivation of differential forms for invariant measures on various manifolds. Although somewhat premature, in case the manifold is a locally compact metric matrix group, the basic idea has already been discussed in Chapter 3, Section 3.6. We discuss several manifolds, the Stiefel and Grassman manifolds, which are not groups, in Chapter 7. These examples require some long detailed calculations and this material has been segregated into a separate chapter. The fact that differential forms are so useful in the derivation of density functions is due to the fact that the manifold of $n \times h$ matrices naturally decomposes into a topological product of manifolds on which groups of transformations act. The theory of Sections 3.3 and 3.4 suggests that the invariant measures should factor. Since Lebesgue measure is absolutely continuous relative to the invariant measures, Lebesgue measure also factors. Chapter 8 discusses the well known methods of decomposing a $n \times h$ matrix. Chapter 9 factors Lebesgue measure over several of these products of manifolds thereby obtaining density functions of several multivariate statistics. In certain parts of inference probability ratios only are needed, not normalized density functions. In these cases the full force of Chapters 6 to 9 is not needed and a more direct approach due to Stein (1956c) may be used to compute the density function of a maximal invariant and the desired probability ratio. See Chapter 10.

Chapter 6 is not totally selfcontained and missing details may be found in Dieudonne (1972) and Helgason (1962). Notably, the manifolds considered here are analytic manifolds, but we do not prove this. We do need the fact that the manifolds considered are $C_2$, but again we do not prove the existence of local coordinates with the necessary differentiability on the overlaps of the charts.

## 6.1. Basic structural definitions and assumptions.

We start with a topological space $\mathfrak{M}$ which is a locally compact Hausdorff space. To each point $x \in \mathfrak{M}$ is assigned an open set $U_x \subset \mathfrak{M}$ such that $x \in U_x$. Associated with the pair $x, U_x$ is a function $\varphi(x, \ ): U_x \to \mathbb{R}_n$ which is a homeomorphism of $U_x$ onto $\mathbb{R}_n$. We assume that the integer $n$ is the same for all $x \in \mathfrak{M}$, and <u>$n$ is called the dimension of the manifold $\mathfrak{M}$</u>. The mapping $\varphi(x, \ )$ is called a coordinate chart at $x$.

The definition of a manifold requires the following consistency relations. If $x \in U_{x_1} \cap U_{x_2}$ then the function

$$\varphi(x_1, \varphi^{-1}(x_2, \ )) \tag{6.1.1}$$

is a homeomorphism of its domain into its range $\mathbb{R}_n$. The differentiability of a manifold is described by specification of the differentiability of the functions (6.1.1). An analytic manifold is one in which the functions (6.1.1) are analytic functions of $n$ real variables. The manifold is said to be $C_k$ if the functions (6.1.1) are k-fold continuously differentiable.

The <u>local coordinates</u> of a point $y \in U_x$ are the numbers

$$\varphi(x,y)^t = (\varphi_1(x,y), \ldots, \varphi_n(x,y)). \tag{6.1.2}$$

A function $f: U_x \to \mathbb{R}$ is called <u>a $C_k$ function locally at $x$</u> if the function

$$f(\varphi^{-1}(x, \ )): \mathbb{R}_n \to \mathbb{R} \tag{6.1.3}$$

is k-fold continuously differentiable. In the sequel we use the <u>local coefficient rings</u> $\mathcal{C}_{x,k}$ of all functions $f:U_x \to \mathbb{R}$ which are $C_k$ functions. The local coefficient rings are the coefficient rings used in the construction of multilinear forms. It should be noted that these rings have multiplicative units, which is important for the construction of canonical bases of modules. Unless otherwise stated we assume that our manifolds are $C_2$ and that the coefficient rings are $\mathcal{C}_{x,2}$, $x \in \mathfrak{M}$.

We list several basic examples of manifolds. $\mathbb{R}_n$ is an n-dimensional analytic manifold using, if $x \in \mathbb{R}_n$ then $U_x = \mathbb{R}_n$ and $\varphi(x,y) = y$, $y \in \mathbb{R}_n$.

The <u>sphere</u> $\{(x_1,\ldots,x_n) | \; x_1^2 + \ldots + x_n^2 = 1\}$ is an (n-1)-dimensional analytic manifold. To each point $x$ let $U_x$ be all points of the sphere making an angle of absolute value $< \pi/2$ with $x$. Project $U_x$ onto $\mathbb{R}_{n-1}$ by taking the (n-1)-dimensional hyperplane tangent at $x$. If $y \in U_x$ then $\varphi(x,y)$ is defined to be the intersection of the hyperplane with the line through the points $0$ and $y$.

<u>$\underline{O}(n)$, the orthogonal group</u>. Here the manifold is the set of $n \times n$ orthogonal matrices with real entries, in the topology of coordinatewise convergence. If we interpret $H \in \underline{O}(n)$ as a point of $\mathbb{R}_{n^2}$ then the map $H \to \det H$ is a continuous function. The nonsingular matrices thus are an open subset of $\mathbb{R}_{n^2}$, and $\underline{O}(n)$ is a closed bounded subset of $\mathbb{R}_{n^2}$, hence a compact topological space.

The condition $H^tH = I_n$ gives $n(n+1)/2$ algebraic relations so one expects the dimension of the manifold to be $(n-1)n/2$. We do not give an explicit construction of local coordinates for this example.

## 6.2. Multilinear forms, algebraic theory.

In developing the algebraic theory we let $\mathcal{C}$ be a coefficient ring with Abelian multiplication and multiplicative unit. We choose a free basis $e_1,\ldots,e_n$ for a n-dimensional free module

(6.2.1) $$E = \{x \mid x = \sum_{i=1}^{n} c_i e_i,\ c_1,\ldots,c_n \in \mathcal{C}\}.$$

We will write $M(E^q,\mathcal{C})$ for the space of multilinear q-forms on $E^q = E \times \ldots \times E$. Of special importance is the fact that since $E$ is a free module, $M(E,\mathcal{C}) = M(E^1,\mathcal{C})$ has a canonical basis $u_1,\ldots,u_n$ defined by the conditions of linearity together with

(6.2.2) $$u_i(e_j) = \delta_{ij} = 1 \quad \text{if} \quad i = j$$
$$= 0 \quad \text{if} \quad i \neq j.$$

Permutations act on $M(E^q,\mathcal{C})$ as follows. Let $\sigma$ be a permutation of $1,\ldots,q$. If $f \in M(E^q,\mathcal{C})$ and $x_1,\ldots,x_q \in E$ then

(6.2.3) $$(\sigma f)(x_1,\ldots,x_q) = f(x_{\sigma(1)},\ldots,x_{\sigma(q)}).$$

Subject to this definition $\sigma$ acts as a linear transformation of $M(E^q,\mathcal{C})$. The product of permutations $\sigma$ and $\tau$ as linear transformations is $\sigma\tau$, that is,

(6.2.4) $$((\sigma\tau)f)(x_1,\ldots,x_q) = (\sigma(\tau(f)))(x_1,\ldots,x_q).$$

The permutation operators are used to define the <u>alternating operators</u>. To make the definition we need the sign function $\epsilon(\sigma)$ of a permutation. We define

(6.2.5) $\epsilon(\sigma) = 1$ if $\sigma$ is the product of an even number of transpositions;

$= -1$ if $\sigma$ is the product of an odd number of transpositions.

We assume that the reader knows that the sign function is well defined and note that it follows from (6.2.5) that if $\sigma$ and $\tau$ are permutations of $1,\dots,q$ then

$$\epsilon(\sigma\tau) = \epsilon(\sigma)\epsilon(\tau). \tag{6.2.6}$$

A linear operator $A$ (we use the same letter for each $q$), $A\colon M(E^q, \mathcal{C}) \to M(E^q, \mathcal{C})$ may be defined by, if $f \in M(E^q, \mathcal{C})$ then

$$(Af) = (q!)^{-1} \Sigma_\sigma\ \epsilon(\sigma)(\sigma f), \tag{6.2.7}$$

and make the definition

$$A^q = \{f \mid f \in M(E^q, \mathcal{C}),\ Af = f\} = A^q(E, \mathcal{C}). \tag{6.2.8}$$

In words, $A^q$ is the set of alternating q-forms.

Lemma 6.2.1. If $f \in M(E^q, \mathcal{C})$ then $A(Af) = Af$.

Proof. $A(Af) = (q!)^{-1} \Sigma_\sigma\ \epsilon(\sigma)\sigma((q!)^{-1} \Sigma_\tau\ \epsilon(\tau)(\tau f))$

$= (q!)^{-2} \Sigma_\sigma \Sigma_\tau\ \epsilon(\sigma\tau)(\sigma\tau)f$

$= (q!)^{-1} \Sigma_\sigma\ Af = Af.$ #

Lemma 6.2.2. $\sigma A = \epsilon(\sigma)A$.

Proof. $(\sigma A)(f) = \sigma(\Sigma_\tau (q!)^{-1}\ \epsilon(\tau)(\tau f)) = \Sigma_\tau (q!)^{-1}\ \epsilon(\tau)(\sigma\tau)(f)$

$= \epsilon(\sigma)\ \Sigma_\tau (q!)^{-1}\ \epsilon(\sigma)\epsilon(\tau)(\sigma\tau)(f) = \epsilon(\sigma)Af.$ #

Definition 6.2.3. If $f \in M(E^q, \mathcal{C})$ and $g \in M(E^r, \mathcal{C})$ then the function $fg$ is defined by

$$(fg)(x_1,\dots,x_{q+r}) = f(x_1,\dots,x_q)g(x_{q+1},\dots,x_{q+r}). \tag{6.2.9}$$

Definition 6.2.4. If $f \in M(E^q, \mathcal{C})$ and $g \in M(E^r, \mathcal{C})$ then $f \wedge g$ is defined by

(6.2.10) $$f \wedge g = A(fg).$$

It is customary to use the same symbols $\wedge$ and $A$ for all choices of $q \geq 1$ and $r \geq 1$. With this understanding it is meaningful to say that $\wedge$ is an associative operation.

<u>Lemma 6.2.5</u>. Let $f \in M(E^q, \mathfrak{C})$, $g \in M(E^r, \mathfrak{C})$ and $h \in M(E^s, \mathfrak{C})$. Then

(6.2.11) $$(f \wedge g) \wedge h = f \wedge (g \wedge h) = A(fgh).$$

<u>Proof</u>. $(f \wedge g)h = (A(fg))h = (\Sigma_\tau((q+r)!)^{-1} \, \epsilon(\tau)\tau(fg))h,$

where $\tau$ runs over all permutations of $1,\ldots,q+r$. Let $\underline{\tau}$ be the permutation of $1,\ldots,q+r+s$ leaving $q+r+1,\ldots,q+r+s$ fixed and acting as $\tau$ on $1,\ldots,q+r$. Then

$$\Sigma_\sigma \ \epsilon(\sigma)\sigma((f \wedge g)h) = \Sigma_\sigma \ \Sigma_\tau \ \epsilon(\sigma)\epsilon(\tau)((q+r)!)^{-1}\sigma(\tau(fg)h) =$$

$$\Sigma_\sigma \ \Sigma_\tau \ \epsilon(\sigma\underline{\tau})((q+r)!)^{-1} \ (\sigma\underline{\tau})(fgh) = (q+r+s)! \ A(fgh).$$

A similar argument will show $f \wedge (g \wedge h) = A(fgh)$. #

<u>Lemma 6.2.6</u>. The operation $\wedge$ on $A^q(E, \mathfrak{C}) \times A^r(E, \mathfrak{C}) \to A^{q+r}(E, \mathfrak{C})$ is a bilinear operation.

<u>Lemma 6.2.7</u>. If $f \in M(E^q, \mathfrak{C})$ and $g \in M(E^r, \mathfrak{C})$ then

(6.2.12) $$f \wedge g = (-1)^{qr} \ (g \wedge f).$$

<u>Proof</u>. Let $\tau$ be the permutation

(6.2.13) $$\tau = \begin{pmatrix} 1 & 2 & \ldots & r & r+1 & \ldots & q+r \\ q+1 & q+2 & \ldots & q+r & 1 & \ldots & q \end{pmatrix}.$$

The proof depends on knowing that $\epsilon(\tau) = (-1)^{qr}$. Note that if $\sigma$ is the cycle $= (1,2,\ldots,q+r)$ then $\tau = \sigma^q$. Then since $\epsilon(\sigma) = (-1)^{q+r-1}$, it follows that $\epsilon(\tau) = (-1)^{(q+r-1)q} = (-1)^{qr}$.

Then

$$(6.2.14)\qquad (q+r)!\,f\wedge g = \Sigma_\sigma\ \epsilon(\sigma)f(x_{\sigma(1)},\ldots,x_{\sigma(q)})g(x_{\sigma(q+1)},\ldots,x_{\sigma(q+r)})$$

$$= \Sigma_\sigma\ \epsilon(\sigma)g(x_{\sigma(q+1)},\ldots,x_{\sigma(q+r)})f(x_{\sigma(1)},\ldots,x_{\sigma(q)})$$

$$= \Sigma_\sigma\ \epsilon(\sigma)g(x_{\sigma\tau(1)},\ldots,x_{\sigma\tau(q+r)})f(x_{\sigma\tau(r+1)},\ldots,x_{\sigma\tau(r+q)})$$

$$= \epsilon(\tau)\ \Sigma_\sigma\ \epsilon(\sigma\tau)((\sigma\tau)(gf)) = (-1)^{qr}\ (q+r)!\ g\wedge f.\ \#$$

Lemma 6.2.8. If $f \in M(E^q, \mathcal{C})$ and $q$ is an odd integer then $f\wedge f = 0$.

Proof. By (6.2.12) $f\wedge f = (-1)^{q^2}\ f\wedge f$. #

Lemma 6.2.9. If $m \geq 1$ is an integer, if $f_1,\ldots,f_m \in M(E, \mathcal{C})$ and $x_1,\ldots,x_m \in E$, then

$$(6.2.15)\qquad (f_1\wedge f_2\wedge \ldots\wedge f_m)(x_1,\ldots,x_m) = (m!)^{-1}\det(f_i(x_j)).$$

Proof. We make an induction on $m$. (6.2.15) is clearly true if $m = 1$. Let a permutation $\sigma_i$ be defined by

$$(6.2.16)\qquad \sigma_i = \begin{pmatrix} 1 & 2 & \ldots & i & i+1 & \ldots & m \\ i & 1 & & i-1 & i+1 & & m \end{pmatrix}.$$

Then $\epsilon(\sigma_i) = (-1)^{i+1}$. We compute

$$(6.2.17)\qquad (m!)(f_1\wedge\ldots\wedge f_m)(x_1,\ldots,x_m)$$

$$= \Sigma_\tau\ \epsilon(\tau)\ f_1(x_{\tau(1)})(f_2\wedge\ldots\wedge f_m)(x_{\tau(2)},\ldots,x_{\tau(m)})$$

$$= \Sigma_i\ f_1(x_i) \sum_{\tau(1)=i} \epsilon(\tau)(\tau\sigma_i^{-1})(f_2\wedge\ldots\wedge f_m)(x_1,\ldots,x_{i-1},x_{i+1},\ldots,x_m).$$

As $\tau$ runs over permutations of $1,\ldots,m$ such that $\tau(1) = i$, we have $\tau\sigma_i^{-1}$ runs over all permutations of $1,\ldots,i-1,\ 1+1,\ldots,m$. Therefore

$$(6.2.18)\quad \sum_{\tau(1)=i} \epsilon(\tau)(\tau\sigma_i^{-1})(f_2\wedge\ldots\wedge f_m)(x_1,\ldots,x_{i-1},x_{i+1},\ldots,x_m)$$

$$= \epsilon(\sigma_i)((m-1)!)(A(f_2\wedge\ldots\wedge f_m))(x_1,\ldots,x_{i-1},x_{i+1},\ldots,x_m)$$

$$= (-1)^{i+1}((m-1)!)(f_2\wedge\ldots\wedge f_m)(x_1,\ldots,x_{i-1},x_{i+1},\ldots,x_m).$$

Apply the inductive hypothesis to the last line of (6.2.18), substitute the result in (6.2.17), and the result follows. #

Lemma 6.2.10. Let $f_1,\ldots,f_m \in M(E,\mathcal{C})$ and $x_{ij} \in \mathcal{C}$, $1 \leq i,j \leq m$. Then

$$(6.2.19)\quad (x_{11}f_1+\ldots+x_{1m}f_m)\wedge(x_{21}f_1+\ldots+x_{2m}f_m)\wedge\ldots\wedge(x_{m1}f_1+\ldots+f_{mm}f_m)$$

$$= (\det(x_{ij}))f_1 \wedge\ldots\wedge f_m.$$

Proof. Expand (6.2.19) to a sum of $m^m$ terms. Using Lemma 6.2.8 and (6.2.12), the only nonzero terms are the terms in which $m$ distinct functions $f_i$ occur. A term

$$x_{1\sigma(1)}x_{2\sigma(2)}\cdots x_{m\sigma(m)}f_{\sigma(1)}\wedge\ldots\wedge f_{\sigma(m)}$$

$$= \epsilon(\sigma)x_{1\sigma(1)}\cdots x_{m\sigma(m)}f_1 \wedge\ldots\wedge f_m$$

by (6.2.12). Summing over these $m!$ terms gives (6.2.19).#

We close Section 6.2 by computing the dimensions of the alternating algebras $A^q(E,\mathcal{C})$. We use the canonical basis of $M(E,\mathcal{C})$ defined in (6.2.2) to construct bases of the various spaces constructed. A linear functional $f \in M(E,\mathcal{C})$ can be expressed as

$$(6.2.20)\qquad f = \sum_{i=1}^{n} c_i u_i, \quad c_1,\ldots,c_n \in \mathcal{C}.$$

Given $f_1,\ldots,f_q$ with $f_j = \sum_{i=1}^{n} c_{ij}u_i$ then

(6.2.21) $$f_1 \cdots f_q = \Sigma_{i_1} \Sigma_{i_2} \cdots \Sigma_{i_q} c_{i_1 1} \cdots c_{i_q q} u_{i_1} \cdots u_{i_q}.$$

Thus the set of terms $u_{i_1} \cdots u_{i_q}$ span $M(E^q, \mathcal{C})$ where repetition of indices is allowed.

To prove linear independence suppose

(6.2.22) $$\Sigma_{i_1} \cdots \Sigma_{i_q} c_{i_1 \cdots i_q} u_{i_1} \cdots u_{i_q} = 0.$$

Evaluate this multilinear form at $e_{j_1}, \ldots, e_{j_q}$. All these terms vanish except for one term so the value is $c_{j_1 \cdots j_q} = 0$. The multiplicative unit of $\mathcal{C}$ is used here.

We summarize the discussion in a Lemma.

Lemma 6.2.11. $M(E^q, \mathcal{C})$ has dimension $n^q$ where $n$ is the dimension of $E$. A canonical basis for this space is the set of $n^q$ terms of the form $u_{i_1} \cdots u_{i_q}$.

In order to obtain a basis of $A^q(E, \mathcal{C})$ we use the following lemma. It is easily verified and a formal proof is omitted. See Lemmas 6.2.1 and 6.2.2.

Lemma 6.2.12. The following identities hold.

(6.2.22) $$A(\sigma f) = \epsilon(\sigma) Af.$$

(6.2.23) $$A(f(Ag)) = f \wedge g.$$

We now compute the dimension of an alternating algebra. The general element of $M(E^q, \mathcal{C})$ can be expresssed as

(6.2.24) $$f = \Sigma \cdots \Sigma\, f(e_{i_1}, \ldots, e_{i_q}) u_{i_1} \cdots u_{i_q}.$$

Using the linearity of $A$ and (6.2.23) it follows that

(6.2.25) $$Af = \Sigma \cdots \Sigma\, f(e_{i_1}, \ldots, e_{i_q}) u_{i_1} \wedge \cdots \wedge u_{i_q}.$$

Then the terms $u_{i_1} \wedge \ldots \wedge u_{i_q}$ span $A^q(E, \mathcal{C})$. Reordering the factors in such a wedge product at most changes the sign so we may suppose $i_1 < i_2 < \ldots < i_q$. Conversely, if

(6.2.26) $$0 = \sum_{i_1 < i_2} \ldots \sum_{< i_q} c_{i_1 \ldots i_q} u_{i_1} \wedge \ldots \wedge u_{i_q} ,$$

then evaluated at $e_{j_1}, \ldots, e_{j_q}$ we find

(6.2.27) if $j_1, \ldots, j_q$ is not a permutation of $i_1, \ldots, i_q$ then $u_{i_1} \wedge \ldots \wedge u_{i_q}(e_{j_1}, \ldots, e_{j_q}) = 0$. If $\sigma$ is a permutation of $1, \ldots, q$ such that $i_{\sigma(1)} = j_1, \ldots, i_{\sigma(q)} = j_q$, then

$$u_{i_1} \wedge \ldots \wedge u_{i_q} (e_{j_1}, \ldots, e_{j_q}) = \epsilon(\sigma).$$

The relations (6.2.27) imply that all coefficients in (6.2.26) vanish. We summarize this in a Theorem.

Theorem 6.2.13. Let $E$ have dimension $n$. The space $A^q(E, \mathcal{C})$ has dimension $\binom{n}{q}$, $1 \leq q \leq n$. If $q > n$ then the space $A^q(E, \mathcal{C}) = \{0\}$. If $1 \leq q \leq n$ a canonical basis for $A^q(E, \mathcal{C})$ is the set of multilinear forms $u_{i_1} \wedge \ldots \wedge u_{i_q}$, $i_1 < i_2 < \ldots < i_q$.

## 6.3. Differential forms and the operator d.

The abstract algebraic theory is now applied when $\mathcal{C}$ is a local coefficient ring $\mathcal{C}_{x,k}$ as described following (6.1.3). In the sequel we suppose to each $x \in \mathfrak{M}$ there is assigned the canonical basis elements of $E_x$, and of $M(E_x, \mathcal{C}_{x,k})$ the canonical basis is

(6.3.1) $$u_1^x, \ldots, u_n^x \quad \text{satisfying} \quad u_i^x(e_j^x) = \delta_{ij}.$$

From Theorem 6.2.13 the terms $u_{i_1}^x \wedge \ldots \wedge u_{i_q}^x$ are a basis of $A^q(E_x, \mathcal{C}_{x,k})$ locally at $x$. We assume $k \geq 2$ but do not otherwise specify k.

We define an operator $d: A^q(E, \mathcal{C}) \to A^{q+1}(E, \mathcal{C})$ simultaneously for all $q \geq 0$ and $\mathcal{C} = \mathcal{C}_{x,k}$, $k \geq 1$.

Definition 6.3.1. If $f \in \mathcal{C}_{x,k}$ then

$$(6.3.2) \qquad df = \sum_{i=1}^{n} \frac{\partial f}{\partial p_i} (\varphi^{-1}(x,p_1,\ldots,p_n))\Big|_{\varphi(x,p)} u_i^x .$$

If $q \geq 1$ and $f\, u_{i_1}^x \wedge \ldots \wedge u_{i_q}^x \in A^q(E_x, \mathcal{C}_{x,1})$ then

$$(6.3.3) \qquad d(f\, u_{i_1}^x \wedge \ldots \wedge u_{i_q}^x) = (df) \wedge u_{i_1}^x \wedge \ldots \wedge u_{i_q}^x .$$

Extend the definition of $d$ by linearity to all of $A^q(E_x, \mathcal{C}_{x,1})$.

Note that by the basis Theorem 6.2.13 the extension of the definition of $d$ by linearity is meaningful. In terms of Definition 6.3.1 we define a relation of equivalence, denoted "$\equiv$" on the intersections $U_x \cap U_y$. If $p = \varphi(x,\varphi^{-1}(y,q))$ we define

Definition 6.3.2.

$$(6.3.4) \qquad \Delta_{ij}(x,y,z) = \frac{\partial p_i}{\partial q_j}\Big|_{\varphi(y,z)} .$$

The function $\Delta_{ij}$ is defined for $z \in U_x \cap U_y$. $p$ as a function of $q \in \mathbb{R}_n$ is $\mathbb{R}_n$ valued so $p^t = (p_1,\ldots,p_n)$. (6.3.4) is to be read as the partial derivative evaluated at $\varphi(y,z) \in \mathbb{R}_n$.

Definition 6.3.3. An element of $A^q(E_x, \mathcal{C}_{x,k})$, $q > 0$, will be called a differential form defined at $x$.

Definition 6.3.4. The equivalence relation $\equiv$ is defined by the following rules. Functions are self equivalent. Next,

$$(6.3.5) \qquad u_i^x \equiv \sum_{j=1}^{n} \Delta_{ij}(x,y,\ )u_j^y .$$

Last, a differential form defined at $x$ is equivalent to a differential form defined at $y$ if upon substitution of $\sum_{j=1}^{n} \Delta_{ij}(x,y,\ )u_j^y$ for $u_i^x$, $1 \leq i \leq n$, the differential form at $x$ is transformed into the differential form at $y$ on the set $U_x \cap U_y$. The coefficients being functions defined on $U_x \cap U_y$ do not require transformation in this definition.

Theorem 6.3.5. "$\equiv$" is an equivalence relation.

Proof. Reflexive. $\varphi(x,\varphi^{-1}(x,\ ))$ is the identity map of $\mathbb{R}_n$ onto $\mathbb{R}_n$. Consequently the functions (6.3.4) and (6.3.5) are

(6.3.6) $$\Delta_{ii}(x,x,z) = 1 \quad \text{for all} \quad z \in U_x;$$
$$\Delta_{ij}(x,x,z) = 0,\ i \neq j, \quad \text{for all} \quad z \in U_x; \quad \text{and}$$
$$u_i^x \equiv u_i^x\ .$$

Substitution into a differential form now clearly shows "$\equiv$" to be a reflexive relation.

Symmetric. To shorten the notations let

(6.3.7) $$\psi(\ ) = \varphi(x,\ \varphi^{-1}(y,\ )).$$

Then clearly

(6.3.8) $$\psi^{-1}(\ ) = \varphi(y,\ \varphi^{-1}(x,\ )).$$

If we write $\psi_{ij}$ for the partial derivative of the i-th component of $\psi$ with respect to the j-th variable of $\psi$, and similarly for $\psi^{-1}$, then the chain rule clearly requires

(6.3.9) $$\delta_{ij} = \sum_{k=1}^{n} \psi_{ik}\psi_{kj}^{-1}\ ,$$

identically on the part of $\mathbb{R}_n$ on which the functions are defined. Making a double substitution, as required by Definition 6.3.4, shows "$\equiv$" to be a symmetric relation.

Transitive. To continue the notation of (6.3.7) we write

(6.3.10) $\psi = \varphi(x,\varphi^{-1}(y,\ )), \overline{\psi} = \varphi(y,\varphi^{-1}(z,\ ))$, and

$$\overline{\overline{\psi}} = \varphi(x,\varphi^{-1}(z,\ )).$$

Here the chain rule clearly requires

(6.3.11) $$\overline{\overline{\psi}}_{ij} = \sum_{k=1}^{n} \psi_{ik}\overline{\psi}_{kj}\ .$$

A double substitution as required by Definition 6.3.4 shows "$\equiv$" to be transitive. #

The definition of "$\equiv$" was made for a single value of $q$, but in the sequel we assume the same symbol of equivalence applies to all the alternating algebras $A^1,\ldots,A^n$ simultaneously, and for all coefficient rings $\mathcal{C}_{x,k}$, $k \geq 1$.

We now develop properties of the operator $d$. It should be noted that from Definitions 6.3.1 and 6.3.4,

(6.3.12) $$u^x \equiv d\varphi(x,\varphi^{-1}(y,\ )),$$

where $d$ is computed locally at $y$ and both sides of (6.3.12) are n-vectors.

Theorem 6.3.6. If $f \in A^q(E_x, \mathcal{C}_{x,2})$ and $g \in A^r(E_x, \mathcal{C}_{x,2})$ then the operator $d$ is a linear operator. The following relations hold.

(6.3.13) $$d(f\wedge g) = (df)\wedge g + (-1)^q f\wedge(dg).$$

(6.4.14) $$d(df) = d^2 f = 0.$$

Proof. It is sufficient to verify (6.3.13) and (6.3.14) on the basis elements. Let $f = f_1\, u^x_{i_1}\wedge\ldots\wedge u^x_{i_q}$ and $g = g_1\, u^x_{j_1}\wedge\ldots\wedge u^x_{j_r}$. As a bilinear form $f\wedge g = f_1 g_1\, u^x_{i_1}\wedge\ldots\wedge u^x_{j_1}\wedge\ldots\wedge u^x_{j_r}$, so that

$$(6.3.15)\qquad d(f\wedge g) = d(f_1g_1)\wedge u^x_{i_1}\wedge\ldots\wedge u^x_{j_r}$$

$$= ((df_1)g_1 + f_1(dg_1))\wedge u^x_{i_1}\wedge\ldots\wedge u^x_{j_r}$$

$$= ((df_1)\wedge u^x_{i_1}\wedge\ldots\wedge u^x_{i_q})\wedge((g_1)\ u^x_{j_1}\wedge\ldots\wedge u^x_{j_r})$$

$$+ (-1)^q((f_1)\ u^x_{i_1}\wedge\ldots\wedge u^x_{i_q})\wedge((dg_1)\wedge u^x_{j_1}\wedge\ldots\wedge u^x_{j_r})$$

$$= (df)\wedge g + (-1)^q\ f\wedge(dg).$$

In (6.3.15) we use the fact that $d(f_1g_1) = (df_1)g_1 + f_1(dg_1) \in A^1$, as follows at once from (6.3.2).

To show $d^2 = 0$, we compute $d^2$ on a basis element.

$$(6.3.16)\qquad d^2(f_1u^x_{i_1}\wedge\ldots\wedge u^x_{i_q}) = d(df_1\wedge u^x_{i_1}\wedge\ldots\wedge u^x_{i_q})$$

$$= (\sum_{j=1}^{n} (d(\frac{\partial f}{\partial x_j}\ u^x_j))\wedge u^x_{i_1}\wedge\ldots\wedge u^x_{i_q})$$

$$= (\sum_{j=1}^{n}\sum_{k=1}^{n} \frac{\partial^2 f}{\partial x_k\partial x_j}\ u^x_k\wedge u^x_j)\wedge u^x_{i_1}\wedge\ldots\wedge u^x_{i_q} = 0.$$

This follows since the terms $u^x_k\wedge u^x_k = 0$ while if $k \neq j$ then $u^x_k\wedge u^x_j = -u^x_j\wedge u^x_k$. Since $f$ is assumed to have continuous second partial derivatives,

$$\frac{\partial^2 f}{\partial x_k\partial x_j} = \frac{\partial^2 f}{\partial x_j\partial x_k}\ .\ \#$$

We state an obvious lemma.

<u>Lemma 6.3.7</u>. If differential forms $\omega_1$ and $\omega_2$ are defined locally at $x$, $\omega_3$ and $\omega_4$ are defined locally at $y$, and $\omega_1 \equiv \omega_3$, $\omega_2 \equiv \omega_4$, then $\omega_1\wedge\omega_2 \equiv \omega_3\wedge\omega_4$.

<u>Theorem 6.3.8</u>. Let $\omega_1$ and $\omega_2$ be differential forms such that $\omega_1$ is defined locally at $x$, and $\omega_2$ is defined locally at $y$,

and $\omega_1 \equiv \omega_2$ (on $U_x \cap U_y$). Then $d\,\omega_1 \equiv d\,\omega_2$.

Proof. We first verify the theorem for zero forms. Thus $\omega$ represents a function defined on $U_x \cap U_y$. Locally at $x$,

$$(6.3.17) \qquad d\omega = \sum_{i=1}^{n} \frac{\partial\omega}{\partial p_1}(\varphi^{-1}(x,\ ))\Big|_{\varphi(x,\ )} u_i^x ,$$

and locally at $y$,

$$(6.3.18) \qquad d\omega = \sum_{i=1}^{n} \frac{\partial\omega}{\partial p_1}(\varphi^{-1}(y,\ ))\Big|_{\varphi(y,\ )} u_i^y .$$

Use of the substitutions (6.3.6) and the chain rule shows that the differential forms exhibited in (6.3.17) and (6.3.18) are equivalent 1-forms.

In order to simplify subsequent computations we take the following observation. Suppose $p_i$ is the i-th component of $\varphi(x,\varphi^{-1}(y,\ ))$. We suppose this is a function of $q \in \mathbb{R}_n$, $q^t = (q_1,\ldots,q_n)$. Then $\frac{\partial p_i}{\partial q_j}\Big|_{\varphi(y,\ )} = \Delta_{ij}(x,y,\ )$ is defined on $U_x \cap U_y$. Locally at $y$ we compute

$$(6.3.19) \qquad d\varphi(x,\varphi^{-1}(y,\varphi(y,\ ))) = \sum_{j=1}^{n} \Delta_{ij}(x,y,\ )\, u_j^y \quad \text{and}$$

$$0 = d^2\varphi(x,\varphi^{-1}(y,\varphi(y,\ ))) = \sum_{j=1}^{n} (d\Delta_{ij}(x,y,\ ))\wedge u_j^y .$$

Given a basis element $f\ u_{i_1}^x \wedge\ldots\wedge u_{i_q}^x$ equivalent to a differential form $\omega$ locally at $y$, we have

$$(6.3.20) \qquad \omega = f\Big(\sum_{j=1}^{n} \Delta_{i_1 j}(x,y,\ )u_j^y\Big)\wedge\ldots\wedge\Big(\sum_{j=1}^{n} \Delta_{i_q j}(x,y,\ )u_j^y\Big).$$

We now use (6.3.13) and (6.3.19) and Lemma 6.3.7. Then, computed locally at $y$,

$$(6.3.21)\quad d\omega = (df)\wedge\left(\sum_{j=1}^{n}\Delta_{i_1 j}(x,y,\ )u_j^y\right)\wedge\ldots\wedge\left(\sum_{j=1}^{n}\Delta_{i_q j}(x,y,\ )u_j^y\right)$$

$$\equiv df\wedge u_{i_1}^x\wedge\ldots\wedge u_{i_q}^x = d(fu_{i_1}^x\wedge\ldots\wedge u_{i_q}^x).$$

The last part of (6.3.21) is computed locally at $x$, finishing the proof. #

Corollary 6.3.9. If $f\colon \mathfrak{M}\to\mathbb{R}$ is a globally defined 0-form which is continuously differentiable then $df$ is a locally defined 1-form, local at $x$ for all $x\in\mathfrak{M}$, representing equivalent forms on the overlaps of charts.

## 6.4. Theory of integration.

Definition 6.4.1. If $\omega_1 = f\, u_1^x\wedge\ldots\wedge u_n^x$ is defined locally at $x$ and if $A\subset U_x$ is a Borel subset of $U_x$ then

$$(6.4.1)\qquad \int_A \omega_1 =_{\mathrm{def}} \int_{\varphi(x,A)} f(\varphi^{-1}(x,(p_1,\ldots,p_n)))dp_1\ldots dp_n.$$

If $\omega$ is a general n-form, since by Theorem 6.2.13 the dimension of $A^n(E,\mathcal{C})$ is one, $\omega$ has the form of $\omega_1$.

Lemma 6.4.2. Let $\omega_1$ and $\omega_2$ be differential n-forms such that $\omega_1$ is defined locally at $x$, and $\omega_2$ is defined locally at $y$, and $\omega_1\equiv\omega_2$. If the determinants $\det(\Delta_{ij}(x,y,z))$ are everywhere positive on $U_x\cap U_y$ and if $C$ is a Borel subset of $U_x\cap U_y$ then

$$(6.4.2)\qquad \int_C \omega_1 = \int_C \omega_2.$$

Proof. We let $\omega_1 = f\, u_1^x\wedge\ldots\wedge u_n^x$. By (6.3.5), the condition of equivalence, and the algebraic relations (6.2.19),

(6.4.3) $$\omega_2 = f(\sum_{j=1}^{n} \Delta_{1j}(x,y,z)u_j^y) \wedge \ldots \wedge (\sum_{j=1}^{n} \Delta_{nj}(x,y,z)u_j^y)$$

$$= f(\det(\Delta_{ij}(x,y,z)))u_1^y \wedge \ldots \wedge u_n^y.$$

Therefore using the notation of (6.3.4) that $p = \varphi(x,\varphi^{-1}(y,q))$, we obtain

(6.4.4) $$\int_C \omega_2 = \int_{\varphi(y,C)} f(\varphi^{-1}(y,q))\det(\frac{\partial p_i}{\partial q_j})dq_1 \ldots dq_n.$$

Since $\varphi^{-1}(x,p) = \varphi^{-1}(x, \varphi(x,\varphi^{-1}(y,q))) = \varphi^{-1}(y,q)$ we obtain from (6.4.4)

(6.4.5) $$\int_C \omega_2 = \int_{\varphi(x,C)} f(\varphi^{-1}(x,p))\ dp_1 \ldots dp_n = \int_C \omega_1. \#$$

Definition 6.4.3. A globally defined differential n-form $\omega = \{\omega_x, x \in \mathfrak{M}\}$ is a set of locally defined differential n-forms satisfying

(6.4.6) if $x,y \in \mathfrak{M}$ then $\omega_x \equiv \omega_y$.

Definition 6.4.4. If $\omega$ is a globally defined n-form then $|\omega|$ is defined locally by $|\omega| = |\omega_x| = |f|\ u_1^x \wedge \ldots \wedge u_n^x$.

Theorem 6.4.5. If $\omega$ is a globally defined n-form on a $C_2$ manifold $\mathfrak{M}$ having countable base for the open sets then $\int_C \omega$ defines a countably additive signed measure which is a regular Borel measure.

Proof. The proof that $\int_C \omega$ defines a measure is an obvious application of Lemma 6.4.2. This application uses the countable cover of $\mathfrak{M}$ to obtain a $\sigma$-finite measure on the manifold. The integral over subsets C will be finite for compact subsets C. It follows from Halmos (1950) that the measure is a regular measure (we assume the manifold is a locally compact Hausdorff space, so that given a countable base for the open sets Baire and Borel subsets are the same.)#

The integrations in subsequent chapters will be purely formal calculations. It is hoped that the machinery of Chapter 6 will provide sufficient theory to justify the applications. In the integrations made a frequent situation is the following. Globally defined functions $a_1,\dots,a_n$ are given and the globally defined differential form is

$$(6.4.7) \qquad \omega = f(a_1,\dots,a_n)\, da_1\wedge\dots\wedge da_n,$$

where on a n-dimensional manifold the differential form (6.4.7) is of maximal degree and hence is integrable locally when $da_1,\dots,da_n$ are computed locally. If we have a Borel subset $C \subset U_x$ let $C'$ be the corresponding set in the range of $(a_1,\dots,a_n)$. Then if we set $b_i(x,p_1,\dots,p_n) = a_i(\varphi^{-1}(x,p_1,\dots,p_n))$, $1 \leq i \leq n$, it follows that

$$(6.4.8) \qquad \int_C \omega = \int_{\varphi(x,C)} f(b_1,\dots,b_n) \det \frac{\partial b_i}{\partial p_j}\, dp_1\dots dp_n$$

$$= \int_{C'} f(a_1,\dots,a_n)\, da_1\dots da_n.$$

In (6.4.8) the variables $a_1,\dots,a_n$ are now formal variables of integration for the n-fold integral over a subset of $R_n$.

In some examples the map $x \longleftrightarrow (a_1(x),\dots,a_n(x))$ is not one to one, this condition failing on a null set $N$ which is seen to satisfy, if $x \in \mathfrak{M}$ then

$$(6.4.9) \qquad \int_{U_x\cap N} u_1\wedge\dots\wedge u_n = 0.$$

In such problems the null set $N$ is usually ignored.

## 6.5. Transformation of manifolds.

We suppose $\mathfrak{M}_1$ and $\mathfrak{M}_2$ are n-dimensional manifolds and that $f: \mathfrak{M}_1 \to \mathfrak{M}_2$ is a homeomorphism. The basic assumption here is that if $g: \mathfrak{M}_2 \to \mathbb{R}$ is a $C_2$ function then the composition $g\circ f: \mathfrak{M}_1 \to \mathbb{R}$

is also a $C_2$ function.

As just noted $f$ induces a map $F$ of $\mathfrak{M}_2$ 0-forms to $\mathfrak{M}_1$ 0-forms. We extend this map to all differential forms subject to the requirement $dF = Fd$, where in each case the operator $d$ is to be computed in the appropriate local coordinates.

If $y = f(x)$ then for $z$ near $x$ we have

$$(6.5.1) \qquad \varphi_2(y, f(z)) \colon \mathfrak{M}_1 \to \mathbb{R}_n$$

gives local coordinates of $f(z)$ on $\mathfrak{M}_2$ near $y$. We want

$$(6.5.2) \qquad \varphi_2(y, f(\varphi_1^{-1}(x,)))$$

to be twice continuously differentiable where defined. If

$$(6.5.3) \qquad (p_1, \ldots, p_n) = \varphi_1(x,z)^t \quad \text{and}$$
$$(q_1, \ldots, q_n) = \varphi_2(y, f(z))^t = \varphi_2(f(x), f(z))^t$$

then $p_1, \ldots, p_n$ are functions defined on $\mathfrak{M}_1$ as are $q_1, \ldots, q_n$. Locally near $x$ we may compute $dq_1, \ldots, dq_n$.

Definition 6.5.1. Define $F$ on 0-forms by

$$(6.5.4) \qquad (Fg)(x) = g(f(x)).$$

Extend $F$ to basis elements of r-forms by the definition

$$(6.5.5) \qquad F(g\, v_{i_1}^y \wedge \ldots \wedge v_{i_r}^y) = (Fg)\, dq_{i_1} \wedge \ldots \wedge dq_{i_r}$$

where $dq_1, \ldots, dq_n$ are computed locally at $x$. Extend $F$ to be a linear transformation of differential forms.

Theorem 6.5.2.

$$(6.5.6) \qquad dF = Fd.$$

Proof. By linearity of $d$ and $F$ it is sufficient to consider basis terms, $g\, v^y_{i_1} \wedge \ldots \wedge v^y_{i_r}$. Using (6.3.13) and (6.3.14),

$$(6.5.7) \qquad d(F(g\, v^y_{i_1} \wedge \ldots \wedge v^y_{i_r})) = d(Fg) \wedge dq_{i_1} \wedge \ldots \wedge dq_{i_r}.$$

Also

$$(6.5.8) \qquad d(Fg) = d(g(f(\ ))) = \sum_{i=1}^{n} \sum_{j=1}^{n} \frac{\partial g}{\partial q_i} \frac{\partial q_i}{\partial p_j} u^x_j$$

$$= \sum_{i=1}^{n} \frac{\partial g}{\partial q_i} dq_i .$$

Also

$$(6.5.9) \qquad F(dg \wedge v^y_{i_1} \wedge \ldots \wedge v^y_{i_r}) = F\Big(\Big(\sum_{i=1}^{n} \frac{\partial g}{\partial q_i} v^y_i\Big) \wedge v^y_{i_1} \wedge \ldots \wedge v^y_{i_r}\Big)$$

$$= \Big(\sum_{i=1}^{n} \frac{\partial g}{\partial q_i} dq_i\Big) \wedge dq_{i_1} \wedge \ldots \wedge dq_{i_r} .$$

Thus (6.5.6) holds. #

The transformation $F$ of differential forms then extends to a transformation of measures obtained by integration of n-forms.

Definition 6.5.3. If $\omega$ is a globally defined n-form on $\mathfrak{M}$, and $\mu$ is defined by $\mu(C) = \int_C \omega$, then $F^{-1}\mu$ is defined by

$$(6.5.10) \qquad (F^{-1}\mu)(C) = \mu(f^{-1}(C)).$$

Theorem 6.5.4. Let $f$ be a transformation of manifolds as described above, $f: \mathfrak{M}_1 \to \mathfrak{M}_2$. Suppose the Jacobians of the transformations (6.5.2) are positive. If $\omega$ is an n-form on $\mathfrak{M}_2$ then

$$(6.5.11) \qquad \int_C F(\omega) = \int_{f(C)} \omega .$$

Proof. Locally for a basis element $g\, v^y_1 \wedge \ldots \wedge v^y_n$ the transformation maps this n-form into $(Fg)\det\left(\frac{\partial q_i}{\partial p_j}\right) u^x_1 \wedge \ldots \wedge u^x_n$. Let $g' = g(\varphi^{-1}(y, \ ))$. Then

$$(6.5.12) \qquad \int_C (Fg) \det(\frac{\partial q_i}{\partial p_j}) dp_1 \cdots dp_n$$

$$= \int_C g'(q_1(p_1,\ldots,p_n),\ldots,q_n(p_1,\ldots,p_n)) \det(\frac{\partial q_i}{\partial p_j}) dp_1 \cdots dp_n$$

$$= \int_{f(C)} g'(q_1,\ldots,q_n)\, dq_1 \cdots dq_n = \int_{f(C)} \omega \ .$$

By linearity the result follows for all local n-forms. Since $\mathfrak{M}_1$ and $\mathfrak{M}_2$ are assumed to be locally compact separable Hausdorff spaces, one may choose a countable cover of sets $U_x$. The Theorem then follows by a countable additivity argument. #

<u>Definition 6.5.5</u>. If $f: \mathfrak{M} \to \mathfrak{M}$ is a transformation of manifolds with induced mapping $F$ of differential forms, then a differential form $\omega = \{\omega_x, x \in \mathfrak{M}\}$ which is globally defined is said to be invariant if,

$$(6.5.13) \qquad \text{if } x \in \mathfrak{M} \text{ then } F(\omega_x) = \omega_{f^{-1}(x)} .$$

<u>Theorem 6.5.6</u>. If $f: \mathfrak{M} \to \mathfrak{M}$ is a transformation of manifolds and the globally defined differential form $\omega$ is invariant then the measure $\mu$ defined by $\mu(C) = \int_C \omega$ satisfies

$$(6.5.14) \qquad \mu = F\mu.$$

<u>Proof</u>. Locally, if $C \subset U_x$ and $y = f(x)$ and $f(C) \subset U_y$, then by (6.5.11)

$$(6.5.15) \qquad \mu(f(C)) = \int_{f(C)} \omega_y = \int_C F(\omega_y) = \int_C F(\omega_{f(x)}) = \int_C \omega_x = \mu(C).$$

Since $f$ is a homeomorphism the sets

$$(6.5.16) \qquad U_x \cap f^{-1}(U_{f(x)})$$

each contain the index point $x$, hence are nonempty open sets and $\mathfrak{M}$ thus has a countable subcover $\{U_{x_i} \cap f^{-1}(U_{f(x_i)}),\ i \geq 1\}$. Thus we may construct a measurable partition $B_i$, $i \geq 1$ of $\mathfrak{M}$ such that if $i \geq 1$ then

(6.5.17) $$B_i \subset U_{x_i}.$$

Given a Borel subset $C$ then (6.5.15) applies to $C \cap B_i$, $i \geq 1$, so that

(6.5.18) $$\mu(f(C)) = \sum_{i=1}^{\infty} \mu(f(C \cap B_i)) = \sum_{i=1}^{\infty} \mu(C \cap B_i) = \mu(C). \ \#$$

## 6.6 A matrix lemma.

Lemma 6.6.1. Let $h$ be a function of $n \times n$ matrices with real entries such that $h(I_n) \neq 0$ and

(6.6.1) for each $A \in GL(n)$, $h(A)$ is a homogeneous polynomial of degree $r$ in the entries of $A$;

(6.6.2) if $A, B \in GL(n)$ then $h(AB) = h(A)h(B)$.

Then $r/n$ is an integer and

(6.6.3) $$h(A) = (\det A)^{(r/n)}.$$

Proof. If $I_n$ is the $n \times n$ identity matrix then $h(I_n) = h(I_n)^2 = h(I_n)^3$ so $h(I_n) = 1$ follows. Then $1 = h(I_n) = h(AA^{-1}) = h(A)\,h(A^{-1})$ so that $h(A) = 1/h(A^{-1})$. Since $h$ is continuous, $h(\underline{O}(n))$ is a compact set of real numbers that is a subgroup. Hence $h(\underline{O}(n)) = \{1, -1\}$, or $h(\underline{O}(n)) = \{1\}$.

Next, take $A_1$ to be the diagonal matrix with $d$ in the (1,1)-position and elsewhere on the diagonal entries $= 1$. Let $A$ be obtained from $A_1$ from permutation of the (1,1)-entry into the (i,i)-entry so that $A_i = QA_1Q^t$ for some permutation matrix $Q$. It follows that $h(A_1) = h(A_i)$, since

$$(6.6.4) \qquad h(A_i) = h(Q)\,h(A_1)\,h(Q^t) = h(A_1).$$

Then the diagonal matrix $dI_n$ factors into $dI_n = A_1 \cdots A_n$ and

$$(6.6.5) \qquad d^r = h(dI_n) = h(A_1 \cdots A_n) = (h(A_1))^n.$$

Since $h(A_1)$ is a polynomial in the variable $d$, it follows that $r/n$ is an integer and

$$(6.6.6) \qquad h(A_1) = d^{r/n}.$$

The identities (6.6.5) and (6.6.6) clearly imply that if $A$ is a diagonal matrix then (6.6.3) holds for $A$.

If $A$ is a symmetric matrix then there exists $U \in \underline{O}(n)$ such that $UAU^t$ is a diagonal matrix, and by the result just obtained, if $A$ is also nonsingular then $h(A) = (\det A)^{r/n}$. For the arbitrary $X \in GL(n)$, we may write $X = AS$ with $A = X(X^tX)^{-1/2} \in \underline{O}(n)$ and $S = (X^tX)^{1/2} \in \underline{S}(n)$. Then the above implies

$$(6.6.7) \qquad h(X) = h(A)\,h(S) = (\pm h(A))(\det X)^{r/n}$$

since $\det(X^tX)^{1/2} = \pm\det X$ depending on the sign choice of the square root.

Since $A \in \underline{O}(n)$ implies $h(A) = \pm 1$, it follows that if $X \in GL(n)$ then $h(X) = \pm(\det X)^{r/n}$. Note that $\{X \mid \det X > 0\}$ is an open subset of $\mathbb{R}_{n^2}$ so one may choose an open neighborhood of $I_n$, call it $U$, such that if $X \in U$ then $|h(X)-(\det X)^{r/n}| < 1/2$ because $|h(X) - 1| < 1/4$ and $|(\det X)^{r/n} - 1| < 1/4$. We use here the continuity of the functions. This clearly implies that if $X \in U$ then $h(X) = (\det X)^{r/n}$. Since these polynomials agree on an open set they are everywhere equal. #

## 6.7. Problems

Problem 6.7.1. If $(p_1,\dots,p_n)$ are local coordinates of $z$ near $x$ and $(q_1,\dots,q_n)$ are local coordinates of $z$ near $y$, so that $z \in U_x \cap U_y$, then

$$f\, u^x_{i_1} \wedge \dots \wedge u^x_{i_r} \quad \text{and} \quad f(dp_{i_1}) \wedge \dots \wedge (dp_{i_r}) \tag{6.7.1}$$

are equivalent r-forms where the latter is computed at $y$.

Problem 6.7.2. Let $u_1 \wedge \dots \wedge \underline{u_i} \wedge \dots \wedge u_n = u_1 \wedge \dots \wedge u_{i-1} \wedge u_{i+1} \wedge \dots \wedge u_n$. The join (wedge product) of n-1 1-forms

$$\left(\sum_{i=1}^{n} b_{1i}u_i\right) \wedge \dots \wedge \left(\sum_{i=1}^{n} b_{(n-1)i}u_i\right) \tag{6.7.2}$$

$$= \sum_{i=1}^{n} \tilde{a}_i\, u_1 \wedge \dots \wedge \underline{u_i} \wedge \dots \wedge u_n,$$

where $\tilde{a}_i$ is given by the following determinant, $1 \leq i \leq n$:

$$\tilde{a}_i = \begin{vmatrix} b_{11} & \dots & b_{1\,i-1} & b_{1\,i+1} & \dots b_{1n} \\ \vdots & & \vdots & \vdots & \vdots \\ b_{n-1\,1} & \dots & b_{n-1\,i-1} & b_{n-1\,i+1} & b_{n-1\,n} \end{vmatrix}. \tag{6.7.3}$$

Problem 6.7.3. Continuation of Problem 6.7.2. Let the $b_{ij}$ be functions of $a_1,\dots,a_n$ such that the matrix

$$\begin{pmatrix} a_1 & a_2 & \dots & a_n \\ b_{11} & b_{12} & \dots & b_{1n} \\ \vdots & \vdots & & \\ b_{n-1\,1} & b_{n-1\,2} & \dots & b_{n-1\,n} \end{pmatrix} \tag{6.7.4}$$

is an orthogonal matrix with determinant $= \epsilon$. Compute $da_1,\dots,da_n$ locally at $p$. Then

(6.7.5) $$\left(\sum_{i=1}^{n} b_{1i} da_i\right) \wedge \ldots \wedge \left(\sum_{i=1}^{n} b_{n-1\, i}\, da_i\right)$$

$$= \epsilon \sum_{i=1}^{n} a_i (-1)^{i+1}\, da_1 \wedge \ldots \wedge \underline{da_i} \wedge \ldots \wedge da_n .$$

Problem 6.7.4. Continue Problem 6.7.3. Since $a_1^2 + \ldots + a_n^2 = 1$, we find

(6.7.6) $$da_n = \sum_{i=1}^{n-1} \frac{-a_i\, da_i}{(1 - a_1^2 - \ldots a_{n-1}^2)^{1/2}}$$

$$= \sum_{i=1}^{n-1} -a_i\, da_i / a_n .$$

Substitution into (6.7.5) shows (6.7.5) to be equal to

(6.7.7) $$\frac{(-1)^{n+1} \epsilon\, da_1 \wedge \ldots \wedge da_{n-1}}{(1 - a_1^2 - \ldots - a_{n-1}^2)^{1/2}} .$$

Problem 6.7.5. Let $\mathfrak{M} = \mathfrak{M}_1 = \mathfrak{M}_2 = \mathbb{R}_{nh}$ with local coordinates given globally by $n \times h$ matrices $X$. Let $A$ be an $n \times h$ matrix and define $f(X) = AX$. Let the canonical ordering of 1-forms $u_{ij}$ be $\bigwedge_{j=1}^{h} \bigwedge_{i=1}^{n} u_{ij}$. Show a nh-form $\omega = g \bigwedge_{j=1}^{h} \bigwedge_{i=1}^{n} u_{ij}$ transforms to

(6.7.8) $$F\omega = (g \circ f) \bigwedge_{j=1}^{h} \bigwedge_{i=1}^{n} \left(\sum_{k=1}^{n} a_{ik} u_{kj}\right) =$$

$$= (\det A)^h (g \circ f) \bigwedge_{j=1}^{h} \bigwedge_{i=1}^{n} u_{ij} .$$

Problem 6.7.6. If $T = \begin{pmatrix} t_{11} & 0 & \ldots & 0 \\ t_{21} & t_{22} & \ldots & 0 \\ t_{n1} & t_{n2} & \ldots & t_{nn} \end{pmatrix}$ and $S = (s_{ij}) = TT^t$,

then wanted is the Jacobian of the substitution $t_{ij} \to s_{ij}$, $1 \leq j \leq i$, $1 \leq i \leq n$.

The value of the Jacobian is not a determinant and this problem is not especially tractibleto exterior algebra. However,

$$s_{nj} = t_{n1}t_{j1} + \dots + t_{nj}t_{jj}, \quad 1 \leq j \leq n, \tag{6.7.9}$$

so for these $n$ variables the Jacobian of the substitution is

$$\frac{\partial(s_{n1},\dots,s_{nn})}{\partial(t_{n1},\dots,t_{nn})} = \det\begin{pmatrix} t_{11} & 0 & \dots & 0 \\ & & & \\ 2t_{n1} & 2t_{n2} & \dots & 2t_{nn}\end{pmatrix} = 2t_{11}t_{22}\dots t_{nn}. \tag{6.7.10}$$

By induction show the required Jacobian is

$$2^n\, t_{11}^n t_{22}^{n-1} \dots t_{nn}. \tag{6.7.11}$$

<u>Problem 6.7.7</u>. Let $S$ be a $n \times n$ symmetric matrix and let $A \in GL(n)$. Set $T = ASA^t$. If the entries of $S$ are differential forms then so are the entries of $T$. Show

$$\bigwedge_{j\leq i} t_{ij} = h(A) \bigwedge_{j\leq i} s_{ij}\,, \tag{6.7.12}$$

where $h(A)$ is a homogeneous polynomial of degree $n(n+1)$ in the variables $(a_{ij})$. Show that $h$ must satisfy $h(AB) = h(A)h(B)$ for all $A, B \in GL(n)$ so that by Lemma 6.6.1, $h(A) = (\det A)^{n+1}$.

<u>Problem 6.7.8</u>. Let $S$ be a $n \times n$ symmetric positive definite matrix and $T = S(I_n + S)^{-1}$. Show that $T$ is a symmetric matrix. Compute

$$d\,T(I_n + S) = dS \tag{6.7.13}$$

and show $I_n = (I_n - T)(I_n + S)$. Then

$$dS = (I_n - T)^{-1}\, dT(I_n - T)^{-1}. \tag{6.7.14}$$

By problem 6.7.7 show

$$(6.7.15)\qquad \bigwedge_{j\le i} ds_{ij} = (\det(I_n - T))^{-(n+1)} \bigwedge_{j\le i} dt_{ij}.$$

Problem 6.7.9. Continue Problems 6.7.7 and 6.7.8. Let $A, T \in \underline{T}(n)$ be nonsingular lower triangular matrices and let $S = AT$. Show the Jacobian of the substitution $t_{ij} \to (AT)_{ij} = s_{ij}$ is

$$(6.7.16)\qquad a_{11}\, a_{22}^2 \cdots a_{nn}^n.$$

Also, note that

$$(6.7.17)\qquad \bigwedge_{j\le i} ds_{ij} = h(A) \bigwedge_{j\le i} dt_{ij}$$

where $h(A) = a_{11}a_{22}^2 \cdots a_{nn}^n$. Thus $h$ is a homogeneous polynomial in the entries of $A$ such that $h(A)h(B) = h(AB)$ but $h(A) \ne (\det A)^r$. Thus Lemma 6.6.1 fails if the function $h$ is defined on a proper subgroup of $GL(n)$.

Problem 6.7.10. Let $f: \mathfrak{M}_1 \to \mathfrak{M}_2$ be a transformation of manifolds with induced mapping $F$ of differential forms. Show

$$(6.7.18)\qquad F(a \wedge b) = (Fa) \wedge (Fb).$$

Hint. Since $F$ is a linear transformation it is sufficient to take $a = g_1\, v_{i_1}^y \wedge \cdots \wedge v_{i_r}^y$ and $b = g_2\, v_{j_1}^y \wedge \cdots \wedge v_{j_s}^y$. #

Problem 6.7.11. Let $\mathfrak{M}_1 = \mathfrak{M}_2 = \underline{S}(n)$ and the transformation of manifolds $f$ be $f(S) = S^{-1}$. Compute the following transformed differential form:

$$(6.7.19)\qquad F\Big(\bigwedge_{j\le i} (dS)_{ij} / (\det S)^{(n+1)/2}\Big).$$

Compare your answer with Problem 7.10.6.

# Chapter 7. Invariant measures on manifolds

## 7.0. Introduction.

In this Chapter we discuss the action of matrix groups on various manifolds. Mostly conclusions will not be stated as formal Theorems except in the last sections of the chapter.

The different manifolds are described below as we treat them. It is our purpose to derive differential forms for regular invariant measures. As suggested in Section 3.4 it is the regular invariant measures which enter into the factorization of measures, and in many examples discussed in Chapter 9 integration out of extra variables is equivalent to integration of the differential form over the entire manifold. The various differential forms introduced in Chapter 3, namely (3.3.18), (3.4.9), (3.4.14), Example 3.5.3 and Section 3.6, are justifiable on the basis of Chapters 6 and 7.

## 7.1. $\mathbb{R}_{nh}$.

We consider the set of $n \times h$ matrices $X$ and let the group action be multiplication on the right by $h \times h$ matrices $A \in GL(h)$. If $Y = XA$ then

$$(7.1.1) \qquad \bigwedge_{j=1}^{h} \bigwedge_{i=1}^{n} dy_{ij} = h(A) \bigwedge_{j=1}^{h} \bigwedge_{i=1}^{n} dx_{ij}$$

where we write $Y = (y_{ij})$, $X = (x_{ij})$ and use the fact that the differential forms are of maximal degree so that space of alternating forms of degree $nh$ has dimension equal one, determining the constant $h(A)$. The function $h$ then satisfies the hypotheses of Lemma 6.6.1. Therefore

$$(7.1.2) \qquad h(A) = (\det A)^{nh/h} = (\det A)^n.$$

In terms of globally defined canonical basis elements $u_{ij}$ for the 1-forms, the differential form

$$(7.1.3) \qquad \omega = (\det X^tX)^{-n/2} \bigwedge_{j=1}^{h} \bigwedge_{i=1}^{n} u_{ij}$$

is an invariant form. The transformation $X \to XA = f(X)$ replaces $(u_{i1},\ldots,u_{ih})$ by $(u_{i1},\ldots,u_{ih})A$ and by (6.2.19) we obtain

$$(7.1.4) \qquad (\sum_{k=1}^{h} u_{ik}a_{k1})\wedge\cdots\wedge(\sum_{k=1}^{h} u_{ik}a_{kh}) = (\det A)u_{i1}\wedge\cdots\wedge u_{ih}.$$

From (7.1.3) and (7.1.4) and Section 6.5 the transformed differential form is

$$(7.1.5) \qquad F\omega = (\det (XA)^t(XA))^{-n/2}(\det A)^n \bigwedge_{j=1}^{h} \bigwedge_{i=1}^{n} u_{ij} = \omega,$$

By Theorem 6.5.6 the differential form $\omega$ defines an invariant measure on the Borel subsets of $\mathbb{R}_{nh}$. The more usual way, of course, of writing the density function of this measure is

$$(7.1.6) \qquad \prod_{j=1}^{h} \prod_{i=1}^{n} dx_{ij}/(\det X^tX)^{n/2}.$$

## 7.2. Lower triangular matrices, left multiplication.

As the Section title implies, right multiplication is a different problem, leading to a different differential form, due to the fact the group $\underline{T}(n)$ is not unimodular.

We consider transformations $T \to AT$, with $A, T \in \underline{T}(n)$. From Problem 6.7.9 with $S = AT$ we find

$$(7.2.1) \qquad \bigwedge_{j\leq i} ds_{ij} = a_{11}a_{22}^2\cdots a_{nn}^n \bigwedge_{j\leq i} dt_{ij}.$$

$\underline{T}(n)$ is an open simply connected subset of $\mathbb{R}_{n(n+1)/2}$ so there exists a one to one $C_\infty$ mapping which maps $\underline{T}(n)$ onto $\mathbb{R}_{n(n+1)/2}$ and this gives a global definition of coordinates, i.e., only one chart function is needed. Consequently in terms of globally defined 1-forms $u_{ij}$ the differential form

(7.2.2) $$\omega = \bigwedge_{j\leq i} u_{ij}/(t_{11}t_{22}^2 \cdots t_{nn}^n)$$

is an invariant differential form. To show this, set $u_{ij} = dt_{ij}$ and obtain

(7.2.3) $$F\omega = (a_{11} \cdots a_{nn}^n) \bigwedge_{j\leq i} dt_{ij}/(a_{11}t_{11}) \cdots (a_{nn}t_{nn})^n = \omega.$$

## 7.3. Lower triangular matrices, right multiplication.

If $S = TA$ then $\bigwedge_{j\leq i} ds_{ij} = a_{11}^n a_{22}^{n-1} \cdots a_{nn} \bigwedge_{j\leq i} dt_{ij}$.

Therefore the differential form (see (7.2.2))

(7.3.1) $$\omega = \bigwedge_{j\leq i} dt_{ij}/(t_{11}^n t_{22}^{n-1} \cdots t_{nn})$$

induces an invariant measure.

## 7.4. The orthogonal group $\underline{O}(n)$

We let $\alpha$ be a generic point of $\underline{O}(n)$ and $a_{ij}(\alpha)$ be the $(i,j)$-entry function. Since $\underline{O}(n)$ is compact local coordinates cannot be globally defined (the continuous image of a compact space is a compact subset of Euclidean space). If we map $\underline{O}(n)$ into $\mathbb{R}_{n(n-1)/2}$ using the functions $\{a_{ij}, 1 \leq j < i \leq n\}$ this map is one to one into and hence is a homeomorphism into $\mathbb{R}_{n(n-1)/2}$. Call this mapping $f$. Clearly there exists an interior point $f(\alpha_0) \subset f(\underline{O}(n))$ and we may choose an open rectangle $U \subset f(\underline{O}(n))$ with $f(\alpha_0) \in U$. Use a $C_\infty$ map of $U$ onto $\mathbb{R}_{n(n-1)/2}$ to obtain a chart function of $\alpha_0$ over the neighborhood $U_{\alpha_0} = f^{-1}(U)$. For the arbitrary point of $\underline{O}(n)$ use translations $U_\alpha = \alpha\,\alpha_0^{-1}(U_{\alpha_0})$ and make the obvious definition of a chart. This makes $\underline{O}(n)$ into a $C_\infty$ manifold.

The question of local coordinates is not treated in more detail here. We suppose it has been shown that the functions $a_{ij}$ are $C_\infty$

functions of the local coordinates. The differential form of interest to us is

$$(7.4.1) \qquad \omega = \bigwedge_{j<i} a_i^t \, da_j$$

where the (column) vectors $a_1,\ldots,a_n$ are the columns of $A$ so that

$$(7.4.2) \qquad A = (a_1,\ldots,a_n) \in \underline{O}(n).$$

The functions $a_1,\ldots,a_n$ are globally defined so that $\omega$ is a globally defined n-form. See Section 6.4.

We begin our calculations by noting that

$$(7.4.3) \qquad I_n = A^tA \quad \text{and} \quad 0 = dI_n = (dA)^tA + A^t(dA).$$

Therefore $A^tdA$ is a skew symmetric matrix.

Let $H \in \underline{O}(n)$ and define a transformation $f:\underline{O}(n) \to \underline{O}(n)$ by $A(f(\alpha)) = H\ A(\alpha)$. Let $F$ be the induced map of differential forms. By Theorem 6.5.2, $dF = Fd$ (when computed locally) so that

$$(7.4.4) \qquad H\ dA = d(HA) = d(F(A)) = F(dA).$$

Therefore

$$(7.4.5) \qquad (FA)^t\ (d(F(A))) = A^tH^t\ H\ dA = A^t\ dA.$$

By Definition 6.5.5 and Theorem 6.5.6 the differential form $\omega$ of (7.4.1) defines a left invariant measure on $\underline{O}(n)$ which by the construction of these measures locally must be of finite mass on $\underline{O}(n)$. The measure is therefore regular, is a Haar measure, and by Chapter 3, is uniquely determined by its total mass and must also be a right invariant measure since $\underline{O}(n)$ is unimodular, see Lemma 3.5.3.

There is a question whether the signed measure so determined can change sign. The positive and negative parts of an invariant measure are easily seen to be invariant, and must be regular measures, hence

Haar measures. Consequently the difference of the positive and negative parts is a Haar measure or is the negative of a Haar measure.

As an exercise we verify the right invariance. Changing our definition, define $f$ by $A(f(\alpha)) = A(\alpha)H$ and let $F$ be the induced map. Then $FA = AH$ and

(7.4.6) $$d(FA) = F(dA) = (dA)H, \quad \text{and}$$

(7.4.7) $$(FA)^t\,(dFA) = H^tA^t\,dA\,H.$$

Write $H = (h_{ij})$ so that the (i,j)-entry of (7.4.7) is

(7.4.8) $$\sum_{k_1=1}^{n}\ \sum_{k_2=1}^{n} h_{k_1 i}\,(a_{k_1}^t\ da_{k_2})h_{k_2 j}.$$

Using the skew symmetry property $a_i^t\,da_j + a_j^t\,da_i = 0$, from (7.4.8) we obtain

(7.4.9) $$\bigwedge_{j<i} (H^tA^t\,dA\,H)_{ij} = f(H) \bigwedge_{j<i} a_i^t\,da_j,$$

where the function $f$ is a polynomial homogeneous of degree $n(n-1)$. (7.4.9) in fact holds for $H \in GL(n)$ rather than just $H \in \underline{O}(n)$. Using successive substitutions we see that $f(H_1H_2) = f(H_1)f(H_2)$ so that $f(H) = (\det H)^{n-1}$, by Lemma 6.6.1. If $H$ is an orthogonal matrix then $f(H) = \pm 1$. Therefore if $n-1$ is odd and $\det H = -1$, right multiplication by $H$ changes orientation and changes the sign of the differential form. The absolute value of the differential form is unchanged.

The original source of this material is James (1954), Theorems 4.2 and 4.3.

## 7.5. Grassman manifolds $G_{k,n-k}$.

The Grassman manifold $G_{k,n-k}$ is defined to be the set of all k-dimensional hyperplanes in $\mathbb{R}_n$ containing 0. No globally defined system of local coordinates suffice to describe the manifold. In these notes we assume local coordinates can be defined but we never explicitly use them.

Given a k-plane $P$ let $a_1,\ldots,a_k$ be an orthonormal basis of $P$, $a_1,\ldots,a_k \in \mathbb{R}_n$. Below we will construct $b_1,\ldots,b_{n-k}$ which are analytic functions of $a_1,\ldots,a_k$ such that the $n \times n$ matrix $a_1,\ldots,a_k, b_1,\ldots,b_{n-k}$ is in $\underline{O}(n)$. We consider the differential form

$$(7.5.1) \qquad \omega = \bigwedge_{j=1}^{n-k} \bigwedge_{i=1}^{k} b_j^t \, da_i .$$

We will show this differential form locally about $P$ is independent of the choices of $a_1,\ldots,a_k,b_1,\ldots,b_{n-k}$ and that the differential form is globally defined. The measure

$$(7.5.2) \qquad \int_C |\omega|$$

is a left invariant measure under the action of $\underline{O}(n)$ on $G_{k,n-k}$.

Let $P_0$ be a k-plane and let the $n \times k$ matrix $X$ give local coordinates of $P$ near $P_0$ with the (i,j)-entry $x_{ij}$ of $X$ analytic in $P$. For ease of discussion we suppose $X_0^t = (X_{01}^t, X_{02}^t)$ represents $P_0$ where $X_{01}$ is a $k \times k$ nonsingular matrix. Similarly we write $X^t = (X_1^t, X_2^t)$ with $X_1$ a $k \times k$ nonsingular matrix. Then $X_1^{-1}$ near $P_0$ is an analytic function of the local coordinates.

From $X$ we pass analytically to $\begin{pmatrix} I_k \\ X_2 X_1^{-1} \end{pmatrix}$ and choose

$$(7.5.3) \qquad Y = \begin{pmatrix} Y_1 \\ I_{n-k} \end{pmatrix}, \quad Y_1 = -(X_1^{-1})^t X_2^t, \quad \text{with } Y_1$$

a $k \times (n-k)$ matrix. This defines $Y$ and $Y_1$ analytically in terms of $X$ and

(7.5.4)
$$Y^t \begin{pmatrix} I_k \\ X_2X_1^{-1} \end{pmatrix} = 0.$$

We take as bases of $P$ and $P^\perp$ the columns of $\begin{pmatrix} I_k \\ X_2X_1^{-1} \end{pmatrix}$ and $Y$.

Apply the Gram-Schmidt process to each matrix. The operations involved are rational or are the taking of square roots, and since lengths are bounded away from zero, the operations will be analytic. The result of the Gram-Schmidt process is an orthonormal set of vectors.

Therefore we may suppose

(7.5.5) $A$ is $n \times k$, $B$ is $n \times (n-k)$, and $(A,B) \in \underline{O}(n)$;
the columns of $A$ are a basis of $P$;
the columns of $A$, $B$ are analytic functions of the local coordinates at $P_0$.

Suppose $\tilde{A}$, $\tilde{B}$ is a second such representation. We show the differential form of (7.5.1) is unchanged. There are $H_1 \in \underline{O}(k)$ and $H_2 \in \underline{O}(n-k)$, $H_1$ and $H_2$ uniquely determined, such that

(7.5.6)
$$\tilde{A} = AH_1 \quad \text{and} \quad \tilde{B} = BH_2.$$

That is

(7.5.7)
$$H_1 = A^t\tilde{A} \quad \text{and} \quad H_2 = B^t\tilde{B}.$$

It is easy to check that $H_1$ and $H_2$ so defined are orthogonal matrices that satisfy (7.5.6) since $AA^t$ and $\tilde{A}\tilde{A}^t$, $BB^t$ and $\tilde{B}\tilde{B}^t$ are orthogonal projections of respective ranges $P$ and $P^\perp$. It follows that the entries of $H_1$ and $H_2$ are analytic functions of the local coordinates at $P_0$. Therefore, passing from $A,B$ to $\tilde{A},B$,

$$(7.5.8)\qquad d\tilde{A} = dA\,H_1 + A\,dH_1 \quad\text{and}\quad B^t\,d\tilde{A} = B^t\,dA\,H_1.$$

Then

$$(7.5.9)\qquad \bigwedge_{i=1}^{k} b_j^t\,d\tilde{a}_i = \bigwedge_{i=1}^{k} (B^t dA)_j H_i = \left(\bigwedge_{i=1}^{k} b_j^t\,da_i\right)\det H_1.$$

In this notation $H_i$ is the i-th column of $H_1$ and $(B^t dA)_j$ is the j-th row of $B^t dA$. By (6.2.19) the last step of (7.5.9) follows. Then

$$(7.5.10)\qquad \bigwedge_{j=1}^{n-k}\bigwedge_{i=1}^{k} b_j^t\,da_i = (\det H_1)^{n-k}\bigwedge_{j=1}^{n-k}\bigwedge_{i=1}^{k} b_j^t\,d\tilde{a}_i.$$

Passing from $\tilde{A}$, B to $\tilde{A},\tilde{B}$, using (7.5.6),

$$(7.5.11)\qquad \tilde{B}^t\,d\tilde{A} = H_2^t\,B^t\,d\tilde{A}.$$

The columns of $BH_2$ are linear combinations of the columns of B. Then

$$(7.5.12)\qquad \bigwedge_{j=1}^{n-k} \tilde{b}_j^t\,d\tilde{a}_i = \bigwedge_{j=1}^{n-k} (H_2^t\,B^t)_j\,d\tilde{a}_i$$

$$= (\det H_2)\bigwedge_{j=1}^{n-k} b_j^t\,d\tilde{a}_i.$$

Therefore

$$(7.5.13)\qquad \bigwedge_{j=1}^{n-k}\bigwedge_{i=1}^{k} \tilde{b}_j^t\,d\tilde{a}_i = (\det H_2)^k\bigwedge_{j=1}^{n-k}\bigwedge_{i=1}^{k} b_j^t\,d\tilde{a}_i$$

$$= (\det H_1)^{n-k}(\det H_2)^k\bigwedge_{j=1}^{n-k}\bigwedge_{i=1}^{k} b_j^t\,da_i.$$

If we suppose the orientation of P and $P^\perp$ is preserved in all representations (A, B) then $\det H_1 = \det H_2 = 1$.

We consider $\omega$ on the overlap of charts locally at $P_0$ and charts locally at $Q_0$. If $A_{P_0}$, $B_{P_0}$ represent P near $P_0$ and $A_{Q_0}$, $B_{Q_0}$ represent P near $Q_0$, then as in (7.5.6),

(7.5.14) $$A_{P_0} = A_{Q_0} H_3 \quad \text{and} \quad B_{P_0} = B_{Q_0} H_4.$$

The uniqueness of $H_3$ and $H_4$ makes it possible to consider these matrices as analytic functions of the local coordinates. Repetition of the above calculations shows $\omega$ to be independent of the use of $A_{P_0}$, $B_{P_0}$ or $A_{Q_0}$, $B_{Q_0}$.

We now consider the action of $\underline{O}(n)$ on $G_{k,n-k}$ by multiplication on the left. Let $P$ locally near $P_0$ be represented by $(A,B)$. Take $H \in \underline{O}(n)$ and let $Q_0$ be the image of $P_0$ under $H$. The point $P_0$ itself is represented by $(A_0,B_0)$, the value of $(A,B)$ at $P = P_0$, so $Q_0$ can be represented by $(HA_0, HB_0)$. The plane $P$ maps to a plane $Q$ which has columns of $HA$ as basis and $Q^{\perp}$ has the columns of $HB$ as basis. The sequence

(7.5.15) $$HA_0 \to A_0 \to A \to HA$$

gives analytic maps whereby the entries of $HA$ and $HB$ can be described analytically in terms of the local coordinates at $Q_0$.

Let $f(P)$ be the image of $P$ under the action of $H$ and let $F$ be the induced mapping of differential forms. Then $Q = f(P)$ and we write $\tilde{A}$ for the matrix representing $Q = f(P)$, with entries $\tilde{a}_{ij}$. Then

(7.5.16) $$F(\tilde{a}_{ij})(P) = \tilde{a}_{ij}(f(P))$$

is the $(i,j)$-entry of $HA$. Similarly

(7.5.17) $$F(\tilde{b}_{ij})(P) = \tilde{b}_{ij}(f(P))$$

is the $(i,j)$-entry of $HB$. Thus

(7.5.18) $$B^t dA = (HB)^t\, d(HA) = (F(\tilde{b}_{ij}))^t\, (dF(\tilde{a}_{ij})) = F(\tilde{B}^t d\tilde{A}).$$

We use (6.7.18) here together with (6.5.6) to evaluate the join of functions (7.5.18). That is, $\omega$ at $P_0$ is the image of $\omega$ at $Q_0$ under F. By Theorem 6.5.6 the measure $\int_C |\omega|$ is a (left) invariant measure under the action of $\underline{O}(n)$ on $G_{k,n-k}$.

This discussion shows $G_{k,n-k}$ to be $\underline{O}(n)$ factored by the compact subgroup $\underline{O}(k) \times \underline{O}(n-k)$ and the action of $\underline{O}(n)$ on $G_{k,n-k}$ is multiplication of cosets on the left. By Theorem 3.3.8 the regular invariant measures are uniquely determined up to multiplicative constant.

## 7.6. Stiefel manifolds $V_{k,n}$

We suppose A is a $n \times k$ matrix such that $A^tA = I_k$. Each such matrix, an "orthonormal k-frame", uniquely represents a point $\alpha$ of $V_{k,n}$. Then the (i,j)-entry of A, say $a_{ij}$, is a function of the points $\alpha \in V_{k,n}$, and this function when evaluated is $a_{ij}(\alpha)$. The functions $a_{ij}$ are globally defined but cannot be used as local coordinates since the ranges $a_{ij}(V_{k,n})$ are compact sets of real numbers. As in the case of $\underline{O}(n)$ the local coordinates do not explicitly enter the calculations. So we do not consider the construction of local coordinates.

The group action considered is $\underline{O}(n)$ acting on the left. We show that if $A \in V_{k,n}$, if $(A,B) \in \underline{O}(n)$, then

$$(7.6.1) \qquad \omega = \bigwedge_{j=1}^{n-k} \bigwedge_{i=1}^{k} b_j^t\, da_i \bigwedge_{i<j} a_j^t\, da_i$$

is the differential form of an invariant measure, where B depends analytically on A. To do this, the entire discussion for Grassman manifolds can be carried over here to show that

(i) the value of $\omega$ is independent of the choice of B except for sign changes due to change of orientation;

(ii) on the overlaps $U_{\alpha_1} \cap U_{\alpha_2}$ the differential form $\omega$ is self-equivalent;

(iii) left invariance follows at once for $\bigwedge_{i<j} a_j^t \, da_i$ as explained in Section 3.6, while left invariance of $\bigwedge_{j=1}^{n-k} \bigwedge_{i=1}^{k} b_j^t \, da_i$ is the left invariance established for Grassman manifolds in Section 7.5.

## 7.7. Total mass on the Stiefel manifold, $k = 1$.

The Stiefel manifold may be thought of as a quotient space $\underline{O}(n)/\underline{O}(n-k)$ since if $(A, B_1) \in \underline{O}(n)$ and if $(A, B_2) \in \underline{O}(n)$ then $\begin{pmatrix} I_k & A^t B_1 \\ B_2^t A, & B_2^t B_1 \end{pmatrix} \in \underline{O}(n)$ which implies $A^t B_1 = 0$ and $B_2^t A = 0$. Thus $B_2^t B_1 \in \underline{O}(n-k)$. By the uniqueness Theorem 3.3.8, the family of regular invariant measures form a 1-dimensional space, i.e., invariant measures on the quotient space $\underline{O}(n)/\underline{O}(n-k)$.

In integration of $\omega = \bigwedge_{j=1}^{n-1} b_j^t \, da$, i.e., the case $k = 1$, we note that $b_1, \ldots, b_{n-1}$ are defined as functions of $a$ which is a function of local coordinates near a given point. But the differential form $\omega$ is globally defined and is independent of the choice of $a$, $b_1, \ldots, b_{n-1}$. It follows from Chapter 6 that if we write $a^t = (a_1, \ldots, a_n)$ and $a_n = (1 - a_1^2 - \ldots - a_{n-1}^2)^{1/2}$ and

(7.7.1) $$\omega = f(a_1, \ldots, a_{n-1}) \, da_1 \wedge \ldots \wedge da_{n-1}$$

then

(7.7.2) $$\int \omega = \int f(a_1, \ldots, a_{n-1}) da_1 \ldots da_{n-1}.$$

See (6.4.8). The elimination of extra variables in $\omega$ follows from Problems 6.7.2, 6.7.3 and 6.7.4, which show that

(7.7.3) $$\omega = (-1)^{n+1} \epsilon (1 - a_1^2 - \dots - a_{n-1}^2)^{-1/2} da_1 \dots da_{n-1}$$

where $\epsilon = \pm 1$. Consequently allowing for $\epsilon = +1$, i.e., $a_n > 0$, and $\epsilon = -1$, i.e., $a_n < 0$,

(7.7.4) $$\int |\omega| = 2 \int \dots \int_{x_1^2 + \dots + x_{n-1}^2 < 1} \frac{dx_1 \dots dx_{n-1}}{(1 - x_1^2 - \dots - x_{n-1}^2)^{1/2}} .$$

Make the change of variable

(7.7.5) $$x_{n-1} = y(1 - x_1^2 - \dots - x_{n-2}^2)^{1/2}$$

to obtain

(7.7.6) $$\int |\omega| = 2 \int_{-\infty}^{\infty} (1-y^2)^{-1/2} dy \int \dots \int_{x_1^2 + \dots + x_{n-2}^2 < 1} dx_1 \dots dx_{n-2}$$

$$= 2\pi \text{ Volume}((n-2)\text{-ball}) = 2\pi^{n/2}/\Gamma(n/2).$$

See Problem 4.2.3. As a notation used in subsequent calculations we define

(7.7.7) $$A_n = \text{area of unit shell in } R_n = 2\pi^{n/2}/\Gamma(n/2) = \int |\omega| .$$

## 7.8. Mass on the Stiefel manifold, general case.

Wanted is $\int |\omega|$, where $\omega$ is given in (7.6.1). Write

(7.8.1) $$\omega = \epsilon_1 \left( \bigwedge_{j=1}^{n-k} \bigwedge_{i=1}^{k-1} b_j^t \, da_i \bigwedge_{j < i \leq k-1} a_j^t \, da_i \right) \bigwedge_{j=1}^{n-k} b_j^t da_k \bigwedge_{i=1}^{k-1} a_i^t \, da_k .$$

$\epsilon_1 = \pm 1$ is a sign introduced by the transpositions made. Except for a possible change of sign, the last part of (7.8.1) is the differential form of the case $k = 1$ treated in Section 7.7, which depends only on $a_k$. The remaining variables $b_1, \dots, b_{n-k}, a_1, \dots, a_{k-1}$ are in $a_k^\perp$ of dimension n-1. We fix $a_k$ and integrate out the remaining variables as follows, to obtain a recursion relation.

Let $H \in \underline{O}(n)$ have $a_k^t$ as the n-th row and transform variables to

$$(7.8.2) \qquad (Hb_1, \ldots, Hb_{n-k}, Ha_1, \ldots, Ha_{n-1}) = (\tilde{b}_1, \ldots, \tilde{b}_{n-k}, \tilde{a}_1, \ldots, \tilde{a}_{k-1}).$$

Each of these vectors is zero in the n-th position and therefore each of these vectors may be considered as being in $\mathbb{R}_{n-1}$. The differential form

$$(7.8.3) \qquad \bigwedge_{j=1}^{n-k} \bigwedge_{i=1}^{k-1} \tilde{b}_j^t \, d\tilde{a}_i \bigwedge_{j<i\leq k-1} \tilde{a}_j^t \, d\tilde{a}_i$$

$$= F\left(\bigwedge_{j=1}^{n-k} \bigwedge_{i=1}^{k-1} b_j^t \, da_i \bigwedge_{j<i\leq k-1} a_j^t \, da_i\right)$$

is invariant under the transformation of manifolds and is the differential form for a measure on $V_{k-1,n-1}$. Therefore

$$(7.8.4) \qquad (\text{mass of } V_{k-1,n-1}) 2\pi^{n/2}/\Gamma(n/2) = \text{mass } V_{k,n}.$$

Induction using (7.7.7) and (7.8.4) gives the following result.

Theorem 7.8.1. If $\omega$ is given in (7.6.1) then

$$(7.8.5) \qquad \text{mass } V_{k,n} = \int_{V_{k,n}} |\omega| = A_n A_{n-1} \cdots A_{n-k+1},$$

where $A_i = 2\pi^{i/2}/\Gamma(i/2)$, $1 \leq i \leq n$. In particular, if $k = n$, and $V_{k,n} = \underline{O}(n)$, then

$$(7.8.6) \qquad \int_{\underline{O}(n)} |\omega| = A_n \cdots A_1.$$

In this formula $A_1 = 2$ is the area of the unit sphere in $R = R_1$. Source, James (1954).

## 7.9. Total mass on the Grassman manifold $G_{n,n-k}$.

The group action considered is $\underline{O}(n)$ and in this section $G_{k,n-k}$ may be thought of as the quotient space

(7.9.1) $$\underline{O}(n)/\underline{O}(k)\underline{O}(n-k).$$

Then Haar measure on $\underline{O}(n)$ induces an invariant measure of finite total mass on $G_{k,n-k}$ which is uniquely determined up to a multiplicative constant. See Theorem 3.3.8.

$G_{k,n-k}$ may also be thought of in terms of $\underline{O}(k)$ acting on $V_{k,n}$, the Stiefel manifold of k-frames by the action $A \to AU$, $U \in \underline{O}(k)$. The orbit of $A$ under this action is the set of all k-frames which generate a fixed hyperplane $P$, so that $G_{k,n-k}$ is a homogeneous space $V_{k,n}/\underline{O}(k)$.

Recall the discussion at (7.5.5). If $P$ is a plane near $P_0$ and $P$ is represented by $(A,B)$ in the local coordinates at $P_0$, then the orbit in $V_{k,n}$ is the set of $AU$, $U \in O(k)$. Therefore the local coordinates in $G_{k,n-k}$ together with local coordinates in $\underline{O}(k)$ give local coordinates in the Stiefel manifold $V_{k,n}$. Therefore if $A \in G_{k,n-k}$ and $H = AU \in V_{k,n}$ then

(7.9.2) $$dH = A(dU) + (dA)U, \quad \text{and}$$
$$B^t(dH) = (B^tA)(dU) + B^t(dA)U = B^t(dA)U.$$

Also

(7.9.3) $$H^t(dH) = U^tA^tA(dU) + U^tA^t(dA)U = U^t(dU) + U^tA^t(dA)U.$$

From (7.9.2) and using (6.2.19) we obtain

(7.9.4) $$\bigwedge_{j=1}^{n-k} \bigwedge_{i=1}^{k} b_j^t\, dh_i = (\det U)^{n-k} \bigwedge_{j=1}^{n-k} \bigwedge_{i=1}^{k} b_j^t\, da_i.$$

Since $H = AU$ we find for the i-th column $h_i$ of $H$ and $u_i$ of $U$ that

(7.9.5) $$dh_i = (dA)u_i + A(du_i) \quad \text{and}$$
$$h_j^t\, dh_i = u_j^t\, A^t((dA)u_i + A(du_i)) = u_j^t\, du_i + u_j^t\, A^t\, dA\, u_i.$$

In forming a join of these expressions we use the observation that

(7.9.6)
$$\left(\bigwedge_{j=1}^{n-k} \bigwedge_{i=1}^{k} b_j^t \, da_i\right) (u_j^t A^t \, dA \, u_i) = 0,$$

since this is a $((n-k)k+1)$-form on a manifold of dimension $(n-k)k$. Consequently from (7.9.4) and (7.9.5) we obtain

(7.9.7)
$$(\det U)^{n-k} \bigwedge_{j=1}^{n-k} \bigwedge_{i=1}^{k} b_j^t \, da_i \bigwedge_{i<j} u_j^t \, du_i$$
$$= \bigwedge_{j=1}^{n-k} \bigwedge_{i=1}^{k} b_j^t \, dh_i \bigwedge_{i<j} h_j^t \, dh_i,$$

where the right side of (7.9.7) is the differential form of an invariant measure on $V_{k,n}$. Integration of the absolute value gives

Theorem 7.9.1.

(7.9.8)
$$A_n \cdots A_{n-k+1} = \left(\int_{G_{k,n-k}} |\omega|\right) A_k \cdots A_1.$$

Source, James (1954).

## 7.10. Problems.

Problem 7.10.1. For the group $\underline{T}(n)$ of nonsingular lower triangular matrices with positive diagonal elements, let $(T^{-1}dT)_{ij}$ be the $(i,j)$-element of $T^{-1}dT$. Then the differential form $\omega = \bigwedge_{j\leq i} (T^{-1}dT)_{ij}$ is an invariant differential form under the transformations $T \to T_0T$, $T_0 \in \underline{T}(n)$. Therefore $\int_C |\omega|$ is a left invariant Haar measure.

Problem 7.10.2. Continue Problem 7.10.1. If the $(i,j)$-element of $((dT)T^{-1})$ is $s_{ij}$ then the differential form $\omega = \bigwedge_{j\leq i} ds_{ij}$ defines a right invariant Haar measure for $\underline{T}(n)$.

Problem 7.10.3. Continue Problem 7.10.2. Use (6.7.10) and (6.7.16) to compute the modular functions for $\underline{T}(n)$.

Problem 7.10.4. For the group $GL(n)$ of nonsingular $n \times n$ matrices $X$, the differential form $\omega = \bigwedge_{i=1}^{n} \bigwedge_{j=1}^{n} ((XX^t)^{-1}X_j)^t dX_i$ is a left invariant differential form, where the j-th column of $X$ is $X_j$.

Problem 7.10.5. Use the decomposition, if $X \in GL(n)$ then $X = TA^t$ with $T \in \underline{T}(n)$, $A \in \underline{O}(n)$ and $T$ and $A$ uniquely determined. See Chapter 8. Then

(7.10.1) $$dX = T(dA^t) + (dT)A^t, \text{ and}$$

$$A^t((XX^t)^{-1}X)^t\,(dX)A = (dA^t)A + T^{-1}(dT).$$

Use the fact that the $(i,j)$-element of $T^{-1}dT$ is zero if $j > i$. Therefore

(7.10.2) $$\bigwedge_{i=1}^{n} \bigwedge_{j=1}^{n} (((XX^t)^{-1}X)^t dX)_{ij} = \epsilon_1 \bigwedge_{i<j} ((dA^t)A)_{ij} \bigwedge_{j\leq i} (T^{-1}(dT))_{ij}.$$

The sign $\epsilon_1$ is determined by the number of transpositions needed to obtain the arrangement on the right side of (7.10.2). Note that since $I_n = A^tA$ that $0 = A^t(dA) + (dA)^tA$. Then

(7.10.3) $$\bigwedge_{i=1}^{n} \bigwedge_{j=1}^{n} (((XX^t)^{-1}X)^t dX)_{ij}$$

$$= (-1)^{n(n-1)/2}\, \epsilon_1 \bigwedge_{i<j} (a_i^t da_j) \bigwedge_{j\leq i} (T^{-1}dT)_{ij}\,.$$

Problem 7.10.6. Let $\mu$ be an invariant regular Borel measure for the Borel subsets of $\underline{S}(n)$. Define $\nu(A) = \mu(\{S \mid S^{-1} \in A\}) = \mu(A^{-1})$. Show that $\nu$ is an invariant measure defined on the Borel subsets of $\underline{S}(n)$. Therefore, by uniqueness there is a number $\phi$ such that $\nu = \phi\mu$. Show that $\phi = 1$. Recall Example 3.3.9 in which the transformations $S \to X^tSX$ are discussed. This is not the action of $GL(n)$ on the coset space $GL(n)/\underline{O}(n)$ but a uniqueness theorem holds. See also Problem 6.7.11.

## Chapter 8. Decompositions of matrices.

### 8.1. The Stiefel manifolds.

In Section 7.9 we constructed local homeomorphisms between $\underline{O}(k) \times G_{k,n-k}$ and $V_{k,n}$. These maps are (locally) analytic and we state

Theorem 8.1.1. $V_{k,n}$ and $O(k) \times G_{k,n-k}$ are isomorphic as analytic manifolds.

### 8.2. Symmetric matrices.

The following decomposition is basic in the treatment of various multivariate problems.

Theorem 8.2.1. Let $A$ be a $n \times n$ symmetric matrix with $(i,j)$-entry $a_{ij}$, $1 \leq i, j \leq n$. If $1 \leq k \leq n$ implies

$$\det \begin{pmatrix} a_{11} & \cdots & a_{1k} \\ a_{k1} & \cdots & a_{kk} \end{pmatrix} > 0 , \tag{8.2.1}$$

then there exists a uniquely determined $T \in \underline{T}(n)$ such that

$$TT^t = A . \tag{8.2.2}$$

Proof. By an induction on the number of rows $n$. The result is obvious for $1 \times 1$ matrices. Let the result hold for $(m-1) \times (m-1)$ matrices and suppose

$$T_{m-1} \in \underline{T}(m-1),\ (T_{m-1}T_{m-1}^t)_{ij} = a_{ij},\ 1 \leq i,j \leq m-1. \tag{8.2.3}$$

We want

$$T = \begin{pmatrix} T_{m-1} & & 0 \\ t_{m1} & \cdots & t_{mm} \end{pmatrix} ,\ (TT^t)_{ij} = a_{ij},\ 1 \leq i,j \leq m . \tag{8.2.4}$$

This gives the equations, if $1 \leq i \leq m$, then

$$(8.2.5) \qquad a_{mi} = t_{m1}t_{i1} + \cdots + t_{mi}t_{ii} .$$

The system of equations, $1 \leq i \leq m-1$ is solvable for $t_{m1}, \ldots, t_{mm-1}$ since by construction $t_{11} > 0, \ldots, t_{m-1\,m-1} > 0$. We write $x = t_{m1}^2 + \cdots + t_{mm}^2$ so that

$$(8.2.6) \qquad TT^t = \begin{pmatrix} a_{11} & \cdots & a_{1m} \\ a_{m1} & \cdots & x \end{pmatrix} .$$

By construction the following relations hold.

$$(8.2.7) \qquad (\det TT^t) = (t_{11} \cdots t_{mm})^2 \quad \text{and}$$

$$(\det T_{m-1}T_{m-1}^t) = (t_{11} \cdots t_{m-1\,m-1})^2 > 0 .$$

Since (8.2.1) holds for the integer $k = m$, there exists a number $t_{mm}$ such that $t_{mm} > 0$ and $\det(TT^t) = \det(a_{ij}) > 0$. Since $\det TT^t$ is by (8.2.6) linear in the variable $x$, there is a unique value of $x$ such that the two sides of (8.2.6) have the same determinant. This $x$ is $x = a_{mm}$. Thus the equation $t_{11}^2 + \cdots + t_{mm}^2 = a_{mm}$ is solvable and $T = T_m$ has been constructed. That completes the inductive step. #

## 8.3. Decompositions of nonsingular $n \times n$ matrices.

The proof of Theorem 8.3.2 requires a lemma which we now state and prove.

Lemma 8.3.1. If $A \in \underline{O}(n) \cap \underline{T}(n)$ then $A = I_n$.

Proof. If $A \in \underline{T}(n)$ and $1 \leq i < j \leq n$ then $a_{ij} = 0$. Since $A \in \underline{O}(n)$ then all entires in the last column are zero except $a_{nn}$ so it follows $a_{nn}^2 = 1$. Since the elements of $\underline{T}(n)$ have positive diagonal elements, $a_{nn} = 1$. This implies the last row of $A$ is

$(0,0,\ldots,1)$. Hence, similarly, $a_{n-1\,n-1} = 1$, etc. The obvious backward induction now shows that $A = I_n$. #

Theorem 8.3.2. If $X \in GL(n)$ then

$$X = A_1T_1 = A_2T_2^t = T_3A_2 = T_4^tA_4, \tag{8.3.1}$$

$$\text{with } A_1,\ldots,A_4 \in \underline{O}(n) \text{ and } T_1,\ldots,T_4 \in \underline{T}(n).$$

Each of the factorizations of $X$ is uniquely determined.

Proof. Given that every $X \in GL(n)$ is expressible as $X = AT$, $A \in \underline{O}(n)$ and $T^t \in \underline{T}(n)$, the other three decompositions follow from consideration of $X^t$, $X^{-1}$ and $(X^t)^{-1}$, each of which maps $GL(n)$ onto $GL(n)$.

The Gram-Schmidt orthogonalization process, see Problem 8.9.1, produces an upper triangular matrix $T$ such that the columns of $XT^{-1}$ are orthonormal, i.e., $XT^{-1} = A \in \underline{O}(n)$ and $X = AT$ with $T^t \in \underline{T}(n)$.

It remains to prove uniqueness. If $A_1T_1 = A_2T_2$ then $A_2^{-1}A_1 = T_2T_1^{-1}$. Since $((T_2T_1)^{-1})^t \in \underline{T}(n)$ and $(A_2^{-1}A_1)^t \in \underline{O}(n)$, by the lemma, Lemma 8.3.1, it follows that $T_1 = T_2$ and $A_1 = A_2$. #

Corollary 8.3.3. If $X \in GL(n)$ there exists $A \in \underline{O}(n)$, $D \in \underline{D}(n)$ (a diagonal matrix) and $T^t \in \underline{T}(n)$ having diagonal elements equal one, such that

$$X = ADT. \tag{8.3.2}$$

## 8.4. Decomposition of $n \times k$ matrices of full rank $k \leq n$.

The uniqueness proof requires a preliminary lemma.

Lemma 8.4.1. Let $D \in \underline{D}(n)$ and suppose the diagonal elements of $D$ are pairwise distinct. If $A$ is $n \times n$ and $DA = AD$, then $A$ is a diagonal matrix.

Proof. If the (i,i)-entry of $D$ is $d_i$ and if $A = (a_{ij})$, then the (i,j)-entry of $AD$ is $a_{ij}d_j$ and the (i,j)-entry of $DA$ is $d_i a_{ij}$. Since $AD = DA$ it follows that $0 = a_{ij}(d_i - d_j)$. By hypothesis, if $i \neq j$, then $d_i \neq d_j$. Hence $a_{ij} = 0$. #

Theorem 8.4.2. Let $X$ be a $n \times k$ matrix of rank $k \leq n$. Assume $X^tX$ has $k$ distinct eigenvalues. There exist matrices $A$ $n \times k$, $D$, and $G$ such that

$$(8.4.1) \qquad A^tA = I_k,\ D \in \underline{D}(k),\ G \in \underline{O}(k),\ \text{and}\ X = ADG.$$

If the diagonal entries of $D_2$ and $D_3$ are positive and in increasing order of magnitude and if $X = A_2D_2G_2 = A_3D_3G_3$ are two such factorizations of $X$ satisfying (8.4.1) then

$$(8.4.2) \qquad D_2 = D_3,\ G_3G_2^t = D_1 \text{ is a diagonal orthogonal matrix, and}$$

$$A_3 = A_2D_1 .$$

Proof. The matrices $X$ and $X^tX$ have the same rank since the matrix entries are real numbers. Thus $X^tX$ is positive definite and

$$(8.4.3) \qquad A = X(X^tX)^{-1/2}$$

is a $n \times k$ matrix satisfying $A^tA = I_k$. Then, choose $G \in O(k)$ such that $D = G(X^tX)^{1/2}G^t$ is a diagonal matrix with the entries in increasing order of magnitude. Then the existence of a factorization follows from

$$(8.4.4) \qquad X = X(X^tX)^{-1/2}G^tG(X^tX)^{1/2}G^tG = ADG .$$

Given two decompositions $X = A_2D_2G_2 = A_3D_3G_3$ satisfying (8.4.1), then

$$(8.4.5) \qquad X^tX = G_2^tD_2^2G_2 = G_3^tD_3^2G_3 .$$

By hypothesis the diagonal entries of the diagonal matrices $D_2^2$, $D_3^2$ are in order of increasing magnitude so that $D_2^2 = D_3^2$ follows. But the diagonal entries of $D_2$ and $D_3$ are the eigenvalues of $(X^tX)^{1/2}$ and are positive. Therefore $D_2 = D_3$. Set $G_3G_2^t = D_1$. This orthogonal matrix clearly satisfies $DD_1 = D_1D$, where from the above, $D = D_2 = D_3$ follows. By Lemma 8.4.1, since the diagonal entries of $D$ are pairwise distinct, it follows that $D_1$ is a diagonal orthogonal matrix. Further $G_3 = D_1G_2$. Thus

$$X = A_2DG_2 = A_3DD_1G_2 , \tag{8.4.6}$$

and since $DD_1 = D_1D$, cancellation of the factor $DG_2$ from both sides yields $A_2 = A_3D_1$. Since $D_1^2 = I_k$ it is also true that $A_2D_1 = A_3$. #

## 8.5. X n x k not necessarily of full rank. Generalized inverses.

In the computation of probability density functions relative to Lebesgue measure, the event (rank X) $<$ k is a set of nk-dimensional Lebesgue measure zero. See Problem 4.2.5. Problem 4.2.6 shows the event that two nonzero eigenvalues of $X^tX$ be equal is an event of Lebesgue measure zero. Thus, except on a set of Lebesgue measure zero, Theorem 8.4.2 is applicable in these problems. Theorem 8.5.1 stated in this Section is primarily of algebraic interest rather than having direct application to the subject of these notes.

Theorem 8.5.1. Let $X$ be a $n \times k$ matrix (of real entries) of rank $r$ and let $X^tX$ have nonzero eigenvalues $\lambda_1 \le \lambda_2 \le \cdots \le \lambda_r$. Then

(8.5.1) $X = ADG$, with $A$ a $n \times r$ matrix, $D \in \underline{D}(r)$, and $G$ a $r \times k$ matrix, such that

$$A^tA = I_r, GG^t = I_r, \text{ and } D = \begin{pmatrix} \sqrt{\lambda_1} & & 0 \\ & \ddots & \\ 0 & & \sqrt{\lambda_r} \end{pmatrix} .$$

Given a second factorization $X = A_1 D_1 G_1$ with the diagonal entries of $D_1$ in increasing order, then $D = D_1$ and there exists $H \in \underline{O}(r)$ such that

$$(8.5.2) \qquad HD = DH \quad \text{and} \quad A_1 H = A,\ HG = G_1 .$$

Remark. Since the diagonal entries of $D$ are not necessarily pairwise distinct, it no longer follows from (8.5.2) that $H$ is a diagonal orthogonal matrix.

Proof. Let $Y_1, \ldots, Y_r$ be an orthonormal set such that $Y_1, \ldots, Y_r \in \mathbb{R}_k$ and

$$(8.5.3) \qquad (X^t X) Y_i = \lambda_i Y_i,\ 1 \leq i \leq r .$$

Then $\sum_{i=1}^{r} Y_i Y_i^t$ is a projection matrix of rank $r$ that maps $\mathbb{R}_k$ onto the range of $X^t X$. It is easy to see that the range of $X^t X$ is the same as the linear span of the rows of $X$ and therefore if $a \in \mathbb{R}_n$ then

$$(8.5.4) \qquad (a^t X) \sum_{i=1}^{r} Y_i Y_i^t = a^t X .$$

In particular

$$(8.5.5) \qquad X = X \sum_{i=1}^{r} Y_i Y_i^t = X \left( \frac{Y_1}{\sqrt{\lambda_1}}, \ldots, \frac{Y_r}{\sqrt{\lambda_r}} \right) \begin{pmatrix} \sqrt{\lambda_1} & \cdots & 0 \\ \vdots & & \vdots \\ 0 & \cdots & \sqrt{\lambda_r} \end{pmatrix} \begin{pmatrix} Y_1^t \\ \vdots \\ Y_r^t \end{pmatrix} .$$

It is easily verified that the matrices

$$(8.5.6) \qquad A = X(\lambda_1^{-1/2} Y_1, \ldots, \lambda_r^{-1/2} Y_r) \text{ and } G^t = (Y_1, \ldots, Y_r)$$

satisfy (8.5.1). We write $X = ADG$.

Given two representations $X = ADG = A_1D_1G_1$ then

(8.5.7) $$X^tX = G_1^tD_1^2G_1 = G^tD^2G, \text{ and}$$

(8.5.8) $$\lambda_iY_i = (X^tX)Y_i = G_1^tD_1^2(G_1Y_i).$$

This implies $G_1Y_i \neq 0$ and since $G_1G_1^t = I_r$, we obtain

(8.5.9) $$\lambda_i(G_1Y_i) = D_1^2(G_1Y_i) .$$

That is, $\lambda_1,\ldots,\lambda_r$ are eigenvalues of $D_1^2$, and since $D_1^2$ is a diagonal matrix with entries in order of increasing magnitude,

(8.5.10) $$D^2 = D_1^2 \text{ and } D = D_1 \text{ follows.}$$

By construction the columns of $G^t$ are nonzero eigenvectors for the nonzero eigenvalues of $X^tX$. By (8.5.7) it follows that the columns of $G_1^t$ likewise are an orthonormal set of eigenvectors for the nonzero eigenvalues of $X^tX$. Therefore there is an orthogonal matrix $H \in \underline{O}(r)$ such that $HG = G_1$. To see this, let $\tilde{G}$ be a $k \times k$ orthogonal matrix whose first $r$ rows are $\tilde{G}$, and $\tilde{G}_1$ be a $k \times k$ orthogonal matrix whose first $r$ rows are $G_1$. Then let $\tilde{H} \in \underline{O}(k)$ satisfy $\tilde{H}\tilde{G} = \tilde{G}_1$. Since

(8.5.11) $$\tilde{G}(X^tX) = \begin{pmatrix} D^2 & 0 \\ 0 & 0 \end{pmatrix}\tilde{G} \quad \text{and} \quad \tilde{G}_1(X^tX) = \begin{pmatrix} D^2 & 0 \\ 0 & 0 \end{pmatrix}\tilde{G}_1$$

if we write

(8.5.12) $$\tilde{H} = \begin{pmatrix} \tilde{H}_{11} & \tilde{H}_{12} \\ \tilde{H}_{21} & \tilde{H}_{22} \end{pmatrix},$$

then

(8.5.13) $$\begin{pmatrix} D^2 & 0 \\ 0 & 0 \end{pmatrix} = \tilde{G}_1(X^tX)\tilde{G}_1^t = (\tilde{H}\tilde{G})(X^tX)(\tilde{H}\tilde{G})^t = \tilde{H}\begin{pmatrix} D^2 & 0 \\ 0 & 0 \end{pmatrix}\tilde{H}^t .$$

Since $D$ is a nonsingular matrix, this implies $\widetilde{H}_{21} = 0$, hence that $\widetilde{H}_{22} = I_{k-r}$, and $\widetilde{H}_{12} = 0$. Therefore $\widetilde{H}_{11}G = G_1$ and to obtain the assertion above we now call $\widetilde{H}_{11}$ the matrix $H \in \underline{O}(r)$.

From (8.5.7) it follows that since $GG^t = I_r$,

(8.5.14) $$X = ADG = A_1DHG \quad \text{and} \quad AD = A_1DH.$$

Thus

(8.5.15) $$D^2 = DA^tAD = H^tDA_1^tA_1DH = H^tD^2H.$$

Hence $H$ commutes with $D^2$, and hence with $D$. Since $D$ is nonsingular, (8.5.14) and (8.5.15) imply

(8.5.16) $$A = A_1H. \ \#$$

<u>Definition 8.5.2.</u> If $X$ is a $n \times k$ matrix the generalized inverse $X^+$ of $X$ is the $k \times n$ matrix

(8.5.17) $$X^+ = G^tD^{-1}A^t ,$$

where $X$ has the factorization $X = ADG$ satisfying (8.5.1).

<u>Theorem 8.5.3. (Penrose (1955))</u>. The generalized inverse $X^+$ satisfies the following properties:

(8.5.18) $X^+X$ is a $k \times k$ orthogonal projection (i.e., is symmetric);

(8.5.19) $XX^+$ is a $n \times n$ orthogonal projection;

(8.5.20) $XX^+X = X$ and $X^+XX^+ = X^+$.

Furthermore, if another $k \times n$ matrix $X^o$ has properties (8.5.18) to (8.5.20) then $X^o = X^+$.

Proof. It is easily verified that the matrices $X = ADG$ and $X^+ = G^tD^{-1}A^t$ satisfy (8.5.18) to (8.5.20) provided $A$, $D$, and $G$ satisfy (8.5.1). Suppose $X^o$ satisfies the hypothesis. Then

$$X^oX = X^o(XX^+X) = (X^oX)(X^+X). \tag{8.5.21}$$

Take transposes of (8.5.21) and use the symmetry of the matrices involved. Then

$$X^oX = (X^oX)^t = (X^+X)^t(X^oX)^t = X^+(XX^oX) = X^+X. \tag{8.5.22}$$

Similarly

$$XX^o = XX^+ . \tag{8.5.23}$$

Then

$$X^+ = X^+(XX^+) = X^+(XX^o) = (X^+X)X^o = (X^oX)X^o = X^o. \ \# \tag{8.5.24}$$

## 8.6. Boundaries of convex sets.

Problems 4.2.4 to 4.2.7 are based on the assumption that the measures involved give zero measure to hyperplanes. To make this a useful hypothesis in subsequent discussions we will need to know that products of nonatomic measures do assign zero mass to hyperplanes. Hyperplanes as the boundaries of half spaces are the boundaries of convex sets. We now state a more general result about convex sets.

Theorem 8.6.1. Let $C$ be a closed convex subset of $\mathbb{R}_n$. Let $\mu_1,\ldots,\mu_n$ be positive $\sigma$-finite nonatomic measures defined on the Borel subsets of $\mathbb{R}_1$. Then

$$\mu_1 \times \cdots \times \mu_n \text{ (boundary of } C) = 0. \tag{8.6.1}$$

Proof. By induction on the dimension $n$. If $n = 1$ the boundary of $C$ contains at most two points and the conclusion follows since $\mu_1$

is a nonatomic measure. We consider sections

$$C_{x_n} = \{t \mid t \in \mathbb{R}_{n-1}, (t, x_n) \in C\} . \tag{8.6.2}$$

Since $C$ is a closed convex set there exists $a \leq b$ such that $C_{x_n} \neq \emptyset$ if and only if $a \leq x_n \leq b$. If $a = b$ then

$$\mu_1 \times \dots \times \mu_n(C) = \mu_1 \times \dots \times \mu_{n-1}(C_a)\mu_n(\{a\}) = 0, \tag{8.6.3}$$

so that in particular the boundary of $C$, being a subset of $C$, has zero mass. We assume in the sequel that $a < b$. If each section $C_{x_n}$, $a < x_n < b$, has void interior then each $C_{x_n}$ is its own boundary and by inductive hypothesis $\mu_1 \times \dots \times \mu_{n-1}(C_{x_n}) = 0$. Then

$$\mu_1 \times \dots \times \mu_n(C) = \mu_1 \times \dots \times \mu_{n-1}(C_a)\mu_n(\{a\}) + \mu_1 \times \dots \times \mu_{n-1}(C_b)\mu_n(\{b\}) + \int_{a+}^{b-} 0\mu_n(dt) = 0. \tag{8.6.4}$$

In (8.6.4) it is understood that if $a = -\infty$ or if $b = \infty$ then that term is zero in value, or in other words, is omitted.

In the remaining case, for some $x_n$, $a < x_n < b$, $C_{x_n}$ has nonvoid interior and hence $C$ has nonvoid interior. This implies that every section $C_{x_n}$, $a < x_n < b$, has nonvoid interior. Then using the inductive hypothesis and Fubini's Theorem,

$$\mu_1 \times \dots \times \mu_n(\text{boundary } C) \leq \int_{a+}^{b-} \mu_1 \times \dots \times \mu_{n-1}(\text{boundary } C_x)\mu_n(dx) + \mu_1 \times \dots \times \mu_{n-1}(C_a)\mu_n(\{a\}) + \mu_1 \times \dots \times \mu_{n-1}(C_b)\mu_n(\{b\}) = 0. \; \# \tag{8.6.5}$$

<u>Corollary 8.6.2</u>. If $C$ is a convex set in $\mathbb{R}_n$ then the boundary of $C$ has zero Lebesgue measure.

## 8.7. A second decomposition of $n \times k$ matrices $X$.

We consider $X$ as a point in $\mathbb{R}_{nk}$. Let $X$ have row vectors $X_1^t, \ldots, X_n^t$. Then Problem 4.2.4 and Theorem 8.6.1 imply that the set

$$N = \{X \mid X_1, \ldots, X_k \text{ are linearly dependent}\} \tag{8.7.1}$$

has Lebesgue measure zero. If $X \notin N$ then the first $k$ rows of $X$ are linearly independent and form a $k \times k$ nonsingular matrix $G^t = (X_1, \ldots, X_k)$. Set

$$Y^t = (I_k, Y_1, \ldots, Y_{n-k}) \text{ and } YG = X. \tag{8.7.2}$$

The matrix $Y$ is uniquely determined.

The null set $N$ is invariant under the action of $GL(k)$ acting as right multipliers of $X$. If $X = YG$ and $G_1 \in GL(k)$ then $XG_1 = Y(GG_1)$. Thus $Y$ represents a maximal invariant under the group action.

## 8.8. Canonical correlations and a decomposition of $n \times k$ matrices.

We begin with a statement of the problem in terms of the population parameters. Later in this section the problem is stated in terms of sample quantities. Assume

$$\underline{X} \text{ is } n \times p, \text{ and } \underline{Y} \text{ is } n \times q, \tag{8.8.1}$$

and that $\underline{X}$ and $\underline{Y}$ are independently distributed random matrices. The assumption about the moments of $\underline{X}$ and $\underline{Y}$ is

$$E\,\underline{X} = E\,\underline{Y} = 0; \tag{8.8.2}$$

$$E\underline{X}^t\underline{X} = \Sigma_{11},\ E\,\underline{Y}^t\underline{Y} = \Sigma_{22}, \text{ and } E\,\underline{X}^t\underline{Y} = \Sigma_{12}\,.$$

The problem is to choose $a \in \mathbb{R}_p$ and $b \in \mathbb{R}_q$ which maximize the correlation between $\underline{X}a$ and $\underline{Y}b$. Since the expectations of $\underline{X}$ and $\underline{Y}$ are zero, the quantity to be maximized is

$$(8.8.3) \qquad a^t \Sigma_{12} b/(a^t\Sigma_{11}a)^{1/2}(b^t\Sigma_{22}b)^{1/2} .$$

The solution to this problem can be obtained from the Cauchy-Schwarz inequality, which we now do. Define

$$(8.8.4) \qquad \alpha = \Sigma_{11}^{1/2}a, \text{ and } \beta = \Sigma_{22}^{1/2}b$$

and normalize $a$ and $b$ so that

$$(8.8.5) \qquad \alpha^t\alpha = 1 \text{ and } \beta^t\beta = 1.$$

Since the expression (8.8.3) for the correlation is homogeneous of degree zero in each of $a$ and $b$, the normalization (8.8.5) can be assumed to hold. Let

$$(8.8.6) \qquad A = \Sigma_{11}^{-1/2}\,\Sigma_{12}\,\Sigma_{22}^{-1/2} .$$

Then the transformed problem is to maximize, subject to (8.8.5), the quantity

$$(8.8.7) \qquad \alpha^t A \beta .$$

By the Cauchy-Schwarz inequality,

$$(8.8.8) \qquad \alpha^tA\beta \leq (\alpha^tAA^t\alpha)^{1/2}(\beta^t\beta)^{1/2} = (\alpha^tAA^t\alpha)^{1/2}$$

with equality if and only if $A^t\alpha$ and $\beta$ are proportional. The condition (8.8.5) then gives

$$(8.8.9) \qquad \beta = A^t\alpha/(\alpha^tAA^t\alpha)^{1/2} \text{ and}$$

$$\alpha^tA\beta = (\alpha^tAA^t\alpha)^{1/2} .$$

Subject to $\alpha^t\alpha = 1$, to maximize

$$(8.8.10) \qquad \alpha^tAA^t\alpha = \alpha^t\Sigma_{11}^{-1/2}\Sigma_{12}\Sigma_{22}^{-1}\Sigma_{12}^t\Sigma_{11}^{-1/2}\alpha$$

is to find the largest eigenvalue and corresponding eigenvector of the indicated matrix,

(8.8.11) $$\Sigma_{11}^{-1/2}\Sigma_{12}\Sigma_{22}^{-1}\Sigma_{12}^{t}\Sigma_{11}^{-1/2} .$$

We let $\lambda_1$ be the largest eigenvalue of the matrix (8.8.11), $\alpha_1$ be the corresponding eigenvector, and $\beta_1 = A^t\alpha_1/(\alpha_1^t AA^t\alpha_1)^{1/2}$ as in (8.8.9). The problem may now be repeated as follows. Choose $a_i \in \mathbb{R}_p$ and $b_i \in \mathbb{R}_q$ so that

(8.8.12) $$E(\underline{X}a_i)^t(\underline{X}a_j) = 0,\ E(\underline{Y}b_i)^t(\underline{Y}b_j) = 0,\ 1 \leq i < j \leq \text{rank } A$$

and subject to (8.8.12) maximize the correlation between $Xa_i$ and $Yb_i$, $1 \leq i \leq$ rank A. As in (8.8.4) and (8.8.6) we set

(8.8.13) $$\alpha_i = \Sigma_{11}^{1/2}a_i,\ \beta_i = \Sigma_{22}^{1/2}b_i \text{ and } A = \Sigma_{11}^{-1/2}\Sigma_{12}\Sigma_{22}^{-1/2},$$

$$1 \leq i \leq \text{rank } A,$$

and suppose the normalization is such that

(8.8.14) $$\alpha_i^t\alpha_i = 1,\ \beta_i^t\beta_i = 1,\ 1 \leq i \leq \text{rank } A.$$

Last, the condition that (8.8.12) holds says that

(8.8.15) $$a_i^t\Sigma_{11}a_j = 0 \text{ and } b_i^t\Sigma_{22}b_j = 0,\ 1 \leq j < i, \text{ or that}$$

(8.8.16) $$\alpha_i^t\alpha_j = 0 \text{ and } \beta_i^t\beta_j = 0,\ 1 \leq j < i.$$

Repetition of the first stage of the argument shows that, by the use of the Cauchy-Schwarz inequality, in order to maximize $\alpha_i^t A\beta_i$ one should take

(8.8.17) $$\beta_i = A^t\alpha_i/(\alpha_i^t AA^t\alpha_i)^{1/2} \text{ and}$$

$$\alpha_i^t A\beta_i = (\alpha_i^t AA^t\alpha_i)^{1/2} ,$$

and seek $\alpha_i$ to maximize (8.8.17). We note that to maximize $\alpha_i^t AA^t \alpha_i$ subject to (8.8.16), where the $\alpha_j$ are eigenvectors of $AA^t$ is equivalent to finding the largest eigenvalue having an eigenvector orthogonal to $\alpha_1,\ldots,\alpha_{i-1}$. If rank $A = r$ and $\lambda_1 \geq \cdots \geq \lambda_r$ are the nonzero eigenvalues of $AA^t$ then $\lambda_i$ is the maximum value subject to the stated conditions and $\alpha_i$ is the corresponding eigenvector. The condition that

$$(8.8.18) \qquad \beta_i^t \beta_j = 0 \text{ is the condition that } \alpha_i^t AA^t \alpha_j = 0$$

which is automatically satisfied.

The numbers $\lambda_1^{1/2},\ldots,\lambda_r^{1/2}$ are the correlation coefficients and have values between zero and one. They are the <u>canonical correlation coefficients</u> and if $\cos \theta_i = \lambda_i^{1/2}$, $1 \leq i \leq r$, then $\theta_1,\ldots,\theta_r$ are the <u>critical angles</u>.

We now give a definition of the corresponding <u>sample quantities</u>. The analogy is to make the correspondences

$$(8.8.19) \qquad \underline{X}^t\underline{X} \text{ to } \Sigma_{11},\ \underline{X}^t\underline{Y} \text{ to } \Sigma_{12}, \text{ and } \underline{Y}^t\underline{Y} \text{ to } \Sigma_{22}.$$

Then

$$(8.8.20) \qquad A = (\underline{X}^t\underline{X})^{-1/2}\underline{X}^t\underline{Y}(\underline{Y}^t\underline{Y})^{-1/2}$$

and

$$(8.8.21) \qquad r = \text{rank } A.$$

Outside an exceptional set of zero Lebesgue measure $\underline{X}$ and $\underline{Y}$ have full rank $p$ and $q$ respectively, and $A$ is well defined. Clearly $r \leq \min(p,q)$. We prove below that $r = \min(p,q)$, except on a set of measure zero.

We seek $a_1,\ldots,a_r$ and $b_1,\ldots,b_r$ satisfying

$$(8.8.22) \qquad a_i^t(\underline{X}^t\underline{X})a_j = 0 \text{ and } b_i^t(\underline{Y}^t\underline{Y})b_j = 0,\ 1 \leq j < i.$$

Define

(8.8.23) $$\alpha_i = (\underline{X}^t\underline{X})^{1/2}a_i \quad \text{and} \quad \beta_i = (\underline{Y}^t\underline{Y})^{1/2}b_i, \; 1 \leq i \leq r,$$

normalized so that

(8.8.24) $$\alpha_i^t\alpha_i = \beta_i^t\beta_i = 1, \; 1 \leq i \leq r.$$

Subject to (8.8.22), (8.8.23) and (8.8.24) choose $\alpha_i$ and $\beta_i$ to maximize

(8.8.25) $$\alpha_i^t A\beta_i, \; 1 \leq i \leq r.$$

The analysis applied to the population quantities may now be applied. Choose

(8.8.26) $$\beta_i = A^t\alpha_i/(\alpha_i^t AA^t\alpha_i)^{1/2}$$

and maximize

(8.8.27) $$\alpha_i^t AA^t\alpha_i$$

by choosing $\alpha_i$ to be the eigenvector corresponding to the i-th largest eigenvalue $\lambda_i$. This guarantees orthogonality of $\alpha_1,\ldots,\alpha_r$ and $\beta_1,\ldots,\beta_r$.

Observe from (8.8.23) that

(8.8.28) $$(\underline{X}a_i)^t(\underline{X}a_j) = a_i^t(\underline{X}^t\underline{X})a_j = ((\underline{X}^t\underline{X})^{1/2}a_i)^t((\underline{X}^t\underline{X})^{1/2}a_j)$$

$$= \alpha_i^t\alpha_j = 0 \text{ if } i \neq j \text{ and } = 1 \text{ if } i = j.$$ Similarly,

$$(\underline{Y}b_i)^t(\underline{Y}b_j) = \beta_i^t\beta_j = 0 \text{ if } i \neq j, \text{ and } = 1 \text{ if } i = j.$$

Therefore

(8.8.29) $\underline{X}a_1,\ldots,\underline{X}a_r$ is an orthonormal set, and

$\underline{Y}b_1,\ldots,\underline{Y}b_r$ is an orthonormal set.

Last, we prove that $r = \min(p,q)$ except for a set of zero Lebesgue measure. Suppose $z^t AA^t z = 0$ and let

$$(8.8.30) \qquad \tilde{z}^t = z^t(\underline{X}^t\underline{X})^{-1/2}\underline{X}^t .$$

Then

$$(8.8.31) \qquad \tilde{z}^t\underline{Y}(\underline{Y}^t\underline{Y})^{-1}\underline{Y}^t\tilde{z} = 0 .$$

The orthogonal projection $\underline{Y}(\underline{Y}^t\underline{Y})^{-1}\underline{Y}^t$ maps $\mathbb{R}_n$ onto the column space of $\underline{Y}$ so (8.8.31) together with (8.8.30) say that

$$(8.8.32) \qquad \tilde{z} \in (\text{column space } \underline{Y})^{\perp} \cap (\text{column space } \underline{X}).$$

Let us assume $p \le q$. Then $p + (n-q) \le q + (n-q) = n$. Given $Y$ fixed we may choose a basis of (column space $\underline{Y}$)$^{\perp}$ and extend it to $n-p$ elements, say $W = W_1,\ldots,W_{n-p}$. Then $\tilde{z} \ne 0$ satisfying (8.8.32) implies

$$(8.8.33) \qquad \det\,(\underline{X}, W) = 0,$$

which by Problem 4.2.4 has $\mathbb{R}_{np}$ dimensional Lebesgue measure zero. This holds for almost all $Y$ so by Fubini's Theorem the set $X$, $Y$ such that

$$(8.8.34) \qquad (\text{column space } Y)^{\perp} \cap (\text{column space } X) \ne \{0\}$$

has $\mathbb{R}_{n(p+q)}$ Lebesgue measure zero.

We summarize the discussion of sample quantities in a Theorem.

Theorem 8.8.1. Let the random $n \times p$ matrix $\underline{X}$ and the random $n \times q$ matrix $\underline{Y}$ have probability density functions relative to Lebesgue measure and suppose $p \le q$. There exists vectors $a_1,\ldots,a_r \in \mathbb{R}_p$ and $b_1,\ldots,b_r \in \mathbb{R}_q$ such that $r = \min(p,q)$, and

(8.8.35) $\underline{X}a_1,\ldots,\underline{X}a_r$ is an orthonormal set;

$\underline{Y}b_1,\ldots,\underline{Y}b_r$ is an orthonormal set; and if $i \neq j$ then

$a_i^t\underline{X}^t\underline{Y}b_j = 0$. #

The <u>canonical</u> <u>correlations</u> defined by

(8.8.36) $$\lambda_i^2 = a_i^t\underline{X}^t\underline{Y}b_i, \; 1 \leq i \leq r,$$

are numbers between zero and one.

(8.8.37) The number of integers $i$ such that $a_i^t\underline{X}^t\underline{Y}b_i = 1$ is $\max(p+q-n, 0)$.

(8.8.38) The projection of $\underline{X}a_i$ on the column space of $\underline{Y}$ is $\underline{Y}(\underline{Y}^t\underline{Y})^{-1}\underline{Y}^t\underline{X}a_i$. The cosine of the angle between $Xa_i$ and its projection is $\lambda_i^{1/2} > 0$. Write $\cos \theta_i = \lambda_i^{1/2}$. $\theta_1,\ldots,\theta_r$ are the critical angles and $\lambda_1^{1/2},\ldots,\lambda_r^{1/2}$ are the canonical correlations.

(8.8.39) The numbers $\lambda_1,\ldots,\lambda_r$ are the nonzero roots of the determinental equation

$$0 = |\lambda\underline{X}^t\underline{X} - (\underline{X}^t\underline{Y})(\underline{Y}^t\underline{Y})^{-1}(\underline{Y}^t\underline{X})|.$$

A few additional properties are stated in Problems 8.9.4 and following, which describe a decomposition of $\underline{X}$. The source of the material presented in this Section is Anderson (1958).

## 8.9. Problems.

### Problem 8.9.1. (Gram-Schmidt Orthogonalization)

Let X be a $n \times k$ matrix of rank h with $h \leq k \leq n$ and suppose the first h columns of X are linearly independent. Then there exists a $h \times h$ matrix $T_1 \in \underline{T}(h)$, a $(k-h) \times h$ matrix $T_2$ and a $n \times h$ matrix A such that if $T^t = (T_1^t, T_2^t)$ then

(8.9.1) $$A^tA = I_h \quad \text{and} \quad X = AT^t .$$

Problem 8.9.2. Continue Problem 8.9.1. Assume rank X = k.

Compute $\bigwedge_{i=1}^{n} \bigwedge_{j=1}^{k} dx_{ij}$ in terms of dA and dT, using 8.9.1.

Augment A so that the matrix (A,B) $\in \underline{O}(n)$ and consider

(8.9.2) $$\binom{A^t}{B^t} dX(T^t)^{-1} = \binom{I_k}{0} dT^t(T^t)^{-1} + \binom{A^t}{B^t} dA .$$

When the indicated join is formed the left side is

(8.9.3) $$\bigwedge_{i=1}^{n} \bigwedge_{j=1}^{k} \left(\binom{A^t}{B^t} dX\, T^{-1}\right)_{ij} = (\det T)^{-n} \bigwedge_{i=1}^{n} \bigwedge_{j=1}^{k} d\, x_{ij}.$$

On the right side of (8.9.2), $dT^t(T^t)^{-1}$ is an upper triangular matrix. This forces the answer on the right side to be of the form

(8.9.4) $$\pm \bigwedge_{1\le i\le j\le k} (dT^t(T^t)^{-1})_{ij} \bigwedge_{1\le j< i\le k} a_i^t da_j \bigwedge_{i=1}^{n-k} \bigwedge_{j=1}^{k} b_i^t da_j .$$

This is the product of two invariant differential forms. Use Problems 6.7.6 and 6.7.7.

Problem 8.9.3. Continue Problem 8.9.2.

Make the change of variable $S = TT^t$ and obtain the Wishart density function. See Chapter 4 and Problem 6.7.6.

Problem 8.9.4. (James (1954)).

Let X be n × p and Y be n × q, each of maximal rank, with $p \le q \le n$. Let $H \in GL(p)$. Then XH makes the same critical angles with Y as does X. Hence the critical angles are an invariant.

Problem 8.9.5. (James (1954)) Continue Problem 8.9.4.

Let Y represent a fixed hyperplane Q and X represent a variable hyperplane P. We are going to decompose X into a product of two Stiefel manifolds and a simplex as follows. (c.f. James, op.

cit., for a treatment of the cases not discussed in this problem.) Refer to Theorem 8.8.1. Project $Xa_i$ on $Q$ and $Q^{\perp}$ as in (8.9.5) below.

(8.9.5) let $\tilde{\alpha}_i = Y(Y^tY)^{-1}Y^tXa_i$, $1 \le i \le p$;

$$\tilde{\beta}_i = (I_n - Y(Y^tY)^{-1}Y^t)Xa_i,\ 1 \le i \le p.$$

Show

(8.9.6) $\tilde{\alpha}_i^t\tilde{\beta}_j = 0$, $1 \le i, j \le p$ and

$$\|\tilde{\alpha}_i\|^2 + \|\tilde{\beta}_i\|^2 = 1,\ 1 \le i \le p;$$

further show that $\tilde{\alpha}_1,\ldots,\tilde{\alpha}_p$ are mutually orthogonal and $\tilde{\beta}_1,\ldots,\tilde{\beta}_p$ are mutually orthogonal.

Write

(8.9.7) $\tilde{\alpha}_i = \alpha_i \cos \theta_i$, $\tilde{\beta}_i = \beta_i \sin \theta_i$, where

$$\|\alpha_i\| = \|\beta_i\| = 1,\ 1 \le i \le p.$$

Then

(8.9.8) $Xa_i = \tilde{\alpha}_i + \tilde{\beta}_i = \alpha_i \cos \theta_i + \beta_i \sin \theta_i$.

If the canonical correlations are distinct real numbers then the vectors $Xa_1,\ldots,Xa_p$, being eigenvectors, are uniquely determined except for sign changes. Fix the signs by requiring

(8.9.9) $\tilde{\tilde{\alpha}}_i = \pm\tilde{\alpha}_i$ such that the first component of $\tilde{\tilde{\alpha}}_i$ is $> 0$.

(8.9.10) $\cos \theta_i > 0$ so $-\pi/2 < \theta_i < \pi/2$ and we eliminate sign changes on $\beta_i$ by requiring $0 \le \theta_i < \pi/2$.

Then show

(8.9.11) Each plane of the Grassman manifold $G_{p,n-p}$

whose critical angles with $Y$ are distinct is uniquely representable by $(\tilde{\tilde{\alpha}}_1,\ldots,\tilde{\tilde{\alpha}}_p)$, $(\tilde{\tilde{\beta}}_1,\ldots,\tilde{\tilde{\beta}}_p)$ and $(\theta_1,\ldots,\theta_p)$ with $0<\theta_1<\ldots<\theta_p$.

(8.9.12) Count dimensions and show that the correct dimension results for $G_{p,n-p}$. That is, $\tilde{\alpha}_1,\ldots,\tilde{\alpha}_p$ generate a p-frame in a q-dimensional space and $\tilde{\beta}_1,\ldots,\tilde{\beta}_p$ generate a p-frame in a (n-q)-dimensional space. Compute the dimensions of $V_{p,q}$ and $V_{p,n-q}$.

(8.9.13) Use Problem 8.9.4 to write X as a product of $H \in GL(k)$ and a point of $G_{p,n-p}$. Check by adding dimensions.

Problem 8.9.6. Let $D$ be a diagonal matrix with pairwise distinct diagonal entries such that $D^{-1}$ exists. If $U \in \underline{O}(n)$ such that $DU$ is symmetric then $U$ is also a diagonal matrix.

Problem 8.9.7. Suppose $S$ is a symmetric positive definite $n \times n$ matrix and $U \in \underline{O}(n)$ such that $US^2U^t = D^2$ is the square of a diagonal matrix $D$. If the entries of $D$ are nonnegative then $USU^t = D$.

Hint. Show $V = D^{-1}USU^t$ is an orthogonal matrix and that $DV$ is a symmetric matrix. Hence $V = I_n$. #

Problem 8.9.8. Suppose $S, T \in \underline{S}(n)$ and $S^2 = T^2$. Then $S = T$.

Hint. Associate to each eigenvalue of $S$ the subspace of eigenvectors. These are the eigenspaces for $S^2$. Therefore $S$ and $T$ have the same eigenspaces and eigenvalues. #

Problem 8.9.9. Let $X$ be a $n \times k$ matrix and the rank $X = k \leq n$. Then there exists a $n \times k$ matrix $A$ and a $k \times k$ matrix $S \in \underline{S}(k)$ such that

(8.9.14) $$X = AS \text{ and } A^tA = I_k .$$

Show that this factorization is unique.

Hint. $X = (X(X^tX)^{-1/2})(X^tX)^{1/2}$. #

Problem 8.9.10. In Theorem 8.2.1 replace (8.2.1) by the hypothesis that

$$(8.9.15) \quad A \text{ has rank } p \le n \text{ and } \det\begin{pmatrix} a_{11} & \cdots & a_{1k} \\ \vdots & & \vdots \\ a_{k1} & \cdots & a_{kk} \end{pmatrix} \ge 0$$

if $1 \le k \le n$ and is greater than zero if $1 \le k \le p$.

Show that there exists a $n \times p$ matrix $T = (t_{ij})$ such that

$$(8.9.16) \quad t_{ij} = 0 \text{ if } j > i,\ TT^t = A, \text{ and } t_{ii} > 0 \text{ if } 1 \le i \le p.$$

Hint. $A + \epsilon I_n \in \underline{S}(n)$. Find $T_\epsilon \in \underline{T}(n)$ such that $T_\epsilon T_\epsilon^t = A + \epsilon I_n$. As $\epsilon \to 0$, $T_\epsilon$ is bounded. Take a convergent subsequence convergent to some $T_0$. Then rank $T_0T_0^t$ = rank $T_0 = p$. The minor

$$\begin{pmatrix} t_{11} & 0 & \cdots & 0 \\ t_{21} & t_{22} & \cdots & 0 \\ \vdots & \vdots & & \\ t_{p1} & t_{p2} & \cdots & t_{pp} \end{pmatrix} = S \text{ must satisfy } SS^t = \begin{pmatrix} a_{11} & \cdots & a_{1p} \\ \vdots & & \vdots \\ a_{p1} & \cdots & a_{pp} \end{pmatrix} > 0 \text{ so}$$

$S$ is nonsingular. This implies the result. #

Problem 8.9.11. (Uniqueness) Under the hypotheses of Problem 8.9.10, if $T_1$ and $T_2$ are $n \times p$ matrices satisfying (8.9.16) then $T_1 = T_2$.

Hint. Write $T_1 = \begin{pmatrix} T_{11} \\ T_{12} \end{pmatrix}$ and $T_2 = \begin{pmatrix} T_{21} \\ T_{22} \end{pmatrix}$ with $T_{11}$ and $T_{21}$ in $\underline{T}(p)$. Then $T_{11} = T_{21}$ and these matrices are nonsingular. Examine $T_1T_1^t = T_2T_2^t$. #

Problem 8.9.12. Let $X$ be a $n \times p$ matrix, $p \le n$, such that the minor $\begin{pmatrix} x_{11} & & x_{1p} \\ \vdots & & \vdots \\ x_{p1} & \cdots & x_{pp} \end{pmatrix}$ is nonsingular. There exist matrices $U \in \underline{O}(p)$ and $T = (t_{ij})$ such that

(8.9.17) $X = TU$ and if $j > i$ then $t_{ij} = 0$.

This factorization is unique.

Hint. $X = \begin{pmatrix} X_1 \\ X_2 \end{pmatrix}$ with $X_1$ a $p \times p$ matrix. Then $X_1 = T_1 U$ with $U \in \underline{O}(p)$ and $T_1 \in \underline{T}(p)$. $T_1$ and $U$ are uniquely determined. Required is $T_2$ solving the equation $T_2 U = X_2$. #

The following result which is stated as a problem has already been used in Problem 4.2.7.

Problem 8.9.13. Let $S_1$ and $S_2$ be symmetric positive semidefinite matrices such that $S_1 + S_2 = S > 0$. There exists $G \in GL(n)$ such that $GSG^t = I_n$ and $GS_1G^t$ is a diagonal matrix. (Uniqueness) If the diagonal entries of $GS_1G^t$ are pairwise distinct and if $G_1 \in GL(n)$ satisfies

(8.9.18) $$G_1 S G_1^t = I_n \quad \text{and} \quad G_1 S_1 G_1^t = GS_1 G^t$$

then $GG_1^{-1} \in \underline{O}(n)$ and is a diagonal matrix.

Problem 8.9.14. Suppose $\begin{pmatrix} A & B \\ B^t & C \end{pmatrix}$ is a $n \times n$ positive definite matrix and that $n = p + q$, $A$ is $p \times p$ and $C$ is $q \times q$. Show

(8.9.19) $$\begin{pmatrix} I_p & 0 \\ -B^tA^{-1} & I_q \end{pmatrix} \begin{pmatrix} A & B \\ B^t & C \end{pmatrix} \begin{pmatrix} I_p & -A^{-1}B \\ 0 & I_q \end{pmatrix} = \begin{pmatrix} A & 0 \\ 0 & C - B^tA^{-1}B \end{pmatrix}.$$

Therefore show

(8.9.20) $$\det \begin{pmatrix} A & B \\ B^t & C \end{pmatrix} = (\det A)(\det (C - B^tA^{-1}B)).$$

## Chapter 9. Examples using differential forms.

### 9.0. Introduction

In this chapter we calculate probability density functions for the canonical correlations (Section 9.1), Hotelling $T^2$ (Section 9.2) and the eigenvalues of the sample covariance matrix (Section 9.3). The calculations of Section 9.1 were stated by James (1954) who did the problem in the central case only. The noncentral case was computed by Constantine (1963). Our derivation differs somewhat from that of Constantine in that we place more emphasis on the use of differential forms. The results of Section 9.3 are taken directly from James (1954). The calculations of Section 9.2 are original to the author and are inserted in order to include this important example. In the problems, Section 9.4, several problems present background material. Problems 9.4.7 and 9.4.8 treat the distribution of correlation coefficients using differential forms.

In reading Chapter 9 keep in mind the existence and uniqueness theorems for invariant measures discussed in Chapter 3, Section 3.3. All the statistics discussed in this chapter are maximal invariants or are closely related to maximal invariants. The use of differential forms allows determination not only of the form of the density but also the normalizations required to give mass one. In Chapter 10 a totally different method of calculation of the density of a maximal invariant is developed. As will be seen on reading of Chapter 10 the method developed there is best suited to determination of ratios of density functions in which normalizations cancel.

A basic concept of the computation methods used in this chapter is as follows. Given are manifolds $\mathfrak{M}_1$ and $\mathfrak{M}_2$ with a transformation $f: \mathfrak{M}_1 \to \mathfrak{M}_2$. Unlike the development given in Chapter 6, $f$ is not assumed to be a homeomorphism onto. We suppose $f$ is continuous and onto. We suppose $\mathfrak{G}$ is a transformation group acting on $\mathfrak{M}_1$ such that

(9.0.1) $$(g,m) \to g(m)$$

is jointly continuous. We suppose $f$ satisfies

(9.0.2) if $g \in \mathcal{G}$, $x,y \in \mathfrak{M}_1$ and $f(x) = f(y)$ then $f(g(x)) = f(g(y))$.

Then $f$ induces a transformation group $\bar{\mathcal{G}}$ on $\mathfrak{M}_2$ by, if $x \in \mathfrak{M}_1$ then

(9.0.3) $$\bar{g}f(x) = f(g(x)).$$

Then clearly by (9.0.2) it follows that $\bar{g}(y)$ is defined uniquely for all $y \in \mathfrak{M}_2$. If $\mu$ is a $\sigma$-finite measure on the Borel subsets of $\mathfrak{M}_1$ then $f$ induces a measure $\bar{\mu}$ on the Borel subsets of $\mathfrak{M}_2$ by

(9.0.4) $$\bar{\mu}(A) = \mu(f^{-1}(A)).$$

Then if $g \in \mathcal{G}$, by (9.0.3) it follows that

(9.0.5) $$\bar{\mu}(\bar{g}^{-1}(A)) = \mu(f^{-1}(\bar{g}^{-1}(A))) = \mu(g^{-1}f^{-1}(A)).$$

In particular, if the measure $\mu$ is an invariant measure for the group $\mathcal{G}$ then the induced measure $\bar{\mu}$ is invariant for the induced group $\bar{\mathcal{G}}$. We state this formally.

Lemma 9.0.1. (James (1954)). If $\mu$ is invariant for $\mathcal{G}$ then the induced measure $\bar{\mu}$ is invariant for the induced group $\bar{\mathcal{G}}$.

We illustrate this discussion by the example of $n \times k$ matrices $X$ as points of $\mathbb{R}_{nk}$, $k \leq n$, rank $X = k$. We let $\mathcal{G}$ be $\underline{O}(n)$ acting as left multipliers of $X$, that is,

(9.0.6) $$g(X) = GX, \quad G \in \underline{O}(n).$$

We let $f(X) \in G_{k,n-k}$ be the hyperplane through $0$ spanned by the columns of $X$. Clearly

(9.0.7) $f(X_1) = f(X_2)$ if and only if there exists $H \in GL(k)$ such that $X_1 H = H_2$.

Then $g(X_1) = GX_1$ and $g(X_2) = GX_2$ and $g(X_1)H = g(X_2)$, that is,

(9.0.8) $$f(g(X_1)) = f(g(X_2)).$$

Therefore the group $\bar{\mathcal{G}}$ of induced transformations is defined. Lebesgue measure on $\mathbb{R}_{nk}$ is invariant under $\mathcal{G}$, but since $\mathbb{R}_{nk}$ is being factored by the noncompact subgroup $GL(k)$ (under the action of $f$) the measure induced on the compact manifold $G_{k,n-k}$ is not regular. Given some other finite invariant measure $\nu$ on $\mathbb{R}_{nk}$ the induced measure $\bar{\nu}$ will be finite and invariant. Since $G_{k,n-k}$ is $\underline{O}(n)/\underline{O}(k)\underline{O}(n-k)$ the invariant measures are uniquely determined up to normalization. See Theorem 3.3.8. Thus if $\nu(\mathbb{R}_{nk}) = 1$ the induced measure $\bar{\nu}$ has mass one and is given by a differential form (c.f. Section 7.5)

(9.0.9) $$K^{-1} \bigwedge_{j=1}^{n-k} \bigwedge_{i=1}^{k} b_j^t \, da_i.$$

This differential form is expected to appear in the integrations. From Section 7.9,

(9.0.10) $$K = A_n \cdots A_{n-k+1}/A_k \cdots A_1.$$

## 9.1. Canonical correlations.

The algebraic definitions were discussed in Section 8.8 and Problems 8.9.4 and 8.9.5. Using Theorem 8.3.2, the $n \times p$ matrix $X = A_1 T_1$ with $T_1^t \in \underline{T}(p)$ and the $n \times p$ matrix $A_1$ satisfying $A_1^t A_1 = I_p$. The p-frame $A_1$ decomposes into a plane $P \in G_{p,n-p}$ and an orientation matrix $U_1 \in \underline{O}(p)$. We write $U_1 T_1 = G_1 \in GL(p)$.

(9.1.1) $$X \sim (P, U_1 T_1) = (P, G_1).$$

We assume $Y$ is $n \times q$ and that $p \leq q$. Our Theorem 9.1.1 will require also $n \geq p+q$. Then

$$(9.1.2) \qquad Y \sim (Q, U_2T_2) = (Q, G_2),$$

where $T_2^t \in \underline{T}(q)$, $U_2 \in \underline{O}(q)$, and $G_2 \in GL(q)$, $G_2 = U_2T_2$, and $Q$ represents a point of $G_{q,n-q}$.

In the plane $P$ we take $a_1,\ldots,a_p$ to be an orthonormal basis. As in Section 7.5 $(a_1,\ldots,a_p)$ represents $P$ locally analytically in terms of local coordinates. In Section 8.8 these vectors were called $Xa_1,\ldots,Xa_p$ while in this section for brevity we call them $a_1,\ldots,a_p$. As in Section 8.8 and Problem 8.9.5, we let $\alpha_1,\ldots,\alpha_p$ be the normalized projections of $a_1,\ldots,a_p$ on $Q$, and $\beta_1,\ldots,\beta_p$ be the normalized projections of $a_1,\ldots,a_p$ on $Q^{\perp}$. It follows from (8.8.20), (8.8.27) and (8.9.5) that

$$(9.1.3) \qquad \alpha_1,\ldots,\alpha_p \text{ are an orthonormal set, and}$$
$$\beta_1,\ldots,\beta_p \text{ are an orthonormal set.}$$

Thus we have the relations

$$(9.1.4) \qquad \alpha_i^t\alpha_j = \delta_{ij},\ \beta_i^t\beta_j = \delta_{ij},\ \alpha_i^t\beta_j = 0, \quad 1 \leq i,j \leq p,$$

and

$$(9.1.5) \qquad \text{if } 1 \leq i \leq p \text{ then } a_i = \alpha_i \cos\theta_i + \beta_i \sin\theta_i,$$

where $\theta_1 \leq \theta_2 \leq \cdots \leq \theta_p$ are the critical angles. We may assume in this construction that $\alpha_1,\ldots,\alpha_p$ have the first component positive thereby uniquely determining $\alpha_1,\ldots,\alpha_p$, and that $0 \leq \theta_1 \leq \theta_p \leq \pi/2$ thereby uniquely fixing $\beta_1,\ldots,\beta_p$. See Problem 8.9.5. Recall that given $P$, $Q$ and the critical angles $\theta_1,\ldots,\theta_p$ the vectors $a_1,\ldots,a_p$ in $P$, which are eigenvectors, are uniquely determined except for sign changes.

The plane $Q$ is determined by a set of $n-q$ linear restrictions

$$(9.1.6) \qquad x^t c_i = 0, \quad i = 1,\ldots,n-q.$$

In the following remember that these conditions require $\beta_1,\ldots,\beta_p$ to be linear combinations of $c_1,\ldots,c_{n-q}$, while $\alpha_1,\ldots,\alpha_p$ are orthogonal to $c_1,\ldots,c_{n-q}$ so that

$$(9.1.7) \qquad \alpha_i^t c_j = 0 \text{ and } (d\alpha_i)^t c_j = 0, \quad 1 \le i \le p,\ 1 \le j \le n-q.$$

Therefore the vectors $d\alpha_1,\ldots,d\alpha_p$ also lie in $Q$ and

$$(9.1.8) \qquad (d\alpha_i)^t \beta_j = 0 \text{ and } (\alpha_i)^t (d\beta_j) = 0, \quad 1 \le i,j \le p.$$

In addition, from (9.1.4) it follows that

$$(9.1.9) \qquad (\alpha_i)^t (d\alpha_i) = 0, \text{ and } (\beta_i)^t (d\beta_i) = 0, \quad 1 \le i \le p;$$

$$(\alpha_i)^t (d\alpha_j) = -(\alpha_j)^t (d\alpha_i), \text{ and}$$

$$(\beta_i)^t (d\beta_j) = -(\beta_j)^t (d\beta_i), \quad 1 \le i \ne j \le p.$$

In using the differential form (7.5.1) the orthonormal set of vectors $b_1,\ldots,b_{n-p}$ are chosen mutually orthogonal to $a_1,\ldots,a_p$ but are otherwise arbitrary, so we make a suitable choice

(9.1.10) if $1 \le i \le p$ then $b_i = -\alpha_i \sin\theta_i + \beta_i \cos\theta_i$; choose $b_{p+1},\ldots,b_q$ in $Q$ orthonormal and orthogonal to $\alpha_1,\ldots,\alpha_p$. And in $Q^\perp$ choose $b_{q+1},\ldots,b_{n-p}$ orthonormal and orthogonal to $\beta_1,\ldots,\beta_p$.

Then clearly the orthonormal set of vectors $b_1,\ldots,b_{n-p}$ are orthogonal to $a_1,\ldots,a_p$. The following relations hold.

$$(9.1.11) \qquad da_i = b_i\, d\theta_i + (d\alpha_i \cos\theta_i + d\beta_i \sin\theta_i), \quad 1 \le i \le p;$$

(9.1.12) $b_i^t \, da_i = d\theta_i, \quad 1 \leq i \leq p;$

(9.1.13) if $i \neq j$ and $1 \leq i,j \leq p$ then

$$b_i^t \, da_j = -\alpha_i^t \, d\alpha_j \sin\theta_i \cos\theta_j + \beta_i^t \, d\beta_j \cos\theta_i \sin\theta_j.$$

By use of these relations we may evaluate the join

$$\bigwedge_{j=1}^{n-p} \bigwedge_{i=1}^{p} b_j^t \, da_i. \tag{9.1.14}$$

At the start we use the symmetries. If $1 \leq i \neq j \leq p$, then

$$-(b_j^t \, da_i) \wedge (b_i^t \, da_j) = (\alpha_j^t \, d\alpha_i) \wedge (\beta_j^t \, d\beta_i)(\cos^2\theta_j - \cos^2\theta_i), \tag{9.1.15}$$

which are obtainable from (9.1.13) together with (9.1.9) and

$$b_i^t \, db_j = -b_j^t \, db_i, \quad 1 \leq i,j \leq n-p. \tag{9.1.16}$$

Using the orthogonality relations, further

(9.1.17) if $i = 1,\ldots,p$ and $j = p+1,\ldots,q$, then

$$b_j^t \, da_i = b_j^t \, d\alpha_i \cos\theta_i; \text{ and}$$

if $i = 1,\ldots,p$ and $j = q+1,\ldots,n-p$, then

$$b_j^t \, da_i = b_j^t \, d\beta_i \sin\theta_i.$$

Use of these relations yields

$$\begin{aligned} \bigwedge_{j=1}^{n-p} \bigwedge_{i=1}^{p} b_j^t \, da_i &= \epsilon \bigwedge_{j<i} \alpha_j^t \, d\alpha_i \bigwedge_{j<i} \beta_j^t \, d\beta_i \bigwedge_{i=1}^{p} d\theta_i \\ &\times \bigwedge_{j=p+1}^{q} \bigwedge_{i=1}^{p} b_j^t \, d\alpha_i \bigwedge_{j=q+1}^{n-p} \bigwedge_{i=1}^{p} b_j^t \, d\beta_i \\ &\times \left(\prod_{i=1}^{p} \cos\theta_i\right)^{q-p} \left(\prod_{i=1}^{p} \sin\theta_i\right)^{n-p-q} \prod_{j<i} (\cos^2\theta_j - \cos^2\theta_i), \end{aligned} \tag{9.1.18}$$

with the sign $\epsilon = \pm 1$ to be determined by the number of interchanges made.

The differential form (9.1.18) is computed for the Grassman manifold containing P, Q being fixed. Since the computed differential form is invariant under left multiplication by $H \in \underline{O}(n)$ we modify the variables as follows. Let $(\tilde{A},\tilde{B})$ represent Q in the sense of Section 7.5. We set

(9.1.19)
$$\alpha_i = \tilde{A}\,\tilde{\alpha}_i, \quad 1 \le i \le p;$$
$$\beta_i = \tilde{B}\,\tilde{\beta}_i, \quad 1 \le i \le p;$$
$$b_i = \tilde{A}\,\tilde{b}_i, \quad p+1 \le i \le q;$$
$$b_i = \tilde{B}\,\tilde{b}_i, \quad q+1 \le i \le n-p.$$

Then

(9.1.20)
$$\text{if } 1 \le i \le p, \quad b_i = -\tilde{A}\,\tilde{\alpha}_i \sin\theta_i + \tilde{B}\,\tilde{\beta}_i \cos\theta_i.$$

We will use the fact that $b_1,\ldots,b_p$ do not appear in the differential form (9.1.18). With $(\tilde{A},\tilde{B})$ fixed, in the new variables $\tilde{\alpha},\tilde{\beta},\tilde{b}$, the differential form (9.1.18) is unchanged except to replace $\alpha,\beta,b$ by $\tilde{\alpha},\tilde{\beta},\tilde{b}$. The representation of P has become

(9.1.21)
$$a_i = \tilde{A}\,\tilde{\alpha}_i \cos\theta_i + \tilde{B}\,\tilde{\beta}_i \sin\theta_i, \quad i = 1,\ldots,p,$$
$$b_i = -\tilde{A}\,\tilde{\alpha}_i \sin\theta_i + \tilde{B}\,\tilde{\beta}_i \cos\theta_i, \quad i = 1,\ldots,p,$$
$$b_i = \tilde{A}\,\tilde{b}_i, \quad i = p+1,\ldots,q:$$
$$b_i = \tilde{B}\,\tilde{b}_i, \quad i = q+1,\ldots,n-p.$$

We now compute in the case of a multivariate normal density function. We assume $\underline{Z} = (\underline{X},\underline{Y})$ has a joint normal density function

(9.1.22)
$$(2\pi)^{-n(p+q)/2}(\det \Sigma)^{-n/2} \operatorname{etr} -1/2\, \Sigma^{-1}Z^tZ,$$

where $\operatorname{etr} A = \exp(\operatorname{tr} A)$, exponential of the trace of the matrix A. Using

$$(9.1.23) \qquad X = AU_1T_1 \quad \text{and} \quad Y = \tilde{A}U_2T_2,$$

we find

$$(9.1.24) \qquad Z^tZ = \begin{pmatrix} X^tX & X^tY \\ Y^tX & Y^tY \end{pmatrix} = \begin{pmatrix} T_1^t\,U_1^t & 0 \\ 0 & T_2^t\,U_2^t \end{pmatrix} \begin{pmatrix} I_p & A^t\tilde{A} \\ \tilde{A}^tA & I_q \end{pmatrix} \begin{pmatrix} U_1T_1 & 0 \\ 0 & U_2T_2 \end{pmatrix} .$$

Let $\tilde{\alpha}$ be the $q \times p$ matrix with $\tilde{\alpha}_i$ as the i-th column. From (9.1.21) if $a_i$ is the i-th column of $A$, we find

$$(9.1.25) \qquad \tilde{A}^tA = \tilde{\alpha} \begin{pmatrix} \cos\theta_1 & & 0 \\ & \ddots & \\ 0 & & \cos\theta_p \end{pmatrix} = \tilde{\alpha}E,$$

where $E$ is the $p \times p$ diagonal matrix in (9.1.25). Then we write

$$(9.1.26) \qquad D^t = (E,0) \quad \text{where } D \text{ is } q \times p.$$

To get back to $X$ and $Y$ we use (9.1.23) and set

$$(9.1.27) \qquad G = U_1T_1 \quad \text{and} \quad \tilde{G} = U_2T_2.$$

Then

$$(9.1.28) \qquad dX = dA\,G + AdG \quad \text{and}$$

$$a_i^t\,dx_j = a_i^t\,dA\,g_j + dg_{ij}, \quad 1 \leq i \leq p, \quad 1 \leq j \leq p,$$

$$b_i^t\,dx_j = b_i^t\,dA\,g_j \qquad 1 \leq i \leq n\text{-}p, \; 1 \leq j \leq p.$$

This yields

$$(9.1.29) \qquad (\det(A,B))^p \bigwedge_{j=1}^{p} \bigwedge_{i=1}^{n} dx_{ij} = \bigwedge_{j=1}^{p} \bigwedge_{i=1}^{p} a_i^t\,dx_j \bigwedge_{i=1}^{n-p} b_i^t\,dx_j$$

$$= \bigwedge_{j=1}^{p} \bigwedge_{i=1}^{p} dg_{ij} \bigwedge_{i=1}^{n-p} b_i^t\,dA\,g_j \quad .$$

On the otherhand

$$(9.1.30) \qquad \bigwedge_{j=1}^{p} (b_i^t\,dA)g_j = (\det G) \bigwedge_{j=1}^{p} b_i^t\,da_j$$

so that

$$(9.1.31)\qquad (\det G)^{n-p} \bigwedge_{j=1}^{p} \bigwedge_{i=1}^{n-p} b_i^t \, da_j = \text{part of } (9.1.29).$$

Therefore

$$(9.1.32)\qquad \bigwedge_{j=1}^{p} \bigwedge_{i=1}^{n} dx_{ij} = (\det G)^{n-p}(\det(A,B))^p \bigwedge_{j=1}^{p} \bigwedge_{i=1}^{p} dg_{ij} \bigwedge_{i=1}^{n-p} b_i^t \, da_j .$$

In the new variables the differential form $\bigwedge_{i=1}^{n-p} \bigwedge_{j=1}^{p} b_i^t \, da_j$ is given by (9.1.18). Similarly for $\bigwedge_{j=1}^{q} \bigwedge_{i=1}^{n} dy_{ij}$.

In the integration, integrate first over the variables $g_{ij}$ and $\tilde{g}_{ij}$. Then one obtains

$$(9.1.33)\qquad \int \operatorname{etr} -1/2\, \Sigma^{-1}\begin{pmatrix} G^t & 0 \\ 0 & \tilde{G}^t \end{pmatrix}\begin{pmatrix} I_p & E^t\tilde{\alpha}^t \\ \tilde{\alpha}E & I_q \end{pmatrix}\begin{pmatrix} G & 0 \\ 0 & \tilde{G} \end{pmatrix} (\det G)^{n-p}(\det \tilde{G})^{n-q}$$

$$\times \prod_{i=1}^{p} \prod_{j=1}^{p} dg_{ij} \prod_{i=1}^{p} \prod_{j=1}^{p} d\tilde{g}_{ij}.$$

Since Haar measure on $GL(q)$ is orthogonally invariant and since there exists $H \in O(q)$ such that $H\alpha = \begin{pmatrix} I_p \\ 0 \end{pmatrix}$,

$$(9.1.34)\qquad (9.1.33) = \int \operatorname{etr} -1/2\, \Sigma^{-1}\begin{pmatrix} G^t & 0 \\ 0 & \tilde{G}^t \end{pmatrix}\begin{pmatrix} I_p & D^t \\ D & I_q \end{pmatrix}\begin{pmatrix} G & 0 \\ 0 & \tilde{G} \end{pmatrix}$$

$$\times (\det G)^{n-p}(\det \tilde{G})^{n-q}$$

$$\times \prod_{i=1}^{p} \prod_{j=1}^{p} dg_{ij} \prod_{i=1}^{p} \prod_{j=1}^{p} d\tilde{g}_{ij} .$$

The remaining variables consist of $2^{-p} V_{p,q}$, the factor $2^{-p}$ resulting from the choice of signs in $\alpha_1,\ldots,\alpha_p$, of $V_{p,n-q}$, and the Grassman manifold $G_{q,n-q}$. See Problem 8.9.5 and refer to Sections 7.8 and 7.9. The resulting normalization from integration over these manifolds is

$$(9.1.35)\quad K = (2\pi)^{n(p+q)/2}(\det \Sigma)^{-n/2}\, 2^{-p}\, A_q \cdots A_{q-p+1} A_{n-q} \cdots A_{n-q-p+1} \times \frac{A_n \cdots A_{n-q+1}}{A_q \cdots A_1}.$$

Theorem 9.1.1. Let $E$ be as in (9.1.25) and $f(E)$ be the function in (9.1.34). Then if $n \geq q \geq p$ and $n \geq p+q$,

$$(9.1.36)\quad K^{-1} f(E) \left(\prod_{i=1}^{p} \cos \theta_i\right)^{q-p} \left(\prod_{i=1}^{p} \sin \theta_i\right)^{n-p-q} \times \prod_{j<i} (\cos^2 \theta_j - \cos^2 \theta_i)$$

is the joint probability density function of the angles $\theta_1 < \ldots < \theta_p$.

The canonical correlations are $\lambda_i = \cos \theta_i$, $i = 1, \ldots, p$. Substitution into (9.1.36) gives the joint probability density function of the correlations. The function $f(E)$ with $E$ as in (9.1.25) may be derived using the methods of Chapter 10 on maximal invariants. However to obtain the normalization $K$ and weight function of $\theta_1, \ldots, \theta_p$, a more precise calculation seems necessary.

In the derivations given we have assumed $q-p \geq 0$. The exponent $n - p - q$ in (9.1.36) cannot be negative. If it is negative then the term $\left(\prod_{i=1}^{p} \sin \theta_i\right)^{n-p-q}$ is missing, corresponding to the non-empty intersection of $P$ and $Q$. For an analysis of this case, see James (1954).

## 9.2. Hotelling $T^2$.

In this section we assume $\underline{X}$ is $n \times k$, that $\underline{Y}$ is $k \times 1$, and that $\underline{X}$ and $\underline{Y}$ are stochastically independent random variables. We will suppose the rows of $\underline{X}$ are independently and identically distributed, each normal $(0, \Sigma)$, and that $\underline{Y}$ is normal $(a, \Sigma)$ with the same covariance matrix $\Sigma$. The problem of inference is to test $a = 0$ against the alternative $a \neq 0$. The problem is invariant under transformations

(9.2.1) $$(X,Y) \to (XG,\ GY)$$

where $G \in GL(k)$. Under the action of $GL(k)$ the maximal invariant constructed from the sufficient statistic

(9.2.2) $$X^tX + YY^t\ ,\ Y$$

is

(9.2.3) $$(Y^t(X^tX + YY^t)^{-1}Y)^{1/2} = \lambda^{1/2}.$$

The number $\lambda$ is the nonzero root of the equation

(9.2.4) $$0 = \det(YY^t - \lambda\ (X^tX + YY^t)).$$

This follows from the general rule that matrices AB and BA have the same eigenvalues.

In the sequel we let $\underline{T}_1$, $\underline{T} \in \underline{T}(k)$ such that

(9.2.5) $$\underline{T}_1\underline{T}_1^t = \underline{X}^t\underline{X} \quad \text{and} \quad \underline{T}\ \underline{T}^t = \underline{X}^t\underline{X} + \underline{Y}\ \underline{Y}^t.$$

The aim of this section is to calculate the probability density function of $\underline{T}^{-1}\underline{Y}$ by use of differential forms. A different evaluation results by use of the ideas of Chapter 10 on maximal invariants.

We note for reference in the sequel that if

(9.2.6) $$S = TT^t = (s_{ij}), \quad \text{then}$$

(9.2.7)
$$\bigwedge_{j\le i} ds_{ij} = 2^k\ t_{11}^k \cdots t_{kk} \bigwedge_{j\le i} dt_{ij}$$
$$= 2^k(\det T)^{k+1}\Big(\bigwedge_{j\le i} dt_{ij}\Big)/(t_{11}\cdots t_{kk}^k)$$
$$= 2^k(\det T)^{k+1} \bigwedge_{j\le i} (T^{-1}dT)_{ij}.$$

See Section 7.2 and Problem 6.7.6.

The joint density function of $\underline{X}$ and $\underline{Y}$ is

$$(9.2.8) \qquad (2\pi)^{-(n+1)k/2}(\det \Sigma)^{-(n+1)/2} \operatorname{etr} -1/2\ \Sigma^{-1}(X^tX+YY^t-2aY^t+aa^t).$$

We now begin the process of changes of variables followed by integrations. Write

$$(9.2.9) \qquad X = AT_1^t \text{ with } A^tA = I_k,\ A \text{ a } n\times k \text{ matrix, and } T_1 \text{ as in } (9.2.5)$$

Let $B$ be $n \times (n-k)$ satisfying

$$(9.2.10) \qquad B^tA = 0, \quad B^tB = I_{n-k}.$$

We assume $B$ is chosen to be an analytic function of the variables $A$ so the differential form for an invariant measure on the Stiefel manifold is

$$(9.2.11) \qquad \bigwedge_{j=1}^{n-k} \bigwedge_{i=1}^{k} b_j^t\, da_i \bigwedge_{j<i} a_j^t\, da_i.$$

See Section 7.6. We compute

$$(9.2.12) \qquad dX = A\, dT_1^t + (dA)T_1^t\ ;$$

$$A^t\, dX(T_1^t)^{-1} = ((dT_1^t)(T_1^t)^{-1}) + A^t\, dA;$$

$$B^t\, dX(T_1^t)^{-1} = B^t\, dA.$$

These relations give

$$(9.2.13) \qquad (\det A,B)^k\ (\det T_1)^{-n} \bigwedge_{j=1}^{k} \bigwedge_{i=1}^{n} dx_{ij}$$

$$= \epsilon \bigwedge_{j=1}^{k} \bigwedge_{i=1}^{n-k} b_i^t\, da_j \bigwedge_{j<i} a_i^t\, da_j \bigwedge_{i\leq j} ((dT_1^t)(T_1^t)^{-1})_{ij}\ .$$

$\epsilon = \pm 1$, depending on the undetermined number of interchanges. Note that

$$(9.2.14) \qquad \bigwedge_{i\leq j} ((dT_1^t)(T_1^t)^{-1})_{ij} = \bigwedge_{i\leq j} ((T_1^{-1}\, dT_1)^t)_{ij} = \bigwedge_{j\leq i} (T_1^{-1}\, dT_1)_{ij}.$$

In the density function (9.2.8), $X^tX = T_1T_1^t$ so that (9.2.8) does not depend on points $A$ of the Stiefel manifold. This integration out over the Stiefel manifold gives

$$(9.2.15)\qquad (2\pi)^{-(n+1)k/2}(\det \Sigma)^{-(n+1)/2}A_n\cdots A_{n-k+1}(\det T_1)^n$$

$$\times \operatorname{etr} - 1/2\, \Sigma^{-1}(T_1T_1^t + YY^t - 2Ya^t + aa^t) \bigwedge_{j\leq i}(T_1^{-1}dT_1)_{ij} \bigwedge_{i=1}^{k} dy_i.$$

Introduce variables $S$ by

$$(9.2.16)\qquad S = T_1T_1^t + YY^t, \quad \text{so that}$$

$$(9.2.17)\qquad \bigwedge_{j\leq i} ds_{ij} \bigwedge_{i=1}^{k} dy_i = 2^k(\det T_1)^{k+1} \bigwedge_{j\leq i}(T_1^{-1}\, dT_1)_{ij} \bigwedge_{i=1}^{k} dy_i .$$

Factor $S$,

$$(9.2.18)\qquad S = TT^t,\ T \in \underline{T}(k), \quad \text{so that}$$

$$(9.2.19)\qquad \bigwedge_{j\leq i} ds_{ij} = 2^k(\det T)^{k+1} \bigwedge_{j\leq i}(T^{-1}dT)_{ij}.$$

The ratio $(\det T/\det T_1)$ may be evaluated as follows.

$$(9.2.20)\qquad T_1T_1^t = TT^t - YY^t = T(I_k - (T^{-1}Y)(T^{-1}Y)^t)T^t$$

$$= T(I_k - WW^t)T^t ,$$

with $W = T^{-1}Y$. Therefore $(\det T_1)^2 = (\det T)^2((\det I_k - WW^t))$. Choose $U \in \underline{O}(k)$ such that $(UW)^t = (\|W\|, 0,\ldots,0)$. Then it follows that

$$(9.2.21)\qquad \det(I_k - WW^t) = 1 - \|W\|^2 \quad \text{and}$$

$$(9.2.22)\qquad (\det T_1/\det T) = (1 - \|W\|^2)^{1/2}.$$

Upon substitution, (9.2.15) becomes

$$(9.2.23)\qquad (2\pi)^{-(n+1)k/2}(\det \Sigma)^{-(n+1)/2}$$

$$\times\ A_n \cdots A_{n-k+1}(\det T)^{n+1}(1 - \|W\|^2)^{\frac{n-k-1}{2}}$$

$$\times\ \operatorname{etr}-1/2\ \Sigma^{-1}(TT^t - 2TWa^t + aa^t) \bigwedge_{j\le i}(T^{-1}dT)_{ij} \bigwedge_{i=1}^{k} dW_i.$$

We use here the relation (think of $T$ as fixed)

$$(9.2.24)\qquad \bigwedge_{i=1}^{k} dy_i = (\det T) \bigwedge_{i=1}^{k} dW_i.$$

Integration on the variables $T$ by the left invariant Haar measure $\bigwedge_{j\le i}(T^{-1}dT)_{ij}$ gives the density function of $\underline{W}$. We let

$$(9.2.25)\qquad \Sigma = \Sigma_1\Sigma_1^t,\quad \Sigma_1 \in \underline{T}(k).$$

Then $\Sigma^{-1} = (\Sigma_1^t)^{-1}\ \Sigma_1^{-1}$ and we use the change of variable $T \to \Sigma_1^{-1}T$. Note that

$$(9.2.26)\qquad \operatorname{tr} \Sigma^{-1}TWa^t = a^t(\Sigma_1{}^t)^{-1}\ \Sigma_1^{-1}\ TW \quad \text{and}$$

$$(9.2.27)\qquad \operatorname{tr} \Sigma^{-1}aa^t = \operatorname{tr}(\Sigma_1^{-1}\ a)(\Sigma_1^{-1}a)^t.$$

The density function of $\underline{W}$ is then

$$(9.2.28)\qquad (2\pi)^{-(n+1)k/2}\ A_n \cdots A_{n-k+1}(1 - \|W\|^2)^{(n-k-1)/2}$$

$$\times \int(\det T)^{n+1}\operatorname{etr} -1/2(TT^t - 2TWb^t + bb^t) \bigwedge_{j\le i}(T^{-1}dT)_{ij}.$$

The new parameter is $b = \Sigma_1^{-1}\ a$.

The statistic often used in inference is $\|W\|^2 = \underline{Y}^t(TT^t)^{-1}\underline{Y} = \underline{Y}^t(\underline{X}^t\underline{X} + \underline{Y}\,\underline{Y}^t)^{-1}\underline{Y}$. Problems 9.4.1 and 9.4.2 suggest the derivation of the noncentral probability density function of $\|W\|^2$. A different derivation will be obtained in Chapter 11 using random variable techniques. The answer (9.2.28) should be compared with Giri, Kiefer, and Stein (1963).

## 9.3. Eigenvalues of the sample covariance matrix $\underline{X}^t\underline{X}$.

We assume $\underline{X}$ is $n \times k$, $n \geq k$. We use the decomposition obtained in Theorem 8.4.2. By Problem 4.2.7 together with Section 8.6, that $X^tX$ has two equal nonzero eigenvalues is an event of zero Lebesgue measure, so, except on this set, we write

$$(9.3.1) \qquad X = ADG^t, \ A \quad n \times k \quad \text{such that} \quad A^tA = I_k, \ D \in \underline{D}(k),$$

and $\qquad G \in \underline{O}(k)$.

This decomposition is unique provided the diagonal entries of $D$ are in order of increasing magnitude, and $G$ is restricted so the entries of the first column of $G$ are all positive. Since

$$(9.3.2) \qquad X^tX = GD^2G^t ,$$

the entries of $D^2$ are the eigenvalues of $X^tX$. We now begin the computation of differential forms. The k-frame $A$ is to be augmented by a $n \times (n-k)$ matrix $B$ such that $(A,B) \in \underline{O}(n)$ and $\det(A,B) = 1$. Then

$$(9.3.3) \qquad dX = (dA)DG^t + A(dD)G^t + AD(dG^t), \quad \text{and}$$

$$B^t dX = B^t(dA)DG^t. \quad \text{Then,}$$

$$(9.3.4) \qquad \left(\bigwedge_{j=1}^{n-k} \bigwedge_{i=1}^{k} b_j^t\, dx_i\right)(\det DG^t)^{-(n-k)} = \bigwedge_{j=1}^{n-k} \bigwedge_{i=1}^{k} b_j^t\, da_i .$$

The remainder of the computation is more complex.

$$(9.3.5) \qquad A^t dX\, G = A^t(dA)D + dD + D(dG^t)G$$

$$= A^t(dA)D + dD - DG^t(dG).$$

We use here the fact that $(dG^t)G + G^t dG = 0$. From the left side of (9.3.5),

$$(9.3.6) \qquad \bigwedge_{j=1}^{k} \bigwedge_{i=1}^{k} (A^t dX\ G)_{ij} = (\det G)^k \bigwedge_{j=1}^{k} \bigwedge_{i=1}^{k} (a_j^t\ dx_i).$$

On the right side of (9.3.5), we write

$$(9.3.7) \qquad D = \begin{pmatrix} c_1 & \cdots & 0 \\ & \ddots & \\ 0 & \cdots & c_k \end{pmatrix}.$$

In forming the join of terms from the matrix on the right side of (9.3.5), the terms $dc_1 \wedge dc_2 \wedge \cdots \wedge dc_k$ are the $(1,1)$, $(2,2),\ldots,(k,k)$ entries in the product since any other entry from $dD$ is zero. If $i \neq j$ then we form the symmetrical product

$$(9.3.8) \qquad (A^t dX\ G)_{ij} \wedge (A^t dX\ G)_{ji}$$

and form a join of these. We use the relations

$$(9.3.9) \qquad a_i^t\ da_j + d(a_i^t) a_j = 0 \quad \text{and} \quad g_i^t\ dg_j + d(g_i^t) g_j = 0.$$

Note that the j-th column of $(dA)D$ is $c_j\ da_j$ and the i-th row of $DG^t$ is $c_i g_i^t$ where $g_i$ is the i-th column of $G$. Thus,

$$(9.3.10) \qquad \text{if } i \neq j \text{ then } (A^t dX\ G)_{ij} = c_j a_i^t\ da_j - c_i\ g_i^t\ dg_j,$$

and the symmetrical product is, if $i \neq j$, then

$$\begin{aligned}(9.3.11) \qquad & (A^t dX\ G)_{ij} \wedge (A^t dX\ G)_{ji} \\ &= (c_j a_i^t da_j - c_i g_i^t\ dg_j) \wedge (c_i a_j^t da_i - c_j g_j^t\ dg_i) \\ &= (c_j^2 - c_i^2)(a_i^t\ da_j) \wedge (g_i^t\ dg_j).\end{aligned}$$

In this calculation we have used (9.3.9) several times. Putting together (9.3.6) and (9.3.11) we obtain

(9.3.12) $$(\det (A,B))^k \bigwedge_{j=1}^{n} \bigwedge_{i=1}^{k} dx_{ij}$$

$$= \pm(\det DG^t)^{n-k}(\det G)^k \bigwedge_{j=1}^{n-k} \bigwedge_{i=1}^{k} b_j^t\, da_i \bigwedge_{i<j} a_i^t da_j \bigwedge_{i<j} g_i^t\, dg_j$$

$$\times \prod_{i<j} (c_j^2 - c_i^2) \bigwedge_{i=1}^{k} dc_i .$$

Let us assume the rows of $\underline{X}$ are independently and identically distributed normal $(0,\Sigma)$ random vectors. Then the joint probability density function is

(9.3.13) $$(2\pi)^{nk/2} (\det \Sigma)^{-n/2} \operatorname{etr} -1/2\, \Sigma^{-1}X^tX$$

$$= (2\pi)^{nk/2} (\det \Sigma)^{-n/2} \operatorname{etr} -1/2\, \Sigma^{-1}GD^2G^t,$$

where $X = ADG^t$. Let $\Lambda$ be the diagonal with entries in increasing order equal to the eigenvalues of $\Sigma$. Integration over $A$ and $G$ gives for the probability density function

(9.3.14) $$(2\pi)^{nk/2}(\det \Sigma)^{-n/2} \frac{A_n \cdots A_{n-k+1}}{2^k}$$

$$\times \int \operatorname{etr} -1/2\, \Lambda^{-1} GD^2G^t \bigwedge_{i<j} g_i^t\, dg_j$$

$$\times \left(\prod_{i=1}^{k} c_i\right)^{n-k} \prod_{i<j} (c_j^2 - c_i^2) \bigwedge_{i=1}^{k} dc_i .$$

Here we assume $c_1 \le c_2 \le \cdots \le c_k$.

## 9.4. Problems.

Problem 9.4.1. Let $X^t = (x_1,\ldots,x_n)$ have differential form $\bigwedge_{i=1}^{n} dx_i$. Let $z = \| X\|$ and $y^t = (y_1,\ldots,y_n) = X^t/z$. Then show

$$(9.4.1) \qquad \bigwedge_{i=1}^{n} dx_i = z^{n-1}\, dz \sum_{i=1}^{n} (-1)^{i+1} y_i\, dy_1 \wedge \ldots \wedge dy_{i-1} \wedge dy_{i+1} \wedge \ldots \wedge dy_n .$$

The sum is the differential form for an invariant measure, c.f. (6.7.5), on a 1-frame and can be written in a coordinate free form $\bigwedge_{i=1}^{n-1} b_i^t\, dy$. Here $B = (b_1,\ldots,b_{n-1})$ satisfies $B^tB = I_{n-1}$ and $B^ty = 0$.

Problem 9.4.2. Use the results of Section 9.2 together with Problem 9.4.1 to obtain the probability density function of the Hotelling $T^2$ statistic, i.e., $\|\underline{W}\|^2$.

Problem 9.4.3. Integrate (9.2.28) over a set $A$ in the W-space which satisfies, if $G \in \underline{O}(n)$, then $GA = A$. Show the resulting function $\beta(b)$ has the property

$$(9.4.2) \qquad \text{if } G \in \underline{O}(n) \text{ then } \beta(Gb) = \beta(b).$$

Problem 9.4.4. Suppose $\underline{X}$ is a $n \times k$ random matrix such that the probability density function of $\underline{X}$ depends on $X$ only through the eigenvalues of $(X^tX)^{1/2}$. If these eigenvalues are $c_1 \leq c_2 \leq \ldots \leq c_k$ then the density function of $c_1,\ldots,c_k$ is

$$(9.4.3) \qquad 2^{-k} A_n \cdots A_{n-k+1} A_k \cdots A_1 f(X) \left(\prod_{i=1}^{k} c_i\right)^{n-k} \prod_{i<j} (c_j^2 - c_i^2) \bigwedge_{i=1}^{k} dc_i .$$

In this formula $f$ is the density function of $\underline{X}$.

Problem 9.4.5. Let $X$ be a $n \times k$ matrix, $A \in \underline{O}(n)$, $G \in \underline{O}(k)$. Then

$$(9.4.4) \qquad \iint f(AXG) \bigwedge_{j<i} a_i^t da_j \bigwedge_{j<i} g_i^t dg_j$$

is a function of the eigenvalues of $(X^tX)^{1/2}$.

Problem 9.4.6. (Read Chapter 5)

Let $\underline{X}$ be a $n \times k$ random matrix with probability density function f. Let $\underline{Y}$ be a $n \times k$ random matrix with probability density function (9.4.4). Then the eigenvalues of $\underline{X}^t\underline{X}$ and the eigenvalues of $\underline{Y}^t\underline{Y}$ have the same probability density function.

Problem 9.4.7. Continue Problem 9.4.5 and Problem 9.4.6. The eigenvalues of $(\underline{X}^t\underline{X})^{1/2}$ have probability density function

$$(9.4.5) \qquad 2^{-k}A_n \cdots A_{n-k+1}A_k \cdots A_1 \left(\iint f(AXG) \bigwedge_{j<i} a_i^t da_j \bigwedge_{j<i} g_i^t dg_j\right)$$

$$\times \left(\prod_{i=1}^{k} c_i\right)^{n-k} \prod_{i<j} (c_j^2 - c_i^2) \bigwedge_{i=1}^{k} dc_i .$$

Problem 9.4.8. Let S be a $k \times k$ positive definite matrix. There exists a unique diagonal matrix D such that

$$(9.4.6) \qquad S = DRD,$$

and the diagonal entries of R are all equal one. (The off diagonal entries of R are the correlations.)

Problem 9.4.9. Continue Problem 9.4.8. Let

$$(9.4.7) \qquad D = \begin{pmatrix} c_1 & & 0 \\ & \ddots & \\ 0 & & c_k \end{pmatrix}.$$

Show

$$(9.4.8) \qquad D^{-1}dS\, D^{-1} = D^{-1}(dD)R + R(dD)D^{-1} + dR, \quad \text{and}$$

$$D^{-1}(dD) = (dD)D^{-1} = \begin{pmatrix} dc_1/c_1 & & 0 \\ & \ddots & \\ 0 & & dc_k/c_k \end{pmatrix}.$$

Since the diagonal of dR is zero, it follows that

$$(9.4.9) \qquad \bigwedge_{j\le i} (D^{-1}(dS)D^{-1})_{ij} = \pm 2^k \bigwedge_{i=1}^{k} (dc_i/c_i) \bigwedge_{j<i} (dR)_{ij}.$$

By Problem 6.7.7, the left side of (9.4.9) is $(\det D)^{-(k+1)} \bigwedge_{j\le i} (dS)_{ij}$. Therefore

$$(9.4.10) \qquad \bigwedge_{j\le i} (dS)_{ij} = \pm(\det D)^k \, 2^k \bigwedge_{i=1}^{k} dc_i \bigwedge_{j<i} (dR)_{ij}.$$

Problem 9.4.10. Continue Problem 9.4.9. If S has a Wishart density function

$$(9.4.11) \quad (2\pi)^{-nk/2}(\det \Sigma)^{-n/2}A_n \cdots A_{n-k+1}(\det S)^{(n-k-1)/2}$$

$$\times \operatorname{etr} -1/2\, \Sigma^{-1}S \bigwedge_{j\le i} (dS)_{ij},$$

then use the substitutions of Problem 9.4.9 to obtain the joint probability density function of the correlation coefficients.

Problem 9.4.11. The factorization of Problem 8.9.8 does not appear to be useful here since one obtains the density function of $(X^tX)^{1/2}$.

## Chapter 10. Cross sections and maximal invariants.

### 10.0 Introduction.

In some problems of statistical inference, as in the construction of Bayes tests, what is needed is a ratio of probability density functions. In the formation of these ratios normalizations cancel. In the case of probability density functions of maximal invariants the normalizations result in part from integration of unwanted variables which lie in a factor of the manifold. In these cases the detailed answer obtained from use of differential forms may be unnecessary.

The method discussed in this Chapter is due to Stein (1956c) and was taught by Karlin (1960). Since then two different developments of Stein's ideas have appeared. Wijsman (1966) develops a theory of cross sections and uses this as a general tool in obtaining a factorization of an invariant measure. This method was further developed by Koehn (1970). The papers of Koehn and Wijsman require differential geometry and Lie group theory. An alternative approach has been developed by Schwartz (1966a), unpublished. This approach notes that in many examples, after discarding an obvious null set, the remainder of the manifold factors. Then the theory of Section 3.4 applies directly. In these notes we develop a methodology in the spirit of Schwartz's development. In Section 10.2 the examples of the sample covariance matrix and the general linear hypothesis are examined in detail. In the summary of examples given in this section these examples are listed as Example 10.0.1 and 10.0.4.

Schwartz's theory is a global type of theory and Example 10.2.1 provides an easy example of the difficulty one may have in verifying the required hypotheses. This difficulty arises in the construction of a suitable maximal invariant, the problem the users of cross sections hope to solve. In the example at hand it seems that one must factor the Stiefel manifold $V_{p,k}$ by the group $\underline{O}(p)$ and from each orbit pick a suitable representative of the point in the Grassman manifold $G_{p,k-p}$. It should be realized that already, in Chapter 7, we have assumed the existence of local cross sections. Were the above correct the existence of local cross sections would be sufficient for the construction of ratios of probability density functions in as much as the ratio as specified in Corollary 10.1.7 and (10.1.39) does not depend on a choice of representation for a maximal invariant. However closer analysis of Example 10.2.1 shows the entire Stiefel manifold does not occur in the answers and that $(X^tX)^{1/2}$ is always a maximal invariant. The global theory still applies.

In the earlier chapters of this book a number of examples already examined in detail are examples of maximal invariants to which the theory of this chapter might be applied. We now list these examples.

Example 10.0.1. The sample covariance matrix $X^tX$, given that $X$ is $n \times k$, is a maximal invariant under the action of $\underline{O}(n)$ as left multipliers of $X$.

Example 10.0.2. The matrix of correlation coefficients $R$, is a maximal invariant under the action $X \to UXD$, where $X$ is $n \times k$, $U \in \underline{O}(n)$, and $D \in \underline{D}(k)$. See Problems 9.4.8, 9.4.9 and 9.4.10.

Example 10.0.3. Hotelling $T^2$ statistic $Y^t(X^tX)^{-1}Y$ discussed in Section 9.2 is a maximal invariant under the action $(X,Y) \to (UXG, GY)$, where $U \in \underline{O}(n)$, $G \in GL(k)$, and $X$ is $n \times k$, $Y$ is $k \times 1$. See Section 9.2.

Example 10.0.4a. The generalization of the Hotelling $T^2$ statistic to the general linear hypothesis is discussed by Lehmann (1959) and will provide the basis of Example 10.1.1 which is further discussed in Section 10.2 as Example 10.2.2. The problem is to find the distribution of the maximal invariant when $X$ is $p \times k$, $Y$ is $q \times k$,

$X \to UXG$, $Y \to VYG$, $U \in \underline{O}(p)$, $V \in \underline{O}(q)$, and $G \in GL(k)$. A maximal invariant is the vector of roots of the equation $0 = \det(X^tX - \lambda Y^tY)$. The exact probability density function has been computed by Constantine (1963).

Example 10.0.4b. Certain tests relating to canonical correlations and discriminant functions can be based on the maximal invariant of $X \to UXT^t$ and $Y \to VYT^t$ where $U \in \underline{O}(p)$, $V \in \underline{O}(q)$ and $T \in \underline{T}(k)$. The invariant here is the $k \times k$ matrix $L$ satisfying $X^tX = TLT^t$, $Y^tY = T(I_k - L)T^t$, with $L \in \underline{S}(k)$ and $T \in \underline{T}(k)$. See Kshirsagar (1961). In case $E\underline{X} = M$ and $E\underline{Y} = 0$ the probability density function of $L$ is the noncentral multivariate beta density function. We further discuss this example in Section 10.3.

Example 10.0.5. The eigenvalues of the sample covariance matrix, discussed in Section 9.3, are a maximal invariant of the action $X \to UXV$, $U \in \underline{O}(n)$, $V \in \underline{O}(k)$ and $X$ a $n \times k$ matrix.

Example 10.0.6. The canonical correlations, discussed in Section 8.8, Problem 8.9.4, Problem 8.9.5 and Section 9.1, are an invariant in that the action $(X,Y) \to (XG_1, YG_2)$ with $X$ a $n \times p$ matrix, $Y$ a $n \times q$ matrix, $G_1 \in GL(p)$, and $G_2 \in GL(q)$, leaves the canonical correlations unchanged. But the definition seems also to involve a minimization in an inherent way.

Example 10.0.7. Aside from the statistics listed above certain of the manifolds listed in these notes may be thought of as maximal invariants. The Stiefel manifold results from the action $X \to XT$, $X$ a $n \times k$ matrix and $T \in \underline{T}(k)$, and the Grassman manifold results from the action $X \to XG$, $X$ a $n \times k$ matrix and $G \in GL(k)$.

## 10.1. Basic Theory.

In order to illustrate the structure needed we choose the example of a restricted version of the general linear hypothesis.

Example 10.1.1. Let the random matrices $\underline{X}$ and $\underline{Y}$ be $p \times k$ and $q \times k$ respectively with $n = p+q$. We assume $\underline{X}$ and $\underline{Y}$ are independently distributed and that the rows of $\underline{X}$ are independently distributed, that the rows of $\underline{Y}$ are independently distributed, that the i-th row of $\underline{X}$ is normal $(a_i, \Sigma)$ and the i-th row of $\underline{Y}$ is normal $(0, \Sigma)$. We write

(10.1.1) $\quad M$ is the $p \times k$ matrix with i-th row $a_i^t$.

Classically the hypothesis to be tested is

$$M = 0,\ \Sigma > 0 \text{ arbitrary}, \tag{10.1.2}$$

against the alternative

$$M \neq 0,\ \Sigma > 0 \text{ arbitrary}. \tag{10.1.3}$$

The problem is left invariant by transformations

$$(X, Y) \to (U_1XG,\ U_2YG) \tag{10.1.4}$$

with $U_1 \in \underline{O}(p)$, $U_2 \in \underline{O}(q)$ and $G \in GL(k)$. That is to say, the random matrices $U_1\underline{X}G$ and $U_2\underline{Y}G$ satisfy the conditions given and the hypothesis parameter sets defined by (10.1.2) and (10.1.3) are invariant sets. We may consider the transformation group to be $\underline{O}(p) \times \underline{O}(q) \times GL(k)$.

Clearly

$$X^tX + Y^tY \text{ and } X^tX,\ Y^tY\ , \tag{10.1.5}$$

are invariants under the action $(X, Y) \to (U_1X, U_2Y)$, with $U_1 \in \underline{O}(p)$ and $U_2 \in \underline{O}(q)$. The action of $G \in GL(k)$ on (10.1.5) becomes

$$X^tX + Y^tY \to G^t(X^tX + Y^tY)G,\quad X^tX \to G^t(X^tX)G. \tag{10.1.6}$$

As suggested in Problem 8.9.12, $G$ may be chosen so that

$$(10.1.7) \qquad I_k = G^t(X^tX + Y^tY)G \quad \text{and} \quad D = G^tX^tXG,$$

with $D \in \underline{D}(k)$ and the diagonal entries of $D$ between 0 and 1. In fact

$$(10.1.8) \qquad \begin{aligned} \text{rank } X &= k - (\text{number of 0's on the diagonal of } D); \\ \text{rank } Y &= k - (\text{number of 1's on the diagonal of } D). \end{aligned}$$

The numbers $c_i$ on the diagonal of $D$ are the roots of the equation

$$(10.1.9) \qquad 0 = \det(c(X^tX + Y^tY) - (X^tX)).$$

By Problem 4.2.7, that $0 < c_i = c_j < 1$, $i \neq j$, is a set of zero Lebesgue measure which is an invariant set under the group action. In the sequel we let $\mathcal{R}$ be the maximal invariant given by (10.1.7), i.e., $\mathcal{R}$ is the set of such diagonal matrices $D$.

The action of the group $\mathcal{G} = \underline{O}(p) \times \underline{O}(q) \times GL(k)$ is not in general one to one and we analyze this question in complete detail in Section 10.2. It suffices here to say that the group $\mathcal{G}$ has a compact subgroup $\mathcal{G}_0$ such that the action $\mathcal{G}/\mathcal{G}_0$ is one to one, provided rank $X = p$, rank $Y = q$ and the roots of (10.1.9) between 0 and 1 are pairwise distinct. The exceptional set $N$ consisting of those $(X,Y)$ such that at least one of

$$(10.1.10) \qquad \text{rank } X < p, \ \text{rank } Y < q \ \text{or} \ 0 < c_i = c_j < 1, \ i \neq j,$$

occurs, is a set of Lebesgue measure zero.

We obtain the following diagram, with $\pi: \mathcal{G} \to \mathcal{G}/\mathcal{G}_0$ the projection map.

$$(10.1.11) \qquad \begin{array}{ccc} h: \mathcal{G} \times \mathcal{R} & \rightarrow & \mathcal{X} - N \\ \pi \times i \downarrow & \nearrow K^{-1} & \\ (\mathcal{G}/\mathcal{G}_0) \times \mathcal{R} & & \end{array}$$

with the relation

(10.1.12) $$K^{-1}(\pi(g),r) = h(g,r).$$

The thing that makes the factorization work, see Lemma 10.1.2, is the uniqueness of invariant measure on $\mathcal{G}/\mathcal{G}_0$ when $\mathcal{G}_0$ is a compact subgroup of $\mathcal{G}$.

In (10.1.11) the manifold to be factored is $\mathcal{X}$. It will appear in our examples that $\pi$ and $h$ are continuous and $K^{-1}$ is a homeomorphism.

Our discussion above has shown that $\mathcal{R}$ is an invariant. In Section 10.2 we reconstruct X, Y from $D \in \mathcal{R}$ thereby showing $\mathcal{R}$ is a maximal invariant. We now begin a formal statement of the results of Schwartz (1966a).

Lemma 10.1.2.

(10.1.13) Suppose $\mathcal{G}$ is a locally compact separable group with a compact subgroup $\mathcal{G}_0$ and factor space $\mathcal{G}/\mathcal{G}_0 = \bar{\mathcal{G}}$ with elements $\bar{g} = \pi(g)$.

Assume

(10.1.14) $\mathcal{X}$ is a separable metric space and $\mu$ is a $\sigma$-finite Borel measure for $\mathcal{X}$.

(10.1.15) That $\mathcal{G}$ acts on $\mathcal{X}$ and the maps $(g,x) \to g(x)$ are jointly measurable.

(10.1.16) That $N \subset \mathcal{X}$ is a Borel set which is $\mathcal{G}$-invariant and $\mu(N) = 0$.

(10.1.17) That $K: \mathcal{X} - N \to \bar{\mathcal{G}} \times \mathcal{R}$ is a one to one onto map.

(10.1.18) $\mathcal{R}$ has a separable locally compact topology.

(10.1.19) $\mathcal{G}$ acts on $\bar{\mathcal{G}} \times \mathcal{R}$ by the action $g_1(\bar{g}_2, r) = (\overline{g_1 g_2}, r)$. Hereafter the same letters $\mathcal{G}$, $g$ will be used for the actions of $\mathcal{G}$ on $\mathcal{X} - N$ and $\mathcal{G}$ on $\bar{\mathcal{G}} \times \mathcal{R}$.

(10.1.20) $Kg = gK$ and hence $K^{-1}g = gK^{-1}$, assuming the notational convention of (10.1.19).

(10.1.21) If $A \subset \mathcal{R}$ and $\bar{B} \subset \bar{\mathcal{J}}$ are compact sets then $K^{-1}(\bar{B} \times A)$ is a compact subset of $\mathcal{X} - N$.

(10.1.22) $\mu$ is a left invariant regular Borel measure for $\mathcal{X} - N$.

(10.1.23) The $\sigma$-algebra of Borel subsets of $\mathcal{R}$ is $\mathcal{C}$. The $\sigma$-algebra of Borel subsets of $\bar{\mathcal{J}}$ is $\mathcal{B}(\bar{\mathcal{J}})$ and of $\mathcal{X}$ is $\mathcal{B}(\mathcal{X})$.

Then there exist regular measures $\bar{\mu}$ defined on $\mathcal{B}(\bar{\mathcal{J}})$ and $\nu$ defined on $\mathcal{C}$ such that if $\bar{A} \in \mathcal{B}(\bar{\mathcal{J}})$ and $B \in \mathcal{C}$ then

(10.1.24) $$\mu K^{-1}(\bar{A} \times B) = \bar{\mu}(\bar{A})\nu(B).$$

The measure $\bar{\mu}$ is left invariant.

<u>Proof</u>. The argument is very similar to that of Section 3.4. Let $\mathcal{C}_0 \subset \mathcal{C}$ be the set of $B$ such that for all $\bar{A} \in \mathcal{B}(\bar{\mathcal{J}})$, $K^{-1}(\bar{A} \times B) \in \mathcal{B}(\mathcal{X})$. We start by showing that if $B$ is a compact set then $B \in \mathcal{C}_0$. Let $\mathcal{B}(\bar{\mathcal{J}}, B)$ be the set of those $\bar{A} \in \mathcal{B}(\bar{\mathcal{J}})$ such that $K^{-1}(\bar{A} \times B) \in \mathcal{B}(\mathcal{X})$. By (10.1.21), since $B$ is a compact subset, if $\bar{A}$ is a compact subset then $K^{-1}(\bar{A} \times B)$ is a compact subset of $\mathcal{X} - N$, and it follows that $K^{-1}(\bar{A} \times B) \in \mathcal{B}(\mathcal{X})$. Further, it is clear that since $K^{-1}$ is a one to one function, the set $\mathcal{B}(\bar{\mathcal{J}}, B)$ is a monotone class. Since $\bar{\mathcal{J}}$ has a separable topology a monotone class containing all the compact sets must contain all the Borel subsets. Therefore $\mathcal{B}(\bar{\mathcal{J}}, B) = \mathcal{B}(\bar{\mathcal{J}})$.

The argument of the last paragraph shows that $\mathcal{C}_0$ must contain all compact subsets of $\mathcal{C}$. Also, $\mathcal{C}_0$ is a monotone class. By (10.1.18) the topology of $\mathcal{R}$ is separable so it follows that $\mathcal{C}_0$ contains all the Borel subsets of $\mathcal{R}$, hence $\mathcal{C}_0 = \mathcal{C}$. Since $K^{-1}(\bar{A} \times B)$ is a Borel subset for every rectangle $\bar{A} \times B$ with measurable factors, and since $K^{-1}$ is one to one, it follows that $K: \mathcal{X} - N \to \bar{\mathcal{G}} \times \mathcal{R}$ is a Borel measurable function.

The measure $\mu K^{-1}$ is therefore defined for all subsets of the product $\sigma$-algebra $\mathcal{B}(\bar{\mathcal{G}}) \times \mathcal{C}$. If $\mu K^{-1} \equiv 0$ then Lemma 10.1.2 follows trivially. Otherwise, since $\bar{\mathcal{G}} \times \mathcal{R}$ is a countable union of open subsets of compact closure it follows that $\mu K^{-1}$ is a $\sigma$-finite measure that is finite valued on compact subsets of $\bar{\mathcal{G}} \times \mathcal{R}$. Hence $\mu K^{-1}$ is a regular Borel measure. In particular there must exist a rectangle $\bar{A} \times B_0$ of compact factors such that $\mu K^{-1}(\bar{A} \times B_0) > 0$, and since $K^{-1}(\bar{A} \times B_0)$ is a compact subset of $\mathcal{X} - N$, $\mu K^{-1}(\bar{A} \times B_0) < \infty$. We use $B_0$ as a reference set and define a measure on $\mathcal{B}(\bar{\mathcal{G}})$ by, if $\bar{A} \in \mathcal{B}(\bar{\mathcal{G}})$ then

$$(10.1.25) \qquad \bar{\mu}(\bar{A}, B_0) = \mu K^{-1}(\bar{A} \times B_0).$$

This is clearly a $\sigma$-finite measure that is finite in value whenever $\bar{A}$ is a compact subset of $\bar{\mathcal{G}}$. Therefore $\bar{\mu}(\ ,B_0)$ is a regular measure. Further,

$$(10.1.26) \qquad \bar{\mu}(g\bar{A}, B_0) = \mu K^{-1}(g\bar{A} \times B_0) = \mu K^{-1} g(\bar{A} \times B_0) = \mu(gK^{-1}(\bar{A} \times B_0))$$

$$= \mu K^{-1}(\bar{A} \times B_0) = \bar{\mu}(\bar{A}, B_0).$$

By uniqueness, Theorem 3.3.8, if $\bar{\mu}_0$ is a nonzero $\sigma$-finite regular left invariant measure for $\bar{\mathcal{G}}$, then there exists a constant $\nu(B_0)$ such that

$$(10.1.27) \qquad \bar{\mu}(\bar{A}, B_0) = \nu(B_0)\bar{\mu}_0(\bar{A}).$$

The equation (10.1.27) holds for all $\bar{A} \in \mathcal{B}(\bar{\mathcal{G}})$ and all compact subsets $B_0 \in \mathcal{C}$. Further

$$(10.1.28) \qquad \mu K^{-1}(\bar{A} \times B_0) = \bar{\mu}(\bar{A}, B_0) = \nu(B_0)\bar{\mu}_0(\bar{A}).$$

If we let $\mathcal{C}_1 \subset \mathcal{C}$ be the collection of subsets B for which if $\bar{A} \in \mathcal{B}(\bar{\mathcal{G}})$ then

(10.1.29) $$\mu K^{-1}(\bar{A} \times B) = \nu(B)\bar{\mu}_0(\bar{A})$$

then $\mathcal{C}_1$ is clearly a monotone class containing all compact subsets of $\mathcal{R}$, hence $\mathcal{C}_1 = \mathcal{C}$. We use here assumption (10.1.23). This implies the result that $\nu$ is a countably additive measure that is finite valued on compact sets, hence that $\nu$ is a regular measure. Finally we take $\bar{\mu} = \bar{\mu}_0$ for the statement (10.1.24). #

Remark 10.1.3. It is conceivable that although $\mu \neq 0$, $\mu K^{-1} \equiv 0$. Clearly if K is continuous then $K^{-1}(\mathcal{B}(\bar{\mathcal{G}}) \times \mathcal{C}) = \mathcal{B}(\mathcal{X} - N)$ and this problem does not occur. More generally, the argument above shows that K is measurable and it has been assumed that K is one to one and onto. Our hypotheses imply that $\bar{\mathcal{G}} \times \mathcal{R}$ is a complete separable metric space. Since $\mathcal{X}$ is a metric space and $\mathcal{X}$ is a countable union of compact sets $\bigcup_{n=1}^{\infty} K^{-1}(C_n)$. On each $K^{-1}(C_n)$, $K^{-1}$ is a homeomorphism and is a Borel measurable function. Hence $K^{-1}$ is a measurable function and $\mathcal{B}(\mathcal{X}-N) = \mathcal{B}(\bar{\mathcal{G}}) \times \mathcal{C}$.

Lemma 10.1.4. Let $\mu$ be a regular left invariant measure defined on $\mathcal{B}(\mathcal{X})$ with the factorization (10.1.24). If $f: \mathcal{X} - N \to \mathbb{R}$ is a $\mu$-integrable function then

(10.1.30) $$\int_{K^{-1}(\bar{A} \times B)} f(x)\mu(dx) = \iint_{\bar{A} \times B} f(K^{-1}(\bar{g}, r))\bar{\mu}(d\bar{g})\nu(dr).$$

Proof. By Lemma 10.1.2, equation (10.1.30) holds if f is the indicator function of the set $K^{-1}(\bar{A}_1 \times B_1)$. Since $K^{-1}$ maps $\mathcal{B}(\bar{\mathcal{G}}) \times \mathcal{C}$ onto $\mathcal{B}(\mathcal{X} - N)$, the equation (10.1.30) holds for all $\mathcal{B}(\mathcal{X} - N)$-measurable simple functions. By taking of monotone limits of sequences of simple functions (10.1.30), then follows for all $\mu$-integrable functions f. #

Remark. In Lemma 10.1.4 and in the following discussion we assume the hypotheses of Lemma 10.1.2 hold.

Lemma 10.1.5. Let $\eta$ be a nonzero left invariant Haar measure for $\mathcal{G}$ and let $m$ be the modular function for $\eta$. Let $f: \mathcal{X} - N \to [0,\infty)$ be a probability density function relative to the regular left invariant measure $\mu$ defined on $\mathcal{B}(\mathcal{X} - N)$. Let $\mu = \overline{\eta}\cdot\nu$ be the factorization of $\mu$ where $\eta$ induces $\overline{\eta}$ on $\mathcal{G}/\mathcal{G}_0$. Then (c.f. (10.1.32))

$$(10.1.31) \qquad m(h)\int_{\mathcal{G}} f(gx)\eta(dg)$$

is the probability density function for the maximal invariant $r \in \mathcal{R}$ relative to the measure $\nu$ on $\mathcal{C}$. In (10.1.31), $h$ is any group element satisfying

$$(10.1.32) \qquad x = K^{-1}(\pi(h), r) = h\, K^{-1}(\overline{e}, r).$$

Proof. The modular function is discussed in Section 3.5. This function is a continuous group homomorphism to the multiplicative group on $(0,\infty)$. Since we assume $\mathcal{G}_0$ is a compact group, $m(\mathcal{G}_0)$ is a compact subgroup of $(0,\infty)$, and hence $m(\mathcal{G}_0) = \{1\}$. In the notation of Lemma 10.1.5, if $\pi(x) = \pi(y)$, i.e., $x\mathcal{G}_0 = y\mathcal{G}_0$, then there exists $g_0 \in \mathcal{G}_0$ such that $m(y) = m(xg_0) = m(x)m(g_0) = m(x)$. Hence $m(x) = \overline{m}(\pi(x))$ defines the function $\overline{m}$ on $\mathcal{G}/\mathcal{G}_0$. If $U \subset \mathbb{R}$ is an open subset then $m^{-1}(U) = \pi^{-1}(\overline{m}^{-1}(U))$ so that since $\pi$ is an open mapping, $\pi(m^{-1}(U)) = \overline{m}^{-1}(U)$ is an open set. Hence $\overline{m}$ is a continuous function.

In our proof $K(x) = (\pi(h), r)$ and if $\pi_1$ is the projection on the first coordinate then $\pi_1(K(x)) = \pi(h)$ defines $\pi(h)$ implicitly as a function of $x$. Thus $m(h) = \overline{m}(\pi(h)) = \overline{m}(\pi_1(K(x)))$ is a measurable function of $x$. It should be noted that if $K(x_1) = (\pi(h), r_1)$ and $K(x_2) = (\pi(h), r_2)$ then

$\overline{m}(\pi_1(K(x_1))) = \overline{m}(\pi_1(K(x_2)))$ independent of $r_1$ and $r_2$. Thus the constant $m(h)$ which occurs in (10.1.35) does not depend on the variable of integration $r$ in (10.1.36).

By (10.1.19) it follows that

(10.1.33) $$g_1 x = g_2 x \text{ if and only if } g_1^{-1} g_2 \in \mathcal{G}_0 .$$

Therefore

(10.1.34) $$\text{if } x = K^{-1}(\pi(h), r) \text{ then } f(gx) = f(ghK^{-1}(\overline{e}, r)).$$

Using Theorem 3.5.1 and (3.5.3),

(10.1.35) $$\begin{aligned}\int_{\mathcal{G}} f(gx)\eta(dg) &= \int_{\mathcal{G}} f(ghK^{-1}(\overline{e}, r))\eta(dg) \\ &= (m(h))^{-1} \int_{\mathcal{G}h} f(K^{-1}(\overline{g}, r))\eta(dg) \\ &= (m(h))^{-1} \int_{\overline{\mathcal{G}}} f(K^{-1}(\overline{g}, r))\overline{\eta}(d\overline{g}) .\end{aligned}$$

The last step of (10.1.35) uses Lemma 3.3.10 to justify introduction of the induced measure $\overline{\eta}$. For a rectangle $\overline{\mathcal{G}} \times B \in \mathcal{B}(\overline{\mathcal{G}}) \times \mathcal{C}$ , from Lemma 10.1.4,

(10.1.36) $$\int_{\overline{\mathcal{G}} \times B} f(K^{-1}(\overline{g}, r))\overline{\eta}(d\overline{g})\nu(dr) = \int_{K^{-1}(\overline{\mathcal{G}} \times B)} f(x)\mu(dx) .$$

Since the most general invariant subset has the form

(10.1.37) $$K^{-1}(\overline{\mathcal{G}} \times B) ,$$

the right side of (10.1.36) is an invariant measure whose density is

(10.1.38) $$\int_{\overline{\mathcal{G}}} f(K^{-1}(\overline{g}, r))\overline{\eta}(d\overline{g}) = m(h) \int_{\mathcal{G}} f(gx)\eta(dg) .$$

The result follows. #

Theorem 10.1.6. Given the hypotheses of Lemma 10.1.2 and Lemma 10.1.5, if $f_1$ and $f_2$ are probability density functions on $\mathcal{X} - N$ relative to the regular left invariant measure $\mu$, then

$$(10.1.39) \qquad \int_{\mathcal{G}} f_2(gx)\nu(dg)/\int_{\mathcal{G}} f_1(gx)\nu(dg)$$

is the ratio of the probability density functions for the maximal invariant.

## 10.2. Examples.

Example 10.2.1. Case I. rank $X = k \leq n$. The sample covariance matrix. This continues the Example 10.0.1. Here $\mathcal{X}$ is the set of $n \times k$ matrices with real entries and we factor $X \in \mathcal{X}$ into $X = AT$, $T \in \underline{T}(k)$ and the k-frame $A$ satisfying $A^tA = I_k$, $A$ a $n \times k$ matrix. Take for $\mathcal{G}$ the matrix group $\underline{O}(n)$. In terms of the diagram (10.1.11), with $A$ $n \times k$, $(A,B) \in \underline{O}(n)$,

$$(10.2.1) \qquad h((A,B),T) = AT.$$

Note that $h$ is an open mapping which implies $K$ is a continuous map. The null set $N$ is the set of $X$ with rank $X < k$, and this set is clearly invariant. By uniqueness

$$(10.2.2) \qquad h((A_1,B_1),T_1) = h((A_2,B_2),T_2) \text{ if and only if}$$
$$A_1 = A_2 \text{ and } T_1 = T_2 \text{ and for some } U \in \underline{O}(n-k)$$
$$B_1U = B_2 .$$

We take $\mathcal{G}_0$ to be the subgroup of matrices $\begin{pmatrix} I_k & 0 \\ 0 & U \end{pmatrix}$ with $U \in \underline{O}(n-k)$ and have that (10.2.2) holds if and only if some $V \in \mathcal{G}_0$, $(A_1,B_1)V = (A_2,B_2)$. Then $\bar{\mathcal{G}} = \mathcal{G}/\mathcal{G}_0$ and

$$(10.2.3) \qquad (\pi \times i)((A,B),T) = ((A,B)\mathcal{G}_0,T).$$

This defines the mapping $K^{-1}$ by

$$K^{-1}((A,B)\mathcal{S}_0,T) = AT, \tag{10.2.4}$$

and then (10.1.12) holds.

To verify the hypotheses of Lemma 10.1.2, $\mathcal{X} - N$ is an open subset of $\mathbb{R}_{nk}$ and $\mu$ = Lebesgue measure is a left invariant measure. Relative to Lebesgue measure $N$ is a null set. $K^{-1}$ is one to one onto $\mathcal{X} - N$. $\mathcal{R} = \underline{T}(k)$ is a locally compact group. If $U \in \underline{O}(n)$ the condition $UK = KU$ requires $U(A,T) = (UA,T)$ in which case (10.1.19) holds and $UK^{-1}((A,B)\mathcal{S}_0,T) = U(AT) = (UA)T = K^{-1}((UA,B)\mathcal{S}_0,T)$ is correct. Since the mapping $K^{-1}$ is continuous, (10.1.21) holds. (10.1.22) holds for $\mu$ as defined and (10.1.23) is a definition.

By Lemma 10.1.2, Lebesgue measure factors into a left invariant measure on the Stiefel manifold $V_{k,n}$ and a measure $\nu$ on $\underline{T}(k)$. By discussing right invariance it is easy to see $\nu$ must be a right invariant Haar measure for $\underline{T}(k)$. The Haar measures $\eta$ on $\underline{O}(n)$ are unimodular. From (10.1.31) we find

$$\int_{\underline{O}(n)} f(gx)\eta(dg) \tag{10.2.5}$$

is the probability density function of the maximal invariant $T$ relative to the measure $\nu$. Compare this result with that of Chapter 5. The factored measure $\nu$ does not, of course, depend on $f$, and may be computed explicitly using differential forms.

<u>Case II. rank $X = n < k$.</u> The decomposition of Case I no longer holds. From Theorem 8.5.1 we may write

$$X = ADG\ , \tag{10.2.6}$$

with $A \in \underline{O}(n)$, $D$ a $n \times n$ diagonal matrix, and $G$ a $n \times k$ matrix such that $GG^t = I_n$. Almost surely the diagonal elements of $D$ are pairwise distinct and we assume the diagonal of $D$ is in increasing order of magnitude, so that (10.2.6) is unique up to sign changes which result from multiplication by diagonal orthogonal matrices. If $U \in \underline{O}(n)$ then

$$UX = (UA)DG \tag{10.2.7}$$

so the maximal invariant under $\underline{O}(n)$ is DG. In this case

$$X^t X = G^t D^2 G \tag{10.2.8}$$

and it follows that the rows of G are eigenvectors for $(X^t X)^{1/2}$ and are (almost surely) uniquely determined except for the sign changes noted. Because of the uniqueness, application of Lemma 10.1.2 is immediate.

Example 10.2.2. The general linear hypothesis, discussed in Examples 10.0.4 and 10.1.1 is continued here. The set up as described in Section 10.1 now fails and some modification is required. We now discuss such a possible modification. See Problem 10.4.6, also.

Modification of the basic theory.

In the example of the general linear hypothesis the group fails to act in the nice way outlined in Section 10.1. We assume, as in Section 10.1, that diagram (10.1.11) is constructible with functions $K$, $h$, $\pi$, $i$ as described there. Our main hypothesis is

Hypothesis 10.2.3. If $r \in \mathfrak{R}$ and $g_1, g_2 \in \mathfrak{G}$ then

$$g_1 K^{-1}(\bar{e}, r_1) = g_2 K^{-1}(\bar{e}, r_2) \text{ if and only if } g_1^{-1} g_2 \in \mathfrak{G}_0 \text{ and } r_1 = r_2 . \tag{10.2.9}$$

Every $x \in \mathfrak{X} - N$ is expressible as $x = gK^{-1}(\bar{e}, r)$.

Definition 10.2.4. The function $L: \mathfrak{G} \times \mathfrak{R} \to \bar{\mathfrak{G}} \times \mathfrak{R}$ is defined by

$$L(g, r) = K(gK^{-1}(\bar{e}, r)). \tag{10.2.10}$$

The group $\bar{\bar{\mathcal{G}}}$ of transformations of $\bar{\mathcal{G}} \times \mathcal{R}$ is defined by

$$(10.2.11) \qquad \bar{\bar{g}}_1 L(g,r) = K(g_1 K^{-1}(L(g,r))).$$

Lemma 10.2.5. $L(g_1,r_1) = L(g_2,r_2)$ if and only if $r_1 = r_2$ and $g_1^{-1}g_2 \in \mathcal{G}_0$. Further,

$$(10.2.12) \qquad \bar{\bar{g}}_1 L(g,r) = L(g_1 g,r).$$

The action $\bar{\bar{g}}$ is thus well defined and

$$(10.2.13) \qquad \bar{\bar{g}}_1 L(g,r) = \bar{\bar{g}}_2 L(g,r) \text{ if and only if } g_1^{-1}g_2 \in \mathcal{G}_0.$$

Proof. By (10.2.10), $L(g_1,r_1) = L(g_2,r_2)$ requires $g_1 K^{-1}(\bar{e},r_1) = g_2 K^{-1}(\bar{e},r_2)$ which by Hypotheses 10.2.3 requires $r_1 = r_2$ and $g_1^{-1}g_2 \in \mathcal{G}_0$. Then, by definition,

$$(10.2.14) \qquad \bar{\bar{g}}_1 L(g,r) = K(g_1 K^{-1}(L(g,r))) = K(g_1 K^{-1}(KgK^{-1}(\bar{e},r)))$$

$$= L(g_1 g,r).$$

Thus $\bar{\bar{g}}_1 L(g,r) = \bar{\bar{g}}_2 L(g,r)$ means $L(g_1 g,r) = L(g_2 g,r)$ and $g_1^{-1}g_2 \in \mathcal{G}_0$. Also, if $L(g_1,r_1) = L(g_2,r_2)$ then $\bar{\bar{g}}L(g_1,r_1) = L(gg_1,r_1)$ and $\bar{\bar{g}}L(g_2,r_2) = L(gg_2,r_2)$ so that $r_1 = r_2$ and $g_1^{-1}g_2 \in \mathcal{G}_0$ follows and $L(gg_1,r_1) = L(gg_2,r_2)$. #

In order to construct invariant measures we prove the following.

Lemma 10.2.6.

$$(10.2.15) \qquad K^{-1}(\bar{\bar{g}}_1(\bar{g},r)) = g_1 K^{-1}(\bar{g},r) .$$

Proof. First we show that $(\bar{g},r) = L(g_2,r)$ for some $g_2 \in \mathcal{G}$. In fact, by definition, $L(g_2,r) = K(g_2 K^{-1}(\bar{e},r))$ so we wish to solve the equation $K^{-1}(\bar{g},r) = g_2 K^{-1}(\bar{e},r)$. By Hypothesis 10.2.3 the element $g_2$ exists. Then

$$(10.2.16)\quad K^{-1}(\bar{\bar{g}}_1(\bar{g},r)) = K^{-1}(\bar{\bar{g}}_1 L(g_2,r)) = K^{-1}L(g_1g_2,r)$$

$$= K^{-1}K(g_1g_2K^{-1}(\bar{e},r)) = g_1K^{-1}K(g_2K^{-1}(\bar{e},r)) = g_1K^{-1}L(g_2,r)$$

$$= g_1K^{-1}(\bar{g},r).\ \#$$

The group $\bar{\bar{\mathcal{G}}}$ will generally be isomorphic to $\mathcal{G}$. In fact if $\bar{\bar{g}}_1 = \bar{\bar{g}}_2$ then $L(g_1g,r) = L(g_2g,r)$ for all $g \in \mathcal{G}$ and $r \in \mathcal{R}$ so that $g^{-1}g_1^{-1}g_2g \in \mathcal{G}_0$ for all $g \in \mathcal{G}$ or $g_1^{-1}g_2 \in \bigcap_g g\,\mathcal{G}_0 g^{-1} = \bar{\mathcal{G}}_0$. This is a subgroup of $\mathcal{G}_0$ which is a normal subgroup of $\mathcal{G}$. Frequently this is $\{e\}$, which would imply $g_1 = g_2$, and $\mathcal{G} \cong \bar{\bar{\mathcal{G}}}$. Otherwise $\bar{\mathcal{G}}_0$ is a proper normal subgroup and

$$(10.2.17)\qquad \bar{\bar{\mathcal{G}}} \cong \mathcal{G}/\bar{\mathcal{G}}_0$$

under a projection map $\pi_0$. We obtain the following diagram.

$$(10.2.18)\qquad \begin{array}{rccc}
 & (g,r) & \longrightarrow & gK^{-1}(\bar{e},r) \\
 & \mathcal{G}\times\mathcal{R} & \longrightarrow & \mathcal{X}-N \\
 & \downarrow & \nearrow\ \bar{h} & \uparrow K^{-1} \\
\pi_0\times i & \bar{\bar{\mathcal{G}}}\times\mathcal{R} & \longrightarrow & \bar{\mathcal{G}}/\bar{\mathcal{G}}_0\times\{(\bar{e},r)\mid r\in\mathcal{R}\} \\
 & (\bar{\bar{g}},r) & \longrightarrow & \bar{\bar{g}}L(e,r) = \bar{\bar{g}}(\bar{e},r)
\end{array}$$

where the function $\bar{h}$ satisfies

$$(10.2.19)\quad \bar{h}(\bar{\bar{g}},r) = gK^{-1}(\bar{e},r) = K^{-1}(L(g,r)) = K^{-1}(\bar{\bar{g}}L(e,r))$$

$$= K^{-1}(\bar{\bar{g}}(\bar{e},r)).$$

Lemma 10.1.2 now applies directly to the factoring of an invariant measure.

Example 10.2.2., continued.

In this example $X$ is a $p \times k$ matrix, $Y$ is a $q \times k$ matrix and we assume $p+q \geq k$ so that $X^tX + Y^tY$ is almost surely nonsingular relative to Lebesgue measure. The transformations allowed are $(X,Y) \to (UXG, VYG)$ with $U \in \underline{O}(p)$, $V \in \underline{O}(q)$, and $G \in GL(k)$. Since

(10.2.20) $$0 = \det(G^tX^tXG - cG^tY^tYG) \text{ if and only if}$$
$$0 = \det(X^tX - cY^tY),$$

the roots of this equation together with their multiplicities are an invariant. We shall see that if $p \geq k$ and $q \geq k$ this invariant is maximal while in the contrary cases one must include points of the relevant Grassman manifold as part of the invariant.

By Problem 8.9.13 the matrix $G \in GL(k)$ can be chosen so that

(10.2.21) $$G^tG = X^tX + Y^tY \text{ and } G^tDG = X^tX,$$

with $D$ a diagonal matrix having entries in order of increasing magnitude. Since $G$ is nonsingular, rank $D$ = rank $X^tX$ = rank $X$ and rank $I_k - D$ = rank $Y^tY$ = rank $Y$. Then by Problem 4.2.7, except on a set of measure zero,

(10.2.22) $D$ has $\max(k-p, 0)$ zero entries;

$D$ has $\max(k-q, 0)$ entries equal one;

$r = k - \max(k-p, 0) - \max(k-q, 0)$ is the number of pairwise distinct diagonal entries between zero and one.

Of course, the diagonal entries of $D$ are directly related to the roots of (10.2.20).

<u>The case $p \geq k$ and $q \geq k$.</u>

In this case a function $\varphi$ which is invariant must satisfy

(10.2.23) $$\varphi(X,Y) = \varphi(U(XG), V(YG))$$
$$= \varphi(U(XG)(G^tX^tXG)^{-1/2}(G^tX^tXG)^{1/2}, V(YG)(G^tY^tYG)^{-1/2}(G^tY^tYG)^{1/2})$$

(10.2.23 cont) $= \varphi(\binom{I_k}{0}(G^tX^tXG)^{1/2}, \binom{I_k}{0}(G^tY^tYG)^{1/2})$

$= \varphi^*(D^{1/2}, (I_k-D)^{1/2}).$

This proves that in the present case $D$ is a maximal invariant. Except on a set of measure zero, $D$ and $I_k-D$ are nonsingular and the diagonal entries of $D$ are pairwise distinct. It follows that if

(10.2.24) $X^tX + Y^tY = G_1^tG_1 = G_2^tG_2$ and $X^tX = G_1^tDG_1 = G_2^tDG_2$

then $G_2G_1^{-1} \in \underline{O}(k)$ and is a diagonal matrix.

We are required to construct a function $K: \mathfrak{X} - N \to \bar{\mathfrak{G}} \times \mathfrak{R}$, where $\mathfrak{R}$ is the maximal invariant. We use the definition

(10.2.25) $K(X,Y) = (X(X^tX)^{-1/2}, Y(Y^tY)^{-1/2}, G(X,Y), D)$

where $G(X,Y)$ is the matrix defined in (10.2.24) as a representative of a coset of $GL(k)$ factored by the $2^k$ sign change matrices, which we call $GL(\pm)$. The group $\underline{O}(p)$ will be factored by the subgroup of matrices of the form $\begin{pmatrix} I_k & 0 \\ 0 & U \end{pmatrix}$ and the group $\underline{O}(q)$ will be factored by the subgroup of matrices having the same form, and we call these factor groups for brevity $\underline{O}(p)/\underline{O}(p-k)$ and $\underline{O}(q)/\underline{O}(q-k)$. Then $\mathfrak{G}$ is factored by $\mathfrak{G}_0 = (\underline{O}(p)/\underline{O}(p-k)) \times (\underline{O}(q)/\underline{O}(q-k)) \times GL(\pm)$. Clearly the function $K$ as defined in (10.2.25) is a one to one function onto $\mathfrak{G}/\mathfrak{G}_0 \times \mathfrak{R} = \bar{\mathfrak{G}} \times \mathfrak{R}$. If given $U \in \underline{O}(p)$ and $V \in \underline{O}(q)$ the first $k$ columns of each matrix are $U_k$ and $V_k$, then the function $K^{-1}$ defined by

(10.2.26) $K^{-1}(U_k, V_k, G, D) = (U_k(G^tDG)^{1/2}, V_k(G^t(I_k-D)G)^{1/2})$

is readily checked to be the inverse function of $K$, the values of this function not depending on the coset representatives used. If, then, $U = \binom{I_k}{0}$, $V = \binom{I_k}{0}$ and $G = I_k$ we obtain for $K^{-1}$ the value

$$(10.2.27)\qquad \left(\begin{matrix} D^{1/2} \\ 0 \end{matrix},\ \begin{matrix} (I_k-D)^{1/2} \\ 0 \end{matrix}\right) = K^{-1}(\bar{e},r)$$

in the notation of Hypothesis 10.2.3. We need to show that given $(X,Y)$ there exist $(U,V,G)$ solving the equations

$$(10.2.28)\qquad (X,Y) = \left(U\begin{pmatrix} D^{1/2}G \\ 0 \end{pmatrix},\ V\begin{pmatrix} (I_k-D)^{1/2}G \\ 0 \end{pmatrix}\right).$$

However it is clear that this set of equations requires

$$(10.2.29)\qquad X^tX = G^tDG \quad\text{and}\quad Y^tY = G^t(I_k-D)G.$$

Equations (10.2.29) are solvable and let $G$ be a solution. Then since $D^{1/2}G$ and $(I_k-D)^{1/2}G$ are nonsingular $k\times k$ matrices we may define solutions

$$(10.2.30)\qquad U_k = X(D^{1/2}G)^{-1} \quad\text{and}\quad V_k = Y((I_k-D)^{1/2}G)^{-1}.$$

Then, for example, using (10.2.29)

$$(10.2.31)\qquad (D^{1/2}G)^t(D^{1/2}G) = X^tX = (D^{1/2}G)^tU_k^tU_k(D^{1/2}G)\ ,$$

and cancellation on both sides gives the result

$$(10.2.32)\qquad I_k = U_k^tU_k\ .$$

Similarly

$$(10.2.33)\qquad V_k^tV_k = I_k.$$

Therefore the onto condition of Hypothesis 10.2.3 is satisfied. Condition (10.2.9) is obvious so that Hypothesis 10.2.3 is verified.

In order to show $K$ is a continuous function we construct the function $h$ of diagram (10.1.11).

$$(10.2.34)\qquad h(U,V,G,D) = (U_k(G^tDG)^{1/2}, V_k(G^t(I_k-D)G)^{1/2}).$$

Then one checks easily that

$$h = K^{-1} \circ (\pi \times i), \tag{10.2.35}$$

as required by the diagram (10.1.11). A compact subset $\bar{C}_1 \times C_2$ of $\bar{\mathcal{G}} \times \mathcal{R}$ is the image $\bar{C}_1 \times C_2 = \pi(C_1) \times C_2$ with $C_1$ a compact subset of $\mathcal{G}$. Since $h$ is a continuous function,

$$K^{-1}(\bar{C}_1 \times C_2) = h(C_1 \times C_2) \tag{10.2.36}$$

is a compact subset of $\mathcal{X} - N$. Thus $K^{-1}$ maps compact subsets to compact subsets, as required by Lemma 10.1.2. In particular then $\mathcal{X} - N$ has lots of open subsets and $K$ is a continuous function, as follows from the facts that $K$ is one to one and $\bar{\mathcal{G}} \times \mathcal{R}$ is a locally compact space.

In order to prove $K^{-1}$ is a homeomorphism we show $K^{-1}$ is also a continuous function. If $U \subset \mathcal{X} - N$ and $U$ is an open set, then $h^{-1}(U)$ is an open subset of $\mathcal{G} \times \mathcal{R}$. Since $\pi \times i$ is an open mapping, $K(U) = K(h(h^{-1}(U))) = (\pi \times i)(h^{-1}(U))$ which is an open set. Therefore $K^{-1}$ is continuous. We summarize the discussion in a lemma.

<u>Lemma 10.2.7</u>. In diagram (10.1.11), if $h$ is a continuous function and the one to one function $K^{-1}$ maps compact sets to compact sets, then $K$ is a homeomorphism of $\mathcal{X} - N$ to $\bar{\mathcal{G}} \times \mathcal{R}$.

In this construction $\bar{\mathcal{G}}_0 = \cap_g g \mathcal{G}_0 g^{-1}$ and hence $\bar{\mathcal{G}}_0$ is a compact normal subgroup. Therefore $\bar{\bar{\mathcal{G}}}$ is homeomorphic to $\mathcal{G}/\bar{\mathcal{G}}_0$, and we suppose $\bar{\bar{\mathcal{G}}}_0$ is the image of the compact subgroup $\bar{\mathcal{G}}_0$ under this mapping. Then, since $\bar{\mathcal{G}}_0 \subset \mathcal{G}_0$, $\bar{\mathcal{G}}/\bar{\mathcal{G}}_0$ is isomorphic and homeomorphic to $\mathcal{G}/\mathcal{G}_0$. In (10.2.18) the map from $\bar{\mathcal{G}} \times \mathcal{R}$ to $\bar{\bar{\mathcal{G}}}/\bar{\bar{\mathcal{G}}}_0 \times \{(\bar{e}, r) \mid r \in \mathcal{R}\}$ is $\pi \times j$ where $j(r) = (\bar{e}, r)$.

The conditions (10.1.13) to (10.1.23) of Lemma 10.1.2 are easily checked with the appropriate substitutions of $\bar{\bar{g}}$ for $\bar{g}$ as necessary. In order to apply Lemma 10.1.2 it is necessary to construct an invariant measure $\mu$ on the Borel subsets of $\mathcal{X} - N$. Clearly the measure determined by the following differential form is an invariant measure.

$$(10.2.37) \qquad \frac{\bigwedge_{i=1}^{p} \bigwedge_{j=1}^{k} dx_{ij} \bigwedge_{i=1}^{q} \bigwedge_{j=1}^{k} dy_{ij}}{(\det(X^tX + Y^tY))^{(p+q)/2}}$$

If $\underline{X}$ and $\underline{Y}$ are normally distributed as explained in Example 10.1.1, then the joint density of $\underline{X}$, $\underline{Y}$ relative to $\mu$ is

$$(10.2.38) \quad f_{M,\Sigma} = (2\pi)^{-(p+q)k/2}(\det \Sigma)^{-(p+q)/2}(\det(X^tX + Y^tY))^{(p+q)/2}$$

$$\operatorname{etr} - 1/2\ \Sigma^{-1}(X^tX + Y^tY - 2X^tM + M^tM).$$

This corresponds to $E\underline{X} = M$, $E\underline{Y} = 0$, independent rows each with covariance matrix $\Sigma$. Integration over $dU$, $dV$, $dG$ for a Haar measure on $\underline{O}(p) \times \underline{O}(q) \times GL(k)$ gives

$$(10.2.39) \quad \int f_{M,\Sigma}(UXG, VYG)\, dU dV dG$$

$$= (2\pi)^{-(p+q)k/2}(\det(X^tX + Y^tY)^{(p+q)/2}) \times$$

$$\int (\det G^tG)^{-k/2} \operatorname{etr} -1/2(G^t(X^tX+Y^tY)G - 2G^tX^tU^tM\Sigma^{-1/2} + M\Sigma^{-1}M^t)\, dU dG.$$

Since the variables of $V$ do not explicitly enter into (10.2.39) the integration $dV$ merely gives a factor of one. By Lemma 10.1.5, after multiplying by the modular function, (10.2.39) becomes the density function of the maximal invariant. In this case the modular function is identically one.

A more general form of the general linear hypothesis was studied by Schwartz (1966a), (1966b), (1967) and (1969).

## 10.3. The noncentral multivariate beta density function.

In this section we describe the problem in random variable terms and in terms of differential forms. Since the quantity in question is a maximal invariant, the application of Sections 10.1 and 10.2 provides a suitable exercise, so stated in the Problems of Section 10.4.

We will suppose $\underline{X}$ is $p \times k$ and $\underline{Y}$ is $q \times k$ such that $\underline{X}$, and $\underline{Y}$ have independently normally distributed rows each with covariance matrix $I_k$. We set

$$(10.3.1) \qquad S = X^t X,\ T = Y^t Y, \text{ and } CC^t = S + T,\ CLC^t = S, \text{ with } C \in T(k) \text{ and } L \in S(k).$$

The problem discussed here and by Kshirsagar (1961) is the probability density function of $L$. It is at once clear that $L$ is a maximal invariant under the group action

$$(10.3.2) \qquad (X, Y) \to (UXG, VYG),\ U \in \underline{O}(p),\ V \in \underline{O}(q),\ G \in \underline{T}(k),$$

provided $p \geq k$ and $q \geq k$.

If we begin with the differential form

$$(10.3.3) \qquad \bigwedge_{j \leq i} ds_{ij} \bigwedge_{j \leq i} dt_{ij}$$

the change of variables involves

$$(10.3.4) \qquad ds_{ij} + dt_{ij} = (dC\ C^t + C\ dC^t)_{ij}, \text{ and}$$

$$ds_{ij} = (dC\ LC^t + C(dL)C^t + CL(dC^t))_{ij}\ .$$

Using the fact that if $\omega$ is a 1-form then $\omega \wedge \omega = 0$, we obtain at once

$$(10.3.5) \quad \bigwedge_{j\le i} dt_{ij} \bigwedge_{j\le i} ds_{ij}$$

$$= \bigwedge_{j\le i} (ds_{ij} + dt_{ij}) \bigwedge_{j\le i} ds_{ij}$$

$$= \bigwedge_{j\le i} (dC\ C^t + C\ dC^t)_{ij} \bigwedge_{j\le i} (dC\ LC^t + C(dL)C^t + CL(dC^t))_{ij}$$

$$= \bigwedge_{j\le i} (dC\ C^t + C\ dC^t)_{ij} (\det C)^{k+1} \bigwedge_{j\le i} (dL)_{ij}$$

$$= 2^k c_{11}^k c_{22}^{k-1} \cdots c_{kk} (\det C)^{k+1} \bigwedge_{j\le i} (dC)_{ij} \bigwedge_{j\le i} (dL)_{ij} .$$

For notational reasons we write $(C)_{ij}$ and $(L)_{ij}$ for the $(i,j)$-entries of the indicated matrices. It follows from (10.3.5) that the underlying joint measure for $C$ and $L$ factors, a fact long known. See for example Anderson (1958).

In the central case, $E\underline{X} = 0$ and $E\underline{Y} = 0$, and the joint probability density function of $S$ and $T$ is (from the Wishart density)

$$(10.3.6) \quad 2^{-(p+q)k/2} (\pi)^{-k(k-1)/2} (\det S)^{(p-k-1)/2} (\det T)^{(q-k-1)/2}$$

$$\times \operatorname{etr} - 1/2(S+T) / \prod_{i=1}^{k} (\Gamma((p+1-i)/2)\Gamma((q+1-i)/2))$$

$$= 2^{-(p+q)k/2} (\pi)^{-k(k-1)/2} (\det CC^t)^{(p+q-2k-2)/2} \operatorname{etr} - 1/2(CC^t)$$

$$\times (\det L)^{(p-k-1)} (\det(I_k - L))^{(q-k-1)/2} / \prod_{i=1}^{k} (\Gamma((p+1-i)/2)\Gamma((q+1-i)/2)).$$

The function (10.3.6) depends only on $E = CC^t$ so we introduce new variables $(E)_{ij}$. Using Problem 6.7.7 and in particular (6.7.11) we find

$$(10.3.7) \quad \bigwedge_{j\le i} (dE)_{ij} = 2^k c_{11}^k c_{22}^{k-1} \cdots c_{kk} \bigwedge_{j\le i} (dC)_{ij} .$$

Substitution into (10.3.5) and integration of (10.3.6) then yields the (central) probability density function of the $(L)_{ij}$ to be

$$(10.3.8)\quad (\pi)^{-k(k-1)/4} \prod_{i=1}^{k} \Gamma((p+q+1-i)/2)/(\Gamma((p+1-i)/2)\Gamma((q+1-i)/2))$$

$$(\det L)^{(p-k-1)/2}(\det(I_k-L))^{(q-k-1)/2} \bigwedge_{j\leq i} d(L)_{ij} .$$

The differential form (10.3.5) and the function (10.3.6) show the well known fact that $\underline{C}$ and $\underline{L}$ are stochastically independent in the central case.

We now consider the noncentral case in which it is assumed that $E\underline{X} = M \neq 0$ and $E\underline{Y} = 0$. In (10.3.6) we must replace the central Wishart probability density function for $S$ by the noncentral Wishart density function for $S$, which is

$$(10.3.9)\quad 2^{-pk/2}(\pi)^{-k(k-1)/4}(\det S)^{(p-k-1)/2}\operatorname{etr} - 1/2(S+M^tM)$$

$$\times \left( \int_{\underline{O}(k)} \operatorname{etr}((UX)^tM)\,dU\right) / \prod_{i=1}^{k} \Gamma((p+1-i)/2).$$

The integral by Haar measure of unit mass on $\underline{O}(k)$ is equal to

$$(10.3.10)\quad \int_{\underline{O}(k)} \operatorname{etr}(U^t(MX^tXM^t)^{1/2})\,dU = \int_{\underline{O}(k)} \operatorname{etr}(U^t(MCLC^tM^t)^{1/2})\,dU.$$

The integral (10.3.10) must be substituted into (10.3.9) which in turn is substituted into (10.3.6) for the integration over the variables $C$. The variables $C$ are to be integrated out, leading to a problem whose answer in closed form is unknown.

The matrix $L$ which is symmetric may be decomposed as

$$(10.3.11)\quad L = WW^t,\ W \in \underline{T}(k).$$

It is known that the random variables on the diagonal of $\underline{W}$, which we call $\underline{w}_{11},\ldots,\underline{w}_{kk}$, are independently distributed (in the central case) beta random variables such that the density function of $\underline{w}_{ii}$ is

$$(10.3.12)\quad w_{ii}^{(p-1-i)/2}(1-w_{ii})^{(q-2)/2}\Gamma((p+q+1-i)/2)/(\Gamma((p+1-i)/2)\Gamma(q/2)).$$

This result is usually obtained by calculation of the joint moments of the random variables. See Anderson (1958). In Chapter 11 of this book we will obtain part of this result from the canonical decomposition of the sample covariance matrix. See Problem 11.11.2.

The Wilk's criterion for independence of a sample $\underline{X}$ from a sample $\underline{Y}$ uses a statistic which computationally, on the null hypothesis, is equivalent to computing $\det L = (w_{11}\cdots w_{kk})^2$. Therefore the null distribution, that of $\det L$ in the central case, is the distribution of a product of independently distributed beta random variables. The corresponding problem for Chi-square random variables was studied by Kullback (1934) and a general formulation of the problem has been made in Mathai and Saxena (1969). The form of the density function is very unpleasant.

We close this section with some observations on moments. If $r \geq 1$ is an integer then

$$(10.3.13)\quad E(\det \underline{L})^r = \text{constant}\int(\det L)^{r+(p-k-1)/2}$$

$$\times\ (\det(I_k-L))^{(q-k-1)/2} \bigwedge_{j\leq i} d(L)_{ij}.$$

From (10.3.8), by normalization of (10.3.13), we find

$$(10.3.14)\quad E(\det \underline{L})^r = \prod_{i=1}^{k} \frac{\Gamma((2r+p+1-i)/2)\Gamma((p+q+1-i)/2)}{\Gamma((2r+p+q+1-i)/2)\Gamma((p+1-i)/2)}.$$

We note that if $S_s$, $T_s$, $L_s$ and $C_s$ are the $s \times s$ principal minors then

(10.3.15) $$S_s + T_s = C_s C_s^t \quad\text{and}\quad S_s = C_s L_s C_s^t,\ 1 \le s \le k.$$

Therefore (10.3.14) gives the value of $E\,(\det \underline{L}_s)^r$ valid for all $p \ge 1$, $q \ge 1$, $r \ge 0$, and $k \ge 1$. More general moment calculations may be found in Anderson, op. cit., Section 9.4.

## 10.4. Problems.

Problem 10.4.1. Examples 10.0.2, 10.0.3 and 10.0.5 can be treated using the theory of Sections 10.1 and 10.2. That is, find the density function of the maximal invariants by the integration process described in those Sections.

Problem 10.4.2. The F-statistic used in the analysis of variance is the maximal invariant in the case $k = 1$ of the general linear hypothesis. In this case $GL(k)$ consists of the multiplicative group on $(0,\infty)$. Write the density function for the maximal invariant, and compare the result with the non-central F-probability density function.

Remark 10.4.3. Don't forget that the measure $\mu$ on the Borel subsets of $\mathfrak{X} - N$ must be a left invariant measure.

Problem 10.4.4. Use the theory of Sections 10.1 and 10.2 to write the density function of $\underline{L}$ which was defined in Section 10.3. As noted in Section 10.3, $\underline{L}$ is a maximal invariant. In particular show that an integral similar to (10.3.10) results in the noncentral problem. Use transformations $(X,Y) \to (UXG^t, VYG^t)$ and Problem 10.4.6.

Problem 10.4.5. Suppose $\underline{U}_1,\ldots,\underline{U}_k$ are mutually independent real valued random variables such that if $1 \le i \le k$ then $\underline{U}_i$ has a gamma density $u^{a_i-1} e^{-u}/\Gamma(a_i)$. Under transformation by scale change $(\underline{U}_1,\ldots,\underline{U}_k) \to (x\underline{U}_1,\ldots,x\underline{U}_k)$ the maximal invariant is a set of random variables satisfying $\underline{S}_i(\underline{U}_1 + \cdots + \underline{U}_k) = \underline{U}_i$, $1 \le i \le k$. Then $\underline{S}_1 + \cdots + \underline{S}_k \equiv 1$. Find the joint density function of $\underline{S}_1,\ldots,\underline{S}_{k-1}$.

(11.2.13) $$\begin{pmatrix} X_1^t X_1 & \cdots & X_1^t X_p \\ \\ X_p^t X_1 & \cdots & X_p^t X_p \end{pmatrix} = (T)_p (T)_p^t .$$

In particular

<u>Theorem 11.2.2.</u>

(11.2.14) $$\frac{\det((T)_{p+1}(T)_{p+1}^t)}{\det((T)_p \ (T)_p^t)} = t_{p+1\ p+1}^2 ,$$

where $t_{ii}$ is the (i,i)-element of $T$. Also from (11.2.12),

(11.2.15) $$\det X^t X = t_{11}^2 \cdots t_{ii}^2 \det Y^t Y$$

where the $(k-i) \times (k-i)$ matrix $\underline{Y}^t\underline{Y}$ has a Wishart distribution with $n-i$ degrees of freedom.

## 11.3. The generalized variance, zero means.

In this section we use (11.2.15) but require a more expressive notation. So we let

(11.3.1) $\underline{S}(k,n)$ be a $k \times k$ random matrix such that $\underline{S}(k,n)$ has a central Wishart probability density function with $n$ degrees of freedom.

Then (11.2.15) may be phrased as

(11.3.2) $$\det S(k,n) = t_{11}^2 \cdots t_{ii}^2 \det S(k-i,n-i);$$

$\underline{t}_{11},\ldots,\underline{t}_{ii}$ and $\underline{S}(k-i,n-i)$ are mutually independent random variables.

Consider the decomposition of the $n \times k$ matrix $X = AT^t$, with $T \in \underline{T}(k)$ and the $n \times k$ matrix $A$ satisfying $A^t A = I_k$. This decomposition arises from use of the Gram-Schmidt orthogonalization process. If the columns of $X$ are $X_1,\ldots,X_k$ and those of $A$ are $A_1,\ldots,A_k$ then from the Gram-Schmidt process it follows that

$$(11.3.3) \qquad (X_1,\ldots,X_{k-1}) = \tilde{X} = (A_1,\ldots,A_{k-1})\tilde{T}^t$$

where $\tilde{T} \in \underline{T}(k-1)$ such that $(\tilde{T})_{ij} = (T)_{ij}$, $1 \leq j \leq i \leq k-1$. Note that the $(k,k)$-entry of $(X^tX)^{-1}$ is given by

$$(11.3.4) \qquad (\det \tilde{X}^t\tilde{X})/(\det X^tX) = ((X^tX)^{-1})_{kk}.$$

We may now state and prove the following theorem.

Theorem 11.3.1. The $(k,k)$-entry of $(\underline{X}^t\underline{X})^{-1}$ is distributed as $1/\chi^2_{n-k+1}$, where $\underline{X}$ is a $n \times k$ matrix consisting of $nk$ mutually independent random variables, each normal $(0,1)$, $k \leq n$.

Proof. $((\underline{X}^t\underline{X})^{-1})_{kk} = \det(\underline{\tilde{X}}^t\underline{\tilde{X}})/\det(\underline{X}^t\underline{X}) = 1/\underline{t}^2_{kk}$.

We use here Theorem 11.2.2 followed by Theorem 11.0.1. #

Definition 11.3.2. If $\underline{X}$ is a $n \times k$ random matrix and $E\underline{X} = 0$ then the quantity $\det \underline{X}^t\underline{X}$ is called the generalized variance.

Theorem 11.3.3. Let $\underline{X}$ be as in Theorem 11.3.1. Then the trace of $\underline{X}^t\underline{X}$ is distributed as a $\chi^2_{nk}$ random variable, and $\det \underline{X}^t\underline{X}$ is a $\chi^2_n\chi^2_{n-1}\cdots\chi^2_{n-k+1}$, a product of $k$ mutually independent chi-square random variables.

## 11.4. Noncentral Wishart, rank one means.

We let $e_n$ be the $n \times 1$ vector having all entries equal one. Suppose the random $n \times k$ matrix $\underline{X}$ satisfies

$$(11.4.1) \qquad E\underline{X} = e_n a^t$$

where $a \in \mathbb{R}_k$. We further assume that $\underline{X} - e_n a^t$ consists of $nk$ mutually independent normal $(0,1)$ entries. Suppose $U \in \underline{O}(n)$ and $(Ue_n)^t = (0,\ldots,0,\sqrt{n})$. Then $\underline{X}^t\underline{X} = (U\underline{X})^t(U\underline{X})$. Let $\underline{Y}$ be the $(n-1) \times k$ matrix of the first $n-1$ rows of $UX$ and $\underline{Z}^t$ be the last row of $U\underline{X}$. Then

$$(11.4.2) \qquad \underline{X}^t\underline{X} = \underline{Y}^t\underline{Y} + \underline{ZZ}^t.$$

Then $\underline{X}^t\underline{X}$ is the sum of two independent Wishart random variables such that $\underline{Y}^t\underline{Y}$ is a central Wishart $n-1,k$ random variable and $\underline{ZZ}^t$ is a noncentral Wishart $1,k$ random variable, with $E\underline{Z}^t = \sqrt{n}\,a^t$.

As an alternative, let $V \in \underline{O}(k)$ and $a^tV = (\|a\|,0,\ldots,0)$. Set

$$(11.4.3) \qquad \underline{X}' = \underline{X}V \text{ so that } E\underline{X}' = e_n(\|a\|,0,\ldots,0).$$

Then the columns of $\underline{X}'$ satisfy

$$(11.4.4) \qquad E(\underline{X}')_i = 0,\ 2 \leq i \leq k,\ E(\underline{X}')_1 = \|a\|\,e_n.$$

We make the same decomposition as in Section 11.2. The random variables $\underline{z}_{n1},\ldots,\underline{z}_{nk}$ obtained at the first step, see (11.2.7), are mutually independent, $z_{n1}^2$ is a noncentral chi-square, $n$ degrees of freedom, and noncentrality parameter $na^ta/2$. $\underline{z}_{n2},\ldots,\underline{z}_{nk}$ are normal $(0,1)$ random variables and the entries of $\underline{Y}$ (not the same as $\underline{Y}$ of the preceeding paragraph), see (11.2.5), are $(n-1)(k-1)$ independent normal $(0,1)$ random variables. Hence, after the first step the decomposition proceeds as before. This process gives a canonical decomposition of $V^t\underline{X}^t\underline{X}V$ as $\underline{TT}^t$ which satisfies,

(11.4.5) $\underline{t}_{11}^2$ is a noncentral $\chi_n^2$ with noncentrality parameter $na^ta/2$, and the other entries of $\underline{T}$ are distributed as described in Theorem 11.0.1.

## 11.5. Hotelling $T^2$ statistic, noncentral case.

We continue the notation of Section 11.4 and let $e = e_n$ be the $n \times 1$ vector having all entries one. We note that

$$(11.5.1) \qquad ee^t/n \quad \text{and} \quad I_n - ee^t/n$$

are $n \times n$ orthogonal projection matrices.

Theorem 11.5.1. Let $\underline{X}$ be a random $n \times k$ matrix with independently distributed rows, each normal $(a, \Sigma)$. Then

$$(11.5.2) \qquad n(n-k)k^{-1}(n^{-1}e^t\underline{X}-a_0^t)(\underline{X}^t(I_n-ee^t/n)\underline{X})^{-1}(n^{-1}\underline{X}^te-a_0)$$

is distributed as a noncentral $F$ statistic $\chi_k^2/\chi_{n-k}^2$ with noncentrality parameter

$$(11.5.3) \qquad (1/2)n(a - a_0)^t\, \Sigma^{-1}(a - a_0).$$

Proof. $\underline{X}^te/n$ is the $k \times 1$ vector of sample means. We show that $\underline{X}^te/n$ and $(I_n - ee^t/n)\underline{X}$ are mutually independent random variables by computing covariances.

$$(11.5.4) \qquad E(I_n - ee^t/n)\underline{X} = (I_n - ee^t/n)ea^t = 0,$$

and if the i-th row of $(I_n - ee^t/n)$ is $(I_n - ee^t/n)_i^t$, then

$$(11.5.5) \qquad E\, \underline{X}^te(I_n - ee^t/n)_i^t\, \underline{X} = 0.$$

That is, the sum of squares has been partitioned into

$$(11.5.6) \qquad X^tX = X^t(ee^t/n)X + X^t(I_n - ee^t/n)X.$$

Using random variable techniques, the following transformations leave the problem invariant. Let $\Omega \in \underline{O}(n)$ satisfy

$$(11.5.7) \qquad e^t\, \Omega/\sqrt{n} = (0,0,\ldots,1).$$

Then

$$(11.5.8)\qquad E\,\Omega^t\underline{X} = \Omega^t e a^t = \begin{pmatrix} 0 \\ 0 \\ \vdots \\ \sqrt{n}\ a^t \end{pmatrix}$$

a $n \times k$ matrix. As usual,

$$(11.5.9)\qquad E(\Omega^t\underline{X} - \Omega^t e a^t)_i(\Omega^t\underline{X} - \Omega^t e a^t)_j^t = \sigma_{ij}\, I_n$$

so the covariances remain unchanged.

Write $\underline{\tilde{X}}$ to be the $(n-1) \times k$ matrix obtained by deletion of the nth row from $\Omega^t\underline{X}$ so that $E\,\underline{\tilde{X}} = 0$. Then

$$(11.5.10)\qquad \tilde{X}^t\tilde{X} = (\Omega^t X)^t\, \Omega^t(I_n - ee^t/n)\Omega(\Omega^t\, X) = X^t(I_n - ee^t/n)X,$$

so the random variables $\tilde{X}^t\tilde{X}$ and $e^t\underline{X}$ are stochastically independent. From (11.5.2) we seek the distribution of

$$(11.5.11)\qquad \underline{S} = n(n-k)k^{-1}\, (n^{-1}e^t\underline{X} - a_0^t)(\underline{\tilde{X}}^t\underline{\tilde{X}})^{-1}\, (n^{-1}\underline{X}^t\, e - a_0),$$

and transform this problem to

$$(11.5.12)\qquad n(n-k)k^{-1}(n^{-1}e^t\underline{X} - a_0^t)\Sigma^{-1/2}\Sigma^{1/2}(\underline{\tilde{X}}^t\underline{\tilde{X}})^{-1}\Sigma^{1/2}\Sigma^{-1/2}(n^{-1}\underline{X}^t e - a_0).$$

The $(n-1) \times k$ random matrix $\underline{\tilde{X}}\,\Sigma^{-1/2}$ has $(n-1)k$ independent normal $(0,1)$ entries and

$$(11.5.13)\qquad E(n^{-1}e^t\underline{X} - a_0^t)\ \Sigma^{-1/2} = (a - a_0)^t\ \Sigma^{-1/2}, \quad \text{and}$$

$$\mathrm{Cov}(n^{-1}e^t\underline{X} - a_0^t\ )\ \Sigma^{-1/2} = n^{-1}I_k.$$

Therefore without loss of generality in the sequel we assume $\Sigma = I_k$ and that the vector of means is $\Sigma^{-1/2}a$. We require the distribution of (11.5.11) under this assumption and the assumption that $n^{-1}e^t\underline{X}$ is independent of $\underline{\tilde{X}}$.

Let $\Omega$ be a random orthogonal matrix such that $\Omega(e^t\underline{X}) \in \underline{O}(k)$ and

(11.5.14) $$(\Omega(e^t\underline{X})(n^{-1}e^t\underline{X}-a_0^t\Sigma^{-1/2}))^t = (0,\ldots,0,\|n^{-1}e^t\underline{X}-a_0^t\Sigma^{-1/2}\|)^t.$$

By Section 11.1, $\tilde{\underline{X}}\,\Omega(e^t\underline{X})$ is distributed as a matrix of $(n-1)k$ independent normal $(0,1)$ random variables. Therefore, if $\underline{S}$ is as in (11.5.11),

(11.5.15) $$n^{-1}(n-k)^{-1}k\ \underline{S} = \|\, n^{-1}e^t\underline{X} - a_0^t\Sigma^{-1/2}\|^2 \times ((\Omega\ \underline{X}^t\ \underline{X}\ \Omega^t)^{-1})_{kk}.$$

Apply Theorem 11.3.1. The random variable

(11.5.16) $$n\|\, n^{-1}e^t\underline{X} - a_0^t\ \Sigma^{-1/2}\|^2 = \|\, n^{-1/2}e^t\underline{X}-n^{1/2}\ a_0^t\Sigma^{-1/2}\|^2 = Y_1$$

is a noncentral Chi-square with $k$ degrees of freedom and noncentrality parameter

(11.5.17) $$(1/2)\ n(a - a_0)^t\Sigma^{-1}(a - a_0).$$

Using Theorem 11.3.1, it follows that the distribution of $\underline{S}$ is

(11.5.18) $$k^{-1}Y_1/(n-k)^{-1}Y_2,$$

with $Y_1$ defined in (11.5.16) and $Y_2$ distributed as a central Chi-square with $n-k$ degrees of freedom, $Y_1$ and $Y_2$ stochastically independent, and the noncentrality parameter being the expression in (11.5.17). #

## 11.6. Generalized variance, nonzero means.

In the situation of Section 11.5, where $E\ \underline{X} = ea^t$, the sample covariance matrix is

(11.6.1) $$(n-1)^{-1}\underline{X}^t(I_n - ee^t/n)\underline{X}.$$

Definition 11.6.1. If the $n \times k$ random matrix $\underline{X}$ has independently distributed rows each normal $(a,\Sigma)$ then $\det(\underline{X}^t(I_n-ee^t/n)\underline{X})/\det\Sigma$ is the generalized variance.

Theorem 11.6.2. The generalized variance is distributed as a product $\chi^2_{n-1} \cdots \chi^2_{n-k}$ of $k$ independently distributed central Chi-square random variables.

Proof. Define $\tilde{X}$ as in (11.5.10) so that the generalized variance is $\det(\tilde{X}^t \tilde{X})$. By Theorem 11.3.3 the result now follows. #

Remark 11.6.3. The distribution of products of independently distributed Chi-square random variables has been studied by Kullback (1934). In this paper (Kullback's) the Fourier transform of a product is computed. Inversion of the transform is done in several cases by use of the residue theorem of complex variable together with knowledge of residues of the gamma function. The answers are not simple and are expressed as infinite series of n-th derivatives (uncomputed). #

## 11.7. Distribution of the sample correlation coefficient.

In Problems 9.4.8, 9.4.9 and 9.4.10 the joint probability density function of the sample correlation coefficients has been described using differential forms. In this section we give a random variable description for a single correlation coefficient. Thus we may assume $X$ is $n \times k$ with $k = 2$. Let $X$ have first column $X_1$ and second column $X_2$. The sample quantity is defined to be

$$(11.7.1) \qquad r = \frac{n^{-1}X_1^t X_2 - (n^{-1}e^t X_1)(n^{-1}e^t X_2)}{(n^{-1}X_1^t X_1 - (n^{-1}e^t X_1)^2)^{1/2}\,(n^{-1}X_2^t X_2 - (n^{-1}e^t X_2)^2)^{1/2}} .$$

The function (11.7.1) is homogeneous of degree 0, so we may assume at the onset by scale changes that

$$(11.7.2) \qquad \Sigma = \begin{pmatrix} 1 & \rho \\ \rho & 1 \end{pmatrix} = \begin{pmatrix} 1 & 0 \\ \rho & (1-\rho^2)^{1/2} \end{pmatrix} \begin{pmatrix} 1 & \rho \\ 0 & (1-\rho^2)^{1/2} \end{pmatrix} .$$

Write

$$(11.7.3) \qquad C = \begin{pmatrix} 1 & 0 \\ \rho & (1-\rho^2)^{1/2} \end{pmatrix} \text{ and } Z = X(C^t)^{-1}.$$

Then the random variable $\underline{Z} = \underline{X}(C^t)^{-1}$ has $I_2$ as covariance matrix.

Apply the canonical decomposition Theorem 11.0.1 to the $2 \times 2$ matrix

$$(11.7.4) \qquad (C^{-1}\underline{X}^t)(I_n - ee^t/n)(C^{-1}\,\underline{X}^t)^t = \underline{T}\,\underline{T}^t$$

with the $2 \times 2$ matrix $T$ given by

$$(11.7.5) \qquad T = \begin{pmatrix} t_{11} & 0 \\ t_{21} & t_{22} \end{pmatrix}.$$

Then

$$(11.7.6) \qquad S = \begin{pmatrix} s_{11} & s_{12} \\ s_{21} & s_{22} \end{pmatrix} \underset{\text{def}}{=} X^t(I_n - ee^t/n)X = (CT)(CT)^t, \text{ and}$$

$$(CT) = \begin{pmatrix} t_{11} & 0 \\ \rho\, t_{11} + (1-\rho^2)^{1/2} t_{21} & (1-\rho^2)^{1/2} t_{22} \end{pmatrix}.$$

From (11.7.1) we find the sample quantity to be

$$(11.7.7) \qquad r = s_{12}/(s_{11}s_{22})^{1/2}.$$

The distribution of $\underline{r}/(1-\underline{r}^2)^{1/2}$ is easier to describe in random variable terms so the distribution of this quantity is computed below. From (11.7.7)

$$(11.7.8) \qquad r/(1-r^2)^{1/2} = s_{12}/(s_{11}s_{22}-s_{12}^2)^{1/2}.$$

From (11.7.6) we compute

$$(11.7.9)\qquad S = (CT)(CT)^t = \begin{pmatrix} t_{11}^2 & \rho t_{11}^2+(1-\rho^2)^{1/2}t_{11}t_{21} \\ \rho t_{11}^2+(1-\rho^2)^{1/2}t_{11}t_{21} & (1-\rho^2)t_{22}^2+(\rho t_{11}+(1-\rho^2)^{1/2}t_{21})^2 \end{pmatrix}.$$

Then

$$(11.7.10)\qquad s_{11}s_{22} - s_{12}^2 = (1 - \rho^2)t_{11}^2t_{22}^2, \quad \text{and}$$

$$(11.7.11)\qquad r/(1-r^2)^{1/2} = \frac{(\rho/(1-\rho^2)^{1/2})t_{11} + t_{21}}{t_{22}} .$$

Theorem 11.7.1. If $\underline{X}$ is a random $n \times 2$ matrix with independently distributed rows each normal $(a,\Sigma)$, then the distribution of $\underline{r}/(1-\underline{r}^2)^{1/2}$ with $\underline{r}$ defined by (11.7.1) is the same as that of

$$(11.7.12)\qquad [(\rho/(1-\rho^2)^{1/2})\underline{t}_{11} + \underline{t}_{21}]/\underline{t}_{22}$$

where $\underline{t}_{11}$, $\underline{t}_{21}$ and $\underline{t}_{22}$ are mutually independent random variables, $\underline{t}_{11}^2$ is $\chi_{n-1}^2$, $\underline{t}_{22}^2$ is $\chi_{n-2}^2$, and $\underline{t}_{21}$ is normal $(0,1)$. The variables $t_{11}$, $t_{21}$, $t_{22}$ are those of $T$ defined in (11.7.4). The population parameter $\rho$ is defined by (11.7.2). The Chi-squares are central Chi-squares.

## 11.8. Multiple correlation, algebraic manipulations.

Multiple correlation, as are the canonical correlations, is defined in terms of a maximization. The multiple correlation is in-fact a canonical correlation when one of the matrices is rank one. See the discussion in Section 8.8.

We suppose $X$ is $n \times k$ and that $k = p+q$. $X$ is partitioned, $X = (X_1,X_2)$ such that $X_1$ is $n \times p$ and $X_2$ is $n \times q$. We assume the rows of $\underline{X}$ are independently distributed normal $(0,\Sigma)$ random vectors and

$$(11.8.1) \qquad \Sigma = \begin{pmatrix} \Sigma_{11} & \Sigma_{12} \\ \Sigma_{21} & \Sigma_{22} \end{pmatrix}, \quad \Sigma_{11} \text{ a } p \times p \text{ matrix.}$$

Then the rows of $\underline{X}_1$ have $\Sigma_{11}$ as covariance matrix.

The multiple correlation problem is to choose a $q \times 1$ vector $a$ such that the correlation between $(\underline{X}_1)_{i1}$ and $((\underline{X}_2)_{11},\ldots,(\underline{X}_2)_{q1})a$ is maximized. Call the maximum value of the correlation $R_{i\cdot p+1,\ldots,p+q}$. Here $(X_1)_{i1}$ is the first component of the i-th column of $X_1$, and $(X_2)_{i1}$ is the first component of the i-th column of $X_2$, etc.

We let $e_i$ be the $p \times 1$ vector of norm one with the i-th position of $e_i$ equal one. Then, since it has been assumed that $E\,\underline{X} = 0$,

$$(11.8.2) \qquad E\, a^t \underline{X}_2^t \underline{X}_2 a = n\, a^t \Sigma_{22} a, \quad \text{and}$$

$$(11.8.3) \qquad E\, e_i^t \underline{X}_1^t \underline{X}_2 a = n\, e_i^t \Sigma_{12} a.$$

The population parameter, the correlation, is

$$(11.8.4) \qquad e_i^t \Sigma_{12} a / \sigma_{1ii} (a^t \Sigma_{22} a)^{1/2},$$

and this is to be maximized through choice of $a$. As in Section 8.8, the problem may be solved by use of the Cauchy-Schwarz inequality. The maximum is

$$(11.8.5) \qquad R_{i\cdot p+1,\ldots,p+q} = \sigma_{1ii}^{-1} (e_i^t \Sigma_{12} \Sigma_{22}^{-1} \Sigma_{21} e_i)^{1/2}.$$

The maximization just described is related to the following minimization problem.

Lemma 11.8.1. Let $\begin{pmatrix} P & Q \\ R & S \end{pmatrix}$ be a symmetric positive definite matrix in which $P$ and $S$ are square matrices. Suppose $P$ is $p \times p$ and $S$ is $q \times q$ and that $A$ is a $p \times q$ matrix. Then using the partial ordering of semidefinite matrices,

$$P - QS^{-1}R \leq P - AR - QA^t + ASA^t. \tag{11.8.6}$$

Equality holds if and only if $A = QS^{-1}$.

Proof. Let

$$A = QS^{-1} + C. \tag{11.8.7}$$

Substitution into the right side of (11.8.6) yields the result that

$$P - AR - QA^t + ASA^t = P - QS^{-1}R + CSC^t. \tag{11.8.8}$$

Since $CSC^t \geq 0$, the result follows. #

Apply Lemma 11.8.1 to the vector $\underline{X}_1 e_i - \underline{X}_2 a$. The components of this vector are independently and identically distributed, each with variance

$$e_i^t \Sigma_{11} e_i - e_i^t \Sigma_{12} a - a^t \Sigma_{21} e_i + a^t \Sigma_{22} a. \tag{11.8.9}$$

We set $P = e_i^t \Sigma_{11} e_i = \sigma_{1ii}^2$, $Q = e_i^t \Sigma_{12}$, and $S = \Sigma_{22}$, and obtain

$$\begin{aligned} e_i^t(\Sigma_{11} - \Sigma_{12}\Sigma_{22}^{-1}\Sigma_{21})e_i &= (\sigma_{1ii}^2 - e_i^t \Sigma_{12}\Sigma_{22}^{-1}\Sigma_{21} e_i) \\ &\leq (11.8.9). \end{aligned} \tag{11.8.10}$$

From (11.8.5) and the Lemma we obtain

Lemma 11.8.2. The vector $a$ which maximizes the correlation between $\underline{X}_1 e_i$ and $\underline{X}_2 a$ also minimizes the variance of $\underline{X}_1 e_i - \underline{X}_2 a$ and the minimum variance is

$$\sigma_{1ii}^2(1 - R_{i.p+1,\ldots,p+q}^2) = e_i^t(\Sigma_{11} - \Sigma_{12}\Sigma_{22}^{-1}\Sigma_{21})e_i . \tag{11.8.11}$$

Remark 11.8.3. The number $e_i^t(\Sigma_{11}-\Sigma_{12}\Sigma_{22}^{-1}\Sigma_{21})e_i$ is the ii entry of the conditional covariance matrix of a row of $\underline{X}_1$ given the same row of $\underline{X}_2$. #

By Problem 8.9.14, in the notation of Lemma 11.8.1,

$$(11.8.12)\qquad \det\begin{pmatrix} P & Q \\ R & S \end{pmatrix} = (\det S)(\det(P - QS^{-1}R)).$$

With the interpretation $P = e_i^t\Sigma_{11}\, e_i$, $Q = e_i^t\Sigma_{12}$ and $S = \Sigma_{22}$, from (11.8.12) it follows that $\det(P - QS^{-1}R) = e_i^t(\Sigma_{11}-\Sigma_{12}\Sigma_{22}^{-1}\Sigma_{21})e_i$ and the following Lemma has been verified.

Lemma 11.8.4. The following holds.

$$(11.8.13)\qquad ((e_i^t\Sigma_{11}e_i)\det \Sigma_{22})(1 - R^2_{i\cdot p+1,\ldots,p+q})$$

$$= \det\begin{pmatrix} e_i^t\Sigma_{11}e_i & e_i^t\Sigma_{12} \\ \Sigma_{21}e_i & \Sigma_{22} \end{pmatrix}.$$

## 11.9. Distribution of the multiple correlation coefficient.

Start with $\underline{X} = (\underline{X}_1, \underline{X}_2)$ a random $(n+1) \times k$ matrix with independently and identically distributed rows, each normal $(a,\Sigma)$. Eliminate the means by considering the "sample" matrix

$$(11.9.1)\qquad \begin{pmatrix} S_{11} & S_{12} \\ S_{21} & S_{22} \end{pmatrix} = \begin{pmatrix} X_1^t \\ X_2^t \end{pmatrix}(I_n - ee^t/n)(X_1,X_2).$$

Then the sample multiple correlation coefficient may be defined by the equation

$$(11.9.2)\qquad (I - \overline{R}^2)(e_i^tS_{11}e_i)(\det S_{22}) = \det\begin{pmatrix} e_i^t\, S_{11}e_i & e_i^t\, S_{12} \\ S_{21}\, e_i & S_{22} \end{pmatrix}.$$

The computation is done with formula (11.9.2). As in Section 11.7, calculate the distribution of $\overline{R}/(1 - \overline{R}^2)^{1/2}$. The general technique is the same as in Section 11.7.

The population and sample multiple correlation coefficients are invariant under addition of an arbitrary $k \times 1$ vector to the mean vector $a$, together with transformation of the random variables. That is, if the transformation is $a \to a + a_1$ and $(\underline{X}_1, \underline{X}_2) \to (\underline{X}_1, \underline{X}_2) + ea_1^t$ then correlations are unchanged. Thus for simplicity one may suppose $a = 0$. Other transformations that leave the correlations invariant are

$$(11.9.3) \qquad (\underline{X}_1, \underline{X}_2) \to (a\underline{X}_1, \underline{X}_2 E),$$

where $a$ is a constant and $E$ is a nonsingular $q \times q$ matrix. For, by (11.9.1), $S_{11}, S_{12}, S_{22}$ transform to $a^2 S_{11}$, $aS_{12}E$ and $E^t S_{22} E$ so that (11.9.2) becomes

$$(11.9.4) \qquad \frac{\det\begin{pmatrix} a & 0 \\ 0 & E \end{pmatrix} \det\begin{pmatrix} e_i^t S_{11} e_i & e_i^t S_{12} \\ S_{21} e_i & S_{22} \end{pmatrix} \det\begin{pmatrix} a & 0 \\ 0 & E \end{pmatrix}}{a^2\, e_i S_{11} e_i \det E^t S_{22} E}.$$

Similarly for (11.8.13). In order to find the distribution of $\overline{R}$ choose $a$ and $E$ so that $\sigma_{1ii} = 1$, $\Sigma_{22} = I_q$, and (11.8.13) now reads

$$(11.9.5) \qquad (1 - R^2_{i\cdot p+1,\ldots,p+q}) = \det\begin{pmatrix} 1 & e_i^t \Sigma_{12} \\ \Sigma_{21} e_i & I_q \end{pmatrix}.$$

The problem may be further transformed by matrices of the form $\begin{pmatrix} 1 & 0 \\ 0 & U \end{pmatrix}$ with $U \in \underline{O}(q)$ and obtain

$$(11.9.6)\qquad (1 - R^2_{i\cdot p+1,\ldots,p+q}) = \det\begin{pmatrix} 1 & 0 & \cdots & 0 & z \\ 0 & 1 & \cdots & 0 & 0 \\ \vdots & & \ddots & \vdots & \vdots \\ z & 0 & \cdots & 0 & 1 \end{pmatrix} = 1 - z^2,$$

where $z = \| e_i^t \Sigma_{12} \|$.

The orthogonal matrix $U$ would be chosen to satisfy $e_i^t \Sigma_{12} U = (0,\ldots,0,z)$. The transformed covariance matrix is then

$$(11.9.7)\qquad \begin{pmatrix} 1 & b^t \\ b & I_q \end{pmatrix}, \quad b^t = (0,\ldots,0,R), \text{ and } R = R_{i\cdot p+1,\ldots,p+q} \geq 0.$$

In order to simplify notations, relable the variables, with $\underline{X} = (\underline{X}_1, \underline{X}_2)$, $\underline{X}_1$ a $n \times q$ matrix, $\underline{X}_2$ a $n \times 1$ matrix, such that the covariance matrix of a row of $\underline{X}$ is the matrix in (11.9.7). In the new notation wanted is the distribution of $\overline{R}_{q+1\cdot 1,\ldots,q}$. Let

$$(11.9.8)\qquad C = \begin{pmatrix} I_q & & 0 \\ & & \vdots \\ R, 0, \ldots, 0, & & (1-R^2)^{1/2} \end{pmatrix},$$

a lower triangular matrix, so that $CC^t = (11.9.7)$. Apply the canonical decomposition to $(\underline{X}(C^{-1})^t)^t(\underline{X}(C^{-1})^t)$ so that

$$(11.9.9)\qquad \underline{T}\,\underline{T}^t = (\underline{X}(C^{-1})^t)^t(\underline{X}(C^{-1})^t).$$

Needed are the entries of $CT$, which are the same as the entries of $T$ except for the last row. The last row of $CT$ is

$$(11.9.10)\qquad \overline{t}_{q+1\ 1} = Rt_{11} + (1-R^2)^{1/2}\, t_{q+1\ 1}, \text{ and if } 2 \leq i \leq q+1,$$

$$\overline{t}_{q+1\ i} = (1-R^2)^{1/2}\, t_{q+1\ i}.$$

Write

$$(11.9.11) \qquad CT = \begin{pmatrix} \overline{T} & 0 \\ \overline{t}^t & \overline{t}_{q+1\ q+1} \end{pmatrix}.$$

Then

$$(11.9.12) \qquad (CT)(CT)^t = \begin{pmatrix} \overline{T}\,\overline{T}^t & \overline{T}\,\overline{t} \\ \overline{t}^t\overline{T}^t & \overline{t}^t\overline{t} + \overline{t}^2_{q+1\ q+1} \end{pmatrix}.$$

Then, the sample quantity satisfies

$$(11.9.13) \qquad 1 - R^2 = \frac{(\det \overline{T})^2\ (\overline{t}_{q+1\ q+1})^2}{(\det \overline{T})^2(\overline{t}^t\overline{t} + \overline{t}^2_{q+1\ q+1})}$$

and

$$(11.9.14) \qquad \overline{R}^2/(1-\overline{R}^2) = \frac{(R\underline{t}_{11} + (1-R^2)^{1/2}\underline{t}_{q+1\ 1})^2 + (1-R^2)\sum_{i=2}^{q}\underline{t}^2_{q+1\ i}}{(1-R^2)\ \underline{t}^2_{q+1\ q+1}}.$$

Theorem 11.9.1. If $\underline{X}$ is a $(n+1) \times k$ random matrix of independently and identically distributed rows such that each row is normal $(a,\Sigma)$, then the distribution of $\overline{R}_{i\cdot p+1,\ldots,p+q}$, transformed to (11.9.14), is the same as the distribution of the right side of (11.9.14), where $\underline{t}_{11}, \underline{t}_{q+1\ 1},\ldots,\underline{t}_{q+1\ q+1}$ are independently distributed random variables, $\underline{t}^2_{11}$ is a central $\chi^2_n$, $\underline{t}^2_{q+1\ q+1}$ is a central $\chi^2_{n-q}$, and $\underline{t}_{q+1\ 1},\ldots,\underline{t}_{q+1\ q}$ are normal $(0,1)$ random variables.

## 11.10. Best linear unbiased estimation.

This long subject is broken into a number of unnumbered subsections.

### BLUE: Best linear unbiased estimation, an algebraic theory.

In this Section, $\underline{X}$ is a $n \times 1$ random vector satisfying

$$(11.10.1) \qquad E\,\underline{X} = B\emptyset \quad \text{and} \quad E(\underline{X} - B\emptyset)(\underline{X} - B\emptyset)^t = \sigma^2 I_n,$$

where $B$ is a $n \times k$ matrix assumed known to the statistician and $\emptyset$, $\sigma$, are unknown parameters, $\emptyset$ a $k \times 1$ vector.

Definition 11.10.1. An estimator which is a function of $\underline{X}$ is a linear estimator if and only if there exists $a \in \mathbb{R}_n$ such that $a^t\underline{X}$ is the estimator.

Remark 11.10.2. A linear estimator $a^t\underline{X}$ is considered to estimate its expectation, that is, $a^t\underline{X}$ estimates $a^tB\emptyset$.

Definition 11.10.3. A linear estimator is a best linear unbiased estimator (BLUE) of $a_0^tB\emptyset$ if and only if

(11.10.2) $$a_0^t\, B\emptyset = a^tB\emptyset \quad \text{for all } \emptyset \in \mathbb{R}_k, \text{ and}$$

$$\operatorname{var}(a_0^t\underline{X}) = \inf_{a\in\mathbb{R}_n,\, Ea^t\underline{X}=a_0^tB\emptyset} \operatorname{var}(a^t\underline{X}) .$$

The questions dealt with in this Section are (i) necessary and sufficient conditions that $a_0^tB\emptyset$ have a best linear unbiased estimator, (ii) if a best linear unbiased estimator does exist, how does one compute it, and (iii) sums of squares.

Lemma 11.10.4. If $a$ and $b \in \mathbb{R}_n$, then

(11.10.3) $$\operatorname{cov}(a^t\underline{X},\, b^t\underline{X}) = \sigma^2a^tb .$$

Proof. $E(a^t\underline{X} - a^tB\emptyset)(b^t\underline{X} - b^tB\emptyset) = E(a^t\underline{X} - a^tB\emptyset)(b^t\underline{X} - b^tB\emptyset)^t$ $= a^t\, E((\underline{X} - B\emptyset)(\underline{X} - B\emptyset)^t)b = a^t(\sigma^2I_n)b = \sigma^2a^tb$. #

Lemma 11.10.5. If $c \in \mathbb{R}_k$ then $c^t\emptyset$ is estimable if and only if

$c^t$ is in the row space of $B$, i.e., $c^t$ is a linear combination of the rows of $B$.

Proof. If $c^t\emptyset$ is estimable then for some $a \in \mathbb{R}_n$,

$$E\, a^t\underline{X} = a^tB\emptyset = c^t\emptyset . \tag{11.10.4}$$

This is to hold for all $\emptyset \in \mathbb{R}_k$ so that $c^t = a^tB$ follows. That is, $c^t$ is a linear combination of rows of $B$. Conversely, if $a \in \mathbb{R}_n$ and $a^tB = c^t$, then $E\, a^tX = a^tB\emptyset = c^t\emptyset$. #

Lemma 11.10.6. In order that every linear combination $c^t\emptyset$ be estimable, $c \in \mathbb{R}_k$, it is necessary and sufficient that rank $B = k$.

Proof. rank $B$ = dimension of the row space of $B$. Use Lemma 11.10.5. #

Lemma 11.10.7. If $B$ is a $n \times k$ matrix with real entries then the following are equal:

(i) rank $B =_{\text{def}}$ dimension of the row space of $B$ = row rank of $B$.

(ii) dimension of the column space of $B$ = column rank of $B$.

(iii) rank $B^tB$.

(iv) rank $BB^t$.

Proof. The row space of $B$ is the set of vectors $\{B^ta \mid a \in \mathbb{R}_n\}$. Two vectors $a_1$ and $a_2$ in $\mathbb{R}_n$ satisfy $B^ta_1 = B^ta_2$ if and only if $B^t(a_1-a_2) = 0$. Therefore

$$\begin{aligned} \text{dim row space } B &= n - \dim(\text{col space } B)^{\perp} \\ &= n - (n - \text{col rank } B). \end{aligned} \tag{11.10.5}$$

Further, if $c \in \mathbb{R}_k$ and

$$B^tBc = 0, \text{ then } c^tB^tBc = 0, \text{ then } (Bc)^t(Bc) = 0, \text{ then } Bc = 0. \tag{11.10.6}$$

Therefore,

(11.10.7) $\quad (\text{row space } B)^{\perp} = (\text{row space } B^tB)^{\perp}$ .

Similarly,

(11.10.8) $\quad (\text{column space } B)^{\perp} = (\text{column space } BB^t)^{\perp}$ . #

<u>Lemma 11.10.8</u>. If $a \in \mathbb{R}_n$ there exists $c \in \mathbb{R}_k$ such that

(11.10.9) $$a^tB = c^tB^tB.$$

Therefore $d \in \mathbb{R}_k$ and $d^t\emptyset$ is estimable if and only if there exists $c \in \mathbb{R}_k$ with $d^t = c^tB^tB$.

<u>Proof</u>. By Lemma 11.10.7, $B$ and $B^tB$ have the same row spaces. Therefore (11.10.9) is correct. By Lemma 11.10.5, $d^t\emptyset$ is estimable if and only if $d^t$ is in the row space of $B$ if and only if there exists $c \in \mathbb{R}_k$ such that $d^t = c^tB^tB$. #

<u>Definition 11.10.9</u>.

(11.10.10) $$H = \{a \mid a \in \mathbb{R}_n \text{ and for some } c \in \mathbb{R}_k,\ a = Bc\}$$

$$N(B) = \{a \mid a \in \mathbb{R}_n \text{ and } a^tB = 0\}.$$

<u>Lemma 11.10.10</u>. The following hold.

(11.10.11) $H^{\perp} = N(B)$ and the dimension of $H$ = rank of $B$;

$H \cap N(B) = \{0\}$;

$\mathbb{R}_n = H \oplus N(B)$.

<u>Proof</u>. Obvious. #

<u>Lemma 11.10.11</u>. If $c \in \mathbb{R}_k$ and $c^t\emptyset$ is estimable then there is a uniquely determined $a \in H$ such that $a^t\underline{X}$ is <u>the</u> best linear unbiased estimator of $c^t\emptyset$. Conversely, if $a \in H$ then $a^tH$ is <u>the</u> best linear unbiased estimator of $a^tB\emptyset$.

Proof. If $c^t\phi$ is estimable then there exists $a \in \mathbb{R}_n$ such that $a^tB = c^t$, and by Lemma 11.10.8, there exists $r \in \mathbb{R}_k$ such that $c^t = a^tB = r^tB^tB$. Then $Br \in H$ and $E(Br)^t\underline{X} = r^tB^tB\phi = c^t\phi$.

To prove uniqueness, if $a_1$ and $a_2$ are in $H$ and if $E\, a_1\underline{X} = E\, a_2\underline{X}$ then $(a_1 - a_2)^tB\phi = 0$ for all $\phi$ in $\mathbb{R}_k$. Therefore $a_1 - a_2 \in H \cap N(B)$ so that by Lemma 11.10.10, $a_1 = a_2$.

To prove minimum variance, suppose $a \in \mathbb{R}_n$ and $a = a_1 + a_2$ with $a_1 \in H$ and $a_2 \in N(B)$. Then $a^tB\phi = E\, a^t\underline{X} = a_1^tB\phi + a_2^tB\phi = a_1^tB\phi$, and by Lemma 11.10.4, $\mathrm{var}(a^t\underline{X}) = \sigma^2(a_1+a_2)^t(a_1+a_2) = \sigma^2(a_1^ta_1 + a_2^ta_2)$, since $a_1^ta_2 = 0$. Clearly, the variance is then minimized by the choice $a_2 = 0$, i.e., by the choice $a \in H$. In particular the best linear unbiased estimator is uniquely determined. #

Gauss Markov equations.

The preceeding results can be summarized in a Theorem.

Theorem 11.10.12. If $c \in \mathbb{R}_k$ then $c^t\phi$ is estimable if and only if $c$ is in the row space of $B$. In this case, there exists $r \in \mathbb{R}_k$ such that $c = B^tBr$ and the best linear unbiased estimator of $c^t\phi$ is $(Br)^t\underline{X}$. The set of vectors $\{a | a \in \mathbb{R}_n$ and $a^tX$ is a BLUE$\}$ = column space of $B = H$.

If $a \in \mathbb{R}_n$ then $a^tB$ is in the row space of $B$ and $a^tB\phi$ is estimable, so that there exists $r \in \mathbb{R}_k$ such that

(11.10.12) $\qquad r^tB^tB = a^tB$, and

(11.10.13) $\qquad r^tB^t\underline{X}$ is the best linear unbiased estimator of $a^tB\phi$.

Closely related to the equations (11.10.12) are the Gauss Markov equations (11.10.14).

(11.10.14) $\qquad B^t\underline{X} = B^tB\underline{t}$.

Clearly if $(B^tB)^{-1}$ exists then $\underline{t} = (B^tB)^{-1}B^t\underline{X}$ solves the equations (11.10.14) and if $c \in \mathbb{R}_k$ then $B(B^tB)^{-1}c$ is in the column space of $B$ so that $(B(B^tB)^{-1}c)^t\underline{X}$ is a best linear unbiased estimator of

$$(11.10.15) \qquad E(B(B^tB)^{-1}c)^t X = c^t\emptyset.$$

In the sequel we will study in detail the relation between being a solution of the Gauss Markov equations (11.10.14) and being a best linear unbiased estimator.

Theorem 11.10.13. If $\underline{X}$ has a joint normal density function and if $E\,\underline{X} = B\emptyset$, $E(\underline{X} - B\emptyset)(\underline{X} - B\emptyset)^t = \sigma^2 I_n$, then the maximum likelihood estimators $\hat{\emptyset}$ of $\emptyset$ are the solutions of the equations (11.10.14).

Proof. The joint density function is

$$(11.10.16) \qquad (2\pi)^{-n/2}\sigma^{-n} \exp - 1/2(X-B\emptyset)^t(X-B\emptyset)\sigma^{-2}.$$

Therefore $\hat{\emptyset}$ satisfies

$$(11.10.17) \qquad (X-B\hat{\emptyset})^t(X-B\hat{\emptyset}) = \inf_{\emptyset\in\mathbb{R}_k} (X-B\emptyset)^t(X-B\emptyset).$$

If $s$ solves (11.10.14) then

$$(11.10.18) \qquad ((X-Bs) + B(s-\emptyset))^t((X-Bs) + B(s-\emptyset))$$
$$= (X-Bs)^t(X-Bs) + (s-\emptyset)^t(B^tB)(s-\emptyset)$$
$$+ (X^t-s^tB^t)B(s-\emptyset) + (s-\emptyset)^tB^t(X-Bs).$$

By (11.10.14) the last two terms of (11.10.18) vanish and the matrix $B^tB$ being positive semi-definite the minimum is attained if $\hat{\emptyset} = s$ is used. Conversely if $\hat{\emptyset}$ is a maximum likelihood estimator then, since $s$ solves (11.10.14),

(11.10.19) $\quad (s-\hat{\emptyset})^t B^t B(s-B\hat{\emptyset}) = 0;$ therefore

$B(s-\hat{\emptyset}) = 0;$ therefore

$B^t B\hat{\emptyset} = B^t Bs = B^t X.$ #

Remark 11.10.14. The minimization of Theorem 11.10.13 implies equations (11.10.14) always have a solution. This fact may be seen directly as follows. $B^t X$ is an element of the row space of $B$, while by Lemma 11.10.8 the row space of $B$ is the set of vectors $\{c \mid \text{for some } t \in \mathbb{R}_k,\ c = B^t Bt\}$. Hence a vector $t \in \mathbb{R}_k$ exists solving the equations. #

Theorem 11.10.15. (Gauss Markov) Suppose $\underline{t}$ solves the Gauss Markov equations. Let $c \in \mathbb{R}_k$ and suppose $c^t\emptyset$ is estimable. Then $c^t\underline{t}$ is the best linear unbiased estimator of $c^t\emptyset$.

Proof. We write $c^t = r^t B^t B$. Then $c^t\underline{t} = r^t B^t B\underline{t} = r^t B^t \underline{X}$. By Theorem 11.10.12 the result now follows. #

Solutions of the Gauss Markov equations.

The solutions of the Gauss Markov equations have a elegant expression in terms of the generalized inverse of $B^t B$. We now develop this theory and refer the reader to Section 8.5 for a definition of the generalized inverse of a matrix. To begin, consider (11.10.15). Here we may write $c^t = r^t B^t B$ so that (11.10.15) becomes

(11.10.19) $\quad r^t B^t (B(B^t B)^{-1} B^t)\underline{X}.$

The essential things here are the projection matrix $B(B^t B)^{-1} B^t = P$, which is a symmetric matrix satisfying $P = P^2$, and that $E\, r^t B^t P\underline{X} = c^t\emptyset$. In the general theory $B^t B$ may be a singular matrix, i.e., rank $B^t B < k$, in which case we consider the projection matrix (as follows directly from Theorem 8.5.3)

(11.10.20) $\quad P = B(B^t B)^+ B^t.$

The superscript "+" is used to denote generalized inverse, and one readily verifies that $P$ is symmetric and $P = P^2$. The basic calculational lemma is

<u>Lemma 11.10.16</u>. The $n \times n$ projection matrix (11.10.20) has as its range the column space of $B$, and hence

$$(B(B^tB)^+B^t)B = B. \qquad (11.10.21)$$

<u>Proof</u>. We first establish some facts about the generalized inverses of symmetric matrices. If $A$ is a square symmetric matrix then the matrix $A^o = (A^+)^t$ satisfies the conditions of Theorem 8.5.3, and hence by uniqueness $(A^+)^t = A^+$. In words, the generalized inverse of a symmetric matrix is a symmetric matrix.

Next, $\{b \mid \text{exists } a \in \mathbb{R}_k \text{ and } b = BB^tBa\}$ = column space of $B$. To see this, the condition $a^tBB^tB = 0$ implies $a^tBB^tBB^ta = 0$ so that $BB^ta = 0$ and $a^tBB^ta = 0$, or $a^tB = 0$. This argument assumes that $B$ has real entries. Therefore the columns of $B$ and the columns of $BB^tB$ have the same orthogonal complement and hence the column spaces are equal.

Next, if $A$ is a square symmetric matrix then by the first part of the proof,

$$AA^+ = (AA^+)^t = (A^+)^tA^t = A^+A. \qquad (11.10.22)$$

Make the interpretation $A = B^tB$ and obtain

$$\begin{aligned}[B(B^tB)^+B^t][BB^tBa] &= B(B^tB)^+ (B^tB)(B^tB)a \\ &= B[(B^tB)(B^tB)^+(B^tB)]a = BB^tBa.\end{aligned} \qquad (11.10.23)$$

Therefore the range of the projection (11.10.20) includes the column space of $B$. But clearly, rank $B(B^tB)^+B^t \le$ rank $B$. Since the rank is the dimension of the column space, the desired conclusion now follows. #

<u>Lemma 11.10.17</u>. The following projection matrices are equal.

$$B(B^tB)^+B^t = (BB^t)(BB^t)^+. \tag{11.10.24}$$

<u>Proof</u>. By Theorem 8.5.3, $(BB^t)(BB^t)^+$ is a $n \times n$ symmetric projection matrix. By the proof of Lemma 11.10.16, the column space of $B$ is representable as $BB^tc$. Thus

$$(BB^t)(BB^t)^+(BB^tc) = BB^tc. \tag{11.10.25}$$

Since rank $(BB^t)(BB^t)^+ \leq$ rank $B$ = dimension of the column space of $B$, it follows that both orthogonal projections of (11.10.24) have the same range, hence are equal as matrices. #

<u>Theorem 11.10.18</u>. The $k \times 1$ vector

$$s = (B^tB)^+B^tX \tag{11.10.26}$$

solves the Gauss Markov equations (11.10.14). If $s_1$ and $s_2$ are solutions of (11.10.14) then $s_1 - s_2$ is an element of (row space $B)^\perp$. Therefore the Gauss Markov equations have a unique solution in the row space of $B$ which is the vector (11.10.26).

<u>Proof</u>. Substitution of (11.10.26) into (11.10.14) gives

$$B^tB(B^tB)^+ (B^tX) = B^tX. \tag{11.10.27}$$

The computation uses Lemma 11.10.16. Given two solutions $s_1$ and $s_2$,

$$B^tB(s_1-s_2) = 0 \text{ so that } B(s_1-s_2) = B(B^tB)^+B^tB(s_1-s_2) = 0. \tag{11.10.28}$$

Thus, $s_1 - s_2$ is in (row space $B)^\perp$. The Theorem now follows. #

<u>Theorem 11.10.19</u>. The matrix

$$P = I_n - B(B^tB)^+B^t \tag{11.10.29}$$

is a $n \times n$ symmetric projection matrix. In addition,

(11.10.30) $\quad E\,P\underline{X} = 0$ and $E\,\underline{X}^t P\underline{X} = (n - \text{rank } B)\sigma^2$.

<u>Proof</u>. From Lemma 11.10.16, $PB = 0$. Thus $E\,P\underline{X} = 0$ and

$$E\,\underline{X}^t P\underline{X} = E(\text{tr } P(\underline{X} - B\emptyset)(\underline{X}-B\emptyset)^t) = \text{tr}(\sigma^2 P I_n) = (n - \text{rank } B)\sigma^2. \ \#$$

<u>Theorem 11.10.20</u>. Let $a \in \mathbb{R}_n$, and let $c \in \mathbb{R}_k$ such that $c^t\emptyset$ is estimable. Let $\underline{t}$ be a solution of the Gauss Markov equations and let $P$ be the projection matrix (11.10.29). Then

(11.10.31) $\quad \text{cov}(c^t\underline{t},\ a^t P\underline{X}) = 0.$

<u>Proof</u>. Write $c^t = r^t B^t B$ so that $c^t\underline{t} = r^t B^t \underline{X}$. Then (11.10.31) becomes

(11.10.32) $\quad E\ r^t B^t \underline{X}\underline{X}^t Pa = \sigma^2 r^t B^t Pa = 0.$ #

<u>Normal theory. Idempotents and Chi-squares.</u>

Throughout this subsection we assume that $\underline{X}$ has a joint normal density function such that $E\,\underline{X} = 0$ and $E\,\underline{X}^t\underline{X} = \sigma^2 I_n$.

<u>Lemma 11.10.21</u>. Let $P$ be a $n \times n$ symmetric projection matrix of rank $r$. Then $\underline{X}^t P\underline{X}$ is a (central) $\chi_r^2$ random variable.

<u>Proof</u>. Take $U \in \underline{O}(n)$ such that $UPU^t = \begin{pmatrix} I_r & 0 \\ 0 & 0 \end{pmatrix}$ and set $\underline{Y} = U\underline{X}$. Then $\underline{Y}$ has a joint normal density function and $E\,\underline{Y} = 0$, $E\,\underline{Y}^t\underline{Y} = \sigma^2 I_n$. We have that

(11.10.33) $$\underline{X}^t P\underline{X} = \underline{Y}^t \begin{pmatrix} I_r & 0 \\ 0 & 0 \end{pmatrix} \underline{Y} = \underline{Y}_1^2 + \dots + \underline{Y}_r^2\ . \ \#$$

<u>Lemma 11.10.22</u>. If $P$ is a $n \times n$ symmetric matrix and $\underline{X}^t P\underline{X}$ has a Chi-square probability density function then $P^2 = P$.

<u>Proof</u>. As in the preceeding Lemma, take $U \in \underline{O}(n)$ so that $UPU^t$ is a diagonal matrix with nonzero diagonal entries $p_1,\dots,p_r$. If

$\underline{Y} = U\underline{X}$ and the entries of the vector are $\underline{Y}_1,\ldots,\underline{Y}_n$ then

(11.10.34) $$X^t P X = \sum_{i=1}^{r} p_i \, \underline{Y}_i^2 ,$$

which is a sum of independently distributed random variables. Therefore the Laplace transform of the random variable (11.10.34) is

(11.10.35) $$\prod_{i=1}^{r} 1/(1 - 2p_i s)^{1/2}.$$

The hypothesis of the Lemma is that for some integer $r' > 0$,

(11.10.36) $$\prod_{i=1}^{r} 1/(1-2p_i s)^{1/2} = 1/(1-2s)^{r'/2} .$$

Since both sides are analytic functions of $s$, their singularities must be the same. Therefore $p_1 = \ldots = p_r = 1$, and $r = r'$ follows. #

Normal Theory in the Analysis of Variance.

In this subsection we assume that the random variable $\underline{X}$ has the joint normal density function (11.10.16).

Lemma 11.10.23 . The random variable

(11.10.37) $$\sigma^{-2} \underline{X}^t (I_n - B(B^t B)^+ B^t)\underline{X}$$

has a central $\chi^2_{n-r}$ probability density function, where $r$ = rank of $B$.

Proof. Since $I_n - B(B^t B)^+ B^t$ is a $n \times n$ orthogonal projection matrix of rank $n - r$, Lemma 11.10.21 applies directly to the random variable $\underline{X} - B\emptyset$. Clearly by Lemma 11.10.16,

(11.10.38) $$(\underline{X} - B\emptyset)^t (I_n - B(B^t B)^+ B^t)(\underline{X} - B\emptyset)$$

$$= \underline{X}^t (I_n - B(B^t B)^+ B^t)\underline{X}. \; \#$$

Remark 11.10.24. The random variable specified in (11.10.37) is called the sum of squares of error, SSE. In the analysis of variance this Chi-square is used in the denominator of the F-statistic. See the next Theorem.

Theorem 11.10.25. In the normal theory BLUE and SSE are stochastically independent random variables.

Proof. If $\underline{t}$ is a solution of the Gauss Markov equations then by Theorem 11.10.20, $c^t\underline{t}$ is stochastically independent of $(I_n - B(B^tB)^+B^t)\underline{X}$, provided $c$ is in the row space of $B$. #

Theorem 11.10.26. If $c$ is in the row space of $B$ and $\underline{t}$ is a solution of the Gauss Markov equations then

$$(11.10.39) \qquad (n-r)(c^t\underline{t}/(\sigma(c^t(B^tB)^+c)^{1/2}))^2/(\text{sum of squares of error})$$

is distributed as a noncentral F-statistic $(n-r)\chi_1^2/\chi_{n-r}^2$ with non-centrality parameter

$$(11.10.40) \qquad (1/2(c^t\emptyset)^2)/(\sigma(c^t(B^tB)^+c)^{1/2})^2 .$$

Proof. We may write $c^t = r^t(B^tB)$. Of the possible choices of $r$ we want $r$ in the row space of $B$. To see that this is possible, since $c$ is in the row space of $B$,

$$(11.10.41) \qquad c^t = c^t(B^tB)^+(B^tB) = r^t(B^tB)(B^tB)^+(B^tB),$$

and $(B^tB)^+(B^tB)r$ is in the row space of $B$ by virtue of Lemmas 11.10.16 and 11.10.17. Then, using Lemmas 11.10.4 and 11.10.16,

$$(11.10.42) \qquad c^t\underline{t} = r^tB^t\underline{X}, \quad \text{and}$$

$$(11.10.43) \qquad \operatorname{var}(c^t\underline{t}) = \sigma^2 r^t(B^tB)r.$$

Since $r$ is in the row space of $B$ we have that

(11.10.44) $$c^t(B^tB)^+ = r^t, \text{ and thus}$$

(11.10.45) $$\mathrm{var}(c^t\underline{t}) = \sigma^2 c^t(B^tB)^+c.$$

Thus the numerator of (11.10.39) has variance one and (11.10.40) is one-half of the expectation squared. By Lemma 11.10.23 and Theorem 11.10.25, the result now follows. #

Lemma 11.10.27. If $c_1$ and $c_2$ are in the row space of $B$ and $\underline{t}$ is a solution of the Gauss Markov equations then

(11.10.46) $$\mathrm{cov}(c_1^t\underline{t},\ c_2^t\underline{t}) = \sigma^2\ c_1^t(B^tB)^+c_2.$$

Proof. As in the proof of Theorem 11.10.26. #

The general setup of the analysis of variance is expressed in the next Theorem. As follows from Lemma 11.10.27 the best linear unbiased estimators indicated in the following statement are stochastically independent and the Theorem follows trivially.

Theorem 11.10.28. Let $S$ be a subspace of the row space of $B$ and let $s = \dim S$. Choose $c_1,\ldots,c_s$ in $S$ such that

(11.10.47) $$\begin{aligned} c_i^t(B^tB)^+c_j &= 1, \quad 1 \le i = j \le s,\\ &= 0, \quad 1 \le i \ne j \le s. \end{aligned}$$

Then

(11.10.48) $$(n-r)\sigma^{-2} \sum_{i=1}^{s} (c_i^t\underline{t})^2/s(\text{sum of squares of error})$$

is a noncentral F-statistic $(n-r)\chi_s^2/s\chi_{n-r}^2$ with noncentrality parameter

(11.10.49) $$(1/2)\sigma^{-2} \sum_{i=1}^{s} (c_i^t\emptyset)^2.$$

Remark 11.10.29. Normally the experimeter has $k \times 1$ vectors $c_1,\ldots,c_s$ which are a basis of $S$ and needs to find a new basis of

$S$ to satisfy the condition of Theorem 11.10.28. Let $C = (c_1, \ldots, c_s)$ be the indicated $k \times s$ matrix and choose a $n \times s$ matrix $A$ so that the columns of $A$ are in the column space of $B$ and

(11.10.50) $$C^t = A^t B.$$

For example it suffices to take

(11.10.51) $$A = B(B^t B)^+ C .$$

Then the requirement that

(11.10.52) $$C^t (B^t B)^+ C = I_s$$

becomes, using (11.10.50),

(11.10.53) $$I_s = C^t (B^t B)^+ C = A^t B(B^t B)^+ B^t A = A^t A .$$

Usually the experimenter's basis $C$ of $S$ does not satisfy (11.10.53). He modifies the basis to

(11.10.54) $$A_1 = A(A^t A)^{-1/2} \text{ and } C_1 = B^t A(A^t A)^{-1/2} = C(A^t A)^{-1/2} ,$$

satisfying

(11.10.55) $$C_1^t (B^t B)^+ C_1 = (A^t A)^{-1/2} A^t B(B^t B)^+ B^t A(A^t A)^{-1/2} = I_s .$$

Then

(11.10.56) $$E\, A_1^t \underline{X} = A_1^t B\emptyset = C_1^t \emptyset ,$$

and the noncentrality parameter becomes

(11.10.57) $$(1/2)\sigma^{-2}\emptyset^t C_1 C_1^t \emptyset = (1/2)\sigma^{-2}\emptyset^t B^t A(A^t A)^{-1} A^t B\emptyset$$
$$= (1/2)\sigma^{-2}\emptyset^t C(C^t (B^t B)^+ C)^{-1} C^t \emptyset .$$

We summarize this discussion in a Theorem.

Theorem 11.10.30. Let $S$ be a subspace of the row space of $B$ and assume the dimension of $S$ is $s$. Let $C$ be a $k \times s$ matrix such that the columns of $C$ span $S$. There exists a noncentral F-statistic $(n-r)\chi_s^2/s\chi_{n-r}^2$ with noncentrality parameter (11.10.57).

Remark 11.10.31. That the noncentrality parameter vanishes means $\emptyset$ is orthogonal to $S$. Therefore the F-test constructed using the F-statistic (11.10.48) is a test that $\emptyset \in S^{\perp}$.

## 11.11. Problems.

Problem 11.11.1. Let $X$ be a $n \times 1$ vector, $X^t = (x_1,\ldots,x_n)$. Define $U(X)$ to be a $n \times n$ matrix such that if $x_1 = 0$ then $U(X) = I_n$ and if $x_1 \neq 0$ then

$$(11.11.1) \qquad U(X) = \begin{pmatrix} x_2(x_1^2+x_2^2)^{-1/2} & -x_1(x_1^2+x_2^2)^{-1/2} & 0,\ldots,0 \\ x_1(x_1^2+x_2^2)^{-1/2} & x_2(x_1^2+x_2^2)^{-1/2} & 0,\ldots,0 \\ 0 & 0 & \\ 0 & 0 & I_{n-2} \end{pmatrix}.$$

In each case the positive square root is to be used. Show that $U(X) \in \underline{O}(n)$, that the entries of $U(X)$ are measurable functions of $x_1,\ldots,x_n$, and that $(U(X)X)^t = (0,\ (x_1^2+x_2^2)^{1/2},\ x_3,\ldots,x_n)$. By induction construct a matrix valued function $U(X)$ with values in $\underline{O}(n)$ such that $(U(X)X)^t = (0,\ldots,\|X\|)$.

Problem 11.11.2. This problem is about the multivariate beta density function. Refer to Section 10.3 for notations. In this problem, $k \times k$ symmetric random matrices $\underline{S}$ and $\underline{T}$ are stochastically independent Wishart (central) random variables with $p$ and $q$ degrees of freedom respectively. Write $(S)_h$ and $(T)_h$ for the principal minors with elements $s_{ij}$, $t_{ij}$, $1 \leq i,j \leq h$. By (10.3.1), $CC^t = S+T$, so that $\underline{C}$ has the canonical distribution specified by Theorem 11.0.1. Determine the parameter values.

Again by (10.3.1), $L = C^{-1}S(C^t)^{-1}$, so that if $D \in \underline{T}(k)$, and $DD^t = S$ then $L = (C^{-1}D)(C^{-1}D)^t$. Show that for the $h \times h$ principal minor the same relations hold, namely,

(11.11.2) $$(L)_h = (C)_h^{-1}(S)_h(C)_h^{-1} = ((C)_h^{-1}(D)_h)((C)_h^{-1}(D)_h)^t, \text{ and}$$

$$(S+T)_h = (C)_h(C)_h^t.$$

Let the diagonal elements of $W = C^{-1}D$ (see (10.3.11)) be $w_{11},\ldots,w_{kk}$. Then show

(11.11.3) $$w_{h+1\ h+1}^2 = \det(L)_{h+1}/\det(L)_h$$

$$= \frac{(\det(D)_{h+1}\ \det(C)_h)^2}{(\det(C)_{h+1}\ \det(D)_h)^2}.$$

Show by use of Theorem 11.2.2 that $\underline{w}_{hh}$ is stochastically independent of $\underline{w}_{11},\ldots,\underline{w}_{h-1\ h-1}$.

Problem 11.11.3. Let the nonnegative real valued random variables $\underline{X}$ and $\underline{Y}$ have probability density functions $f$ and $g$ respectively. Then the random variable $\underline{W} = \underline{X}/(\underline{X} + \underline{Y})$ has density function

(11.11.4) $$(1-w)^{-2} \int_0^\infty y\ f(wy/(1-w))\ g(y)dy.$$

Problem 11.11.4. Continue Problem 11.11.3. Suppose

(11.11.5) $$f(x) = x^{n-1}\ e^{-x}/\Gamma(n) \text{ and } g(y) = y^{m-1}\ e^{-y}/\Gamma(m).$$

Then the density function of $\underline{X}/(\underline{X} + \underline{Y})$ is the beta density function

(11.11.6) $$h(w) = w^{n-1}(1-w)^{m-1}\ \Gamma(m+n)/\Gamma(m)\Gamma(n).$$

Problem 11.11.5. Continue Problem 11.11.4. Calculate the r-th moment of the density function $h$ and show the answer is

(11.11.7) $$\frac{\Gamma(m+n)\Gamma(n+r)}{\Gamma(m+n+r)\Gamma(n)}.$$

Problem 11.11.6. Using (10.3.14) and (10.3.15), together with (11.11.7), show that

(11.11.8) $$\frac{E(\det(\underline{L})_{h+1})^r}{E(\det(\underline{L})_h)^r} = \frac{\Gamma(r+(p-h)/2)\Gamma((p+q-h)/2)}{\Gamma(r+(p+q-h)/2)\Gamma((p-h)/2)}.$$

In (11.11.8) make the identification

(11.11.9) $$n = (p-h)/2 \quad \text{and} \quad m = q/2.$$

Refer to (10.3.11) and conclude that $\underline{w}_{h+1\ h+1}$ has a beta probability density function. What Theorem about moments are you using?

Problem 11.11.7. (unsolved) Obtain the result of Problem 11.11.6 that $\underline{w}_{h+1\ h+1}$ has a beta probability density function by random variable techniques.

Problems on conditional distributions.

Problem 11.11.8. Suppose $\underline{X} = \underline{X}_1, \underline{X}_2$ is a $n \times k$ random matrix such that the rows of $\underline{X}$ are independent normal $(0, \boldsymbol{\Sigma})$ random vectors. Let $\underline{X}_1$ be $n \times p$ and $\underline{X}_2$ be $n \times q$, and partition the covariance matrix

(11.11.10) $$\boldsymbol{\Sigma} = \begin{pmatrix} \Sigma_{11} & \Sigma_{12} \\ \Sigma_{21} & \Sigma_{22} \end{pmatrix}, \quad \Sigma_{11} \text{ a } p \times p \text{ matrix.}$$

Let

(11.11.11) $$C = \begin{pmatrix} I_p & -\Sigma_{12}\Sigma_{22}^{-1} \\ 0 & I_q \end{pmatrix}.$$

Show that the covariance matrix of a row of $\underline{X}(C^t)$ is

$$(11.11.12) \qquad \begin{pmatrix} \Sigma_{11} - \Sigma_{12}\Sigma_{22}^{-1}\Sigma_{21} & 0 \\ 0 & \Sigma_{22} \end{pmatrix}.$$

Therefore $\underline{X}_2$ and $\underline{X}_1 - \underline{X}_2\Sigma_{22}^{-1}\Sigma_{21}$ have zero covariance.

Problem 11.11.9. Continue Problem 11.11.8. The conditional distribution of the first row of $\underline{X}_1$ given the first row of $\underline{X}_2$ is normal $(\underline{X}_2\Sigma_{22}^{-1}\Sigma_{21}, \Sigma_{11} - \Sigma_{12}\Sigma_{22}^{-1}\Sigma_{21})$.

Problem 11.11.10. Continue Problem 11.11.9. We have assumed that $E\,\underline{X} = 0$. Let $\underline{S}_{ij} = \underline{X}_i^t\underline{X}_j$, $1 \le i,j \le 2$. The $p \times p$ random matrix

$$(11.11.13) \qquad \underline{S}_{11} - \underline{S}_{12}\underline{S}_{22}^{-1}\underline{S}_{21}$$

has a Wishart density with parameters $n-q, p, \Sigma_{11}-\Sigma_{12}\Sigma_{22}^{-1}\Sigma_{21}$.

Hint. The quantity (11.11.13) is

$$(11.11.14) \qquad \underline{X}_1^t\underline{X}_1 - \underline{X}_1^t\underline{X}_2(\underline{X}_2^t\underline{X}_2)^{-1}\underline{X}_2^t\underline{X}_1 = \underline{X}_1^t(I_n - \underline{X}_2(\underline{X}_2^t\underline{X}_2)^{-1}\underline{X}_2^t)\underline{X}_1.$$

The $n \times n$ matrix in parentheses is a $n \times n$ orthogonal projection. Choose $U = U(\underline{X}_2)$ a random orthogonal $n \times n$ matrix such that

$$(11.11.15) \qquad U(I_n - \underline{X}_2(\underline{X}_2^t\underline{X}_2)^{-1}\underline{X}_2)U^t = \begin{pmatrix} I_{n-q} & 0 \\ 0 & 0 \end{pmatrix}.$$

The random matrices $\underline{X}_1$ and $U(\underline{X}_2)\underline{X}_1$ have the same distribution. Therefore (11.11.13) has the same distribution as

$$(11.11.16) \qquad \underline{X}_1^t \begin{pmatrix} I_{n-q} & 0 \\ 0 & 0 \end{pmatrix} \underline{X}_1. \quad \#$$

Problem 11.11.11. Continue Problem 11.11.10. The conditional distribution of $\underline{S}_{11} - \underline{S}_{12}\underline{S}_{22}^{-1}\underline{S}_{21}$ given $\underline{X}_2$ is a Wishart density with parameters $n-q$, $p$, $\Sigma_{11} - \Sigma_{12}\Sigma_{22}^{-1}\Sigma_{21}$.

Hint. Use the choice of $U$ in (11.11.15). As in (11.11.16),

$$\underline{S}_{11} - \underline{S}_{12}\underline{S}_{22}^{-1}\underline{S}_{21} = (U(\underline{X}_2)\underline{X}_1)^t \begin{pmatrix} I_{n-q} & 0 \\ 0 & 0 \end{pmatrix} (U(\underline{X}_2)\underline{X}_1).$$

Show that the conditional means of $U(\underline{X}_2)\underline{X}_1$ are zero by using Problem 11.11.9 and (11.11.15). #

Problem 11.11.12. Continue Problem 11.11.10. If $\Sigma_{12} = 0$ then the conditional distribution of $\underline{S}_{12}\underline{S}_{22}^{-1}\underline{S}_{21}$ given $\underline{X}_2$ is Wishart with parameters $p$, $q$, $\Sigma_{11}$. The quantities $\underline{S}_{11} - \underline{S}_{12}\,\underline{S}_{22}^{-1}\underline{S}_{21}$ and $\underline{S}_{12}\underline{S}_{22}^{-1}\underline{S}_{21}$ are stochastically independent.

Hint. In terms of $U(\underline{X}_2)$ the random variables are

$$(U(\underline{X}_2)\underline{X}_1)^t \begin{pmatrix} 0 & 0 \\ 0 & I_q \end{pmatrix} (U(\underline{X}_2)\underline{X}_1) \quad \text{and} \quad (U(\underline{X}_2)\underline{X}_1)^t \begin{pmatrix} I_{n-q} & 0 \\ 0 & 0 \end{pmatrix} (U(\underline{X}_2)\underline{X}_1).$$

Since $\Sigma_{12} = 0$ the conditional means are zero. The result now follows since $U(X_2)X_1$ and $X_1$ have the same conditional distribution. #

Problems on the analysis of variance.

Problem 11.11.13. Suppose $A$ is a $n \times s$ matrix and $\sigma^{-2}\,\underline{X}^tAA^t\underline{X}$ is a noncentral $\chi_s^2$ random variable. Suppose $\underline{X}$ has a joint normal density function with $E\,\underline{X} = B\emptyset$ and $E(X - B\emptyset)(X - B\emptyset)^t = \sigma^2 I_n$. Then show $AA^t$ is a $n \times n$ projection matrix.

Hint. See Lemma 11.10.22 and Chapter 2. #

Problem 11.11.14. Continue Problem 11.11.13. Show the noncentrality parameter is

$$(11.11.17) \quad (1/2)\sigma^{-2}\|\,A^tB\emptyset\|^2 = (1/2)\sigma^{-2}\,\emptyset^tB^t(AA^t)B\emptyset\,.$$

More generally, if $P$ is an $n \times n$ orthogonal projection matrix then the noncentral Chi-square $\sigma^{-2}\ \underline{X}^t P\underline{X}$ has noncentrality parameter $\sigma^{-2}(B\emptyset)^t P(B\emptyset)/2$. #

Problem 11.11.15. Let $A_1$ be a $n \times s$ matrix and $A_2$ a $n \times r$ matrix and suppose the columns of $A_1$ and $A_2$ are in the column space of B. Let $\sigma^{-2}\underline{X}^t A_1 A_1^t \underline{X}$ and $\sigma^{-2}\underline{X}^t A_2 A_2^t \underline{X}$ be noncentral Chi-square random variables. Suppose $0 = \emptyset^t B^t (A_1 A_1^t) B\emptyset$ implies $0 = \emptyset^t B^t (A_2 A_2^t) B\emptyset$. Then show $A_1 A_1^t \geq A_2 A_2^t$ in the partial ordering of semidefinite matrices.

Problem 11.11.16. Continuation of Problem 11.11.15. The power function of an analysis of variance test is a strictly increasing function of the noncentrality parameter. Let two analysis of variance tests have power functions $\beta_1$ and $\beta_2$ and assume each F-statistic is a function of the $n \times 1$ random vector $\underline{X}$. Let the numerators of the two F-statistics be $\sigma^{-2}\underline{X}^t A_1 A_1^t \underline{X}$ and $\sigma^{-2}\underline{X}^t A_2 A_2^t \underline{X}$ with noncentrality parameters $\| A_1^t B\emptyset\|^2/2\sigma^2$ and $\| A_2^t B\emptyset\|^2/2\sigma^2$ respectively. Let S be a s dimensional subspace of the row space of B and suppose $\emptyset \in S^\perp$ if and only if $\| A_1^t B\emptyset\| = 0$. See Remark 11.10.31. If both tests are similar size $\alpha$ for the null hypotheses $\emptyset \in S^\perp$, show that if $\emptyset \in \mathbb{R}_k$ then

(11.11.18) $$\| A_1^t B\emptyset\| > \| A_2^t B\emptyset\| .$$

Problem 11.11.17. (Kiefer (1958))

Let $n_1 \leq n_1'$ and $n_1 + n_2 \geq n_1' + n_2'$ with at least one strict inequality. Let noncentral F-tests with power functions $\beta_{n_1 n_2}(\lambda, \alpha)$ and $\beta_{n_1' n_2'}(\lambda, \alpha)$ be given with size $\alpha$ and noncentrality parameter $\lambda$. Then if $\lambda > 0$ and $0 < \alpha < 1$,

(11.11.19) $$\beta_{n_1 n_2}(\lambda, \alpha) > \beta_{n_1' n_2'}(\lambda, \alpha).$$

<u>Hint</u>. Given four stochastically independent Chi-squares $\chi^2_{m_1}$, $\chi^2_{m_2}$, $\chi^2_{m_1'}$, $\chi^2_{m_2'}$ with the interpretations $m_1 = n_1$, $m_2 = n_2'$, $m_1' = n_1' - n_1$ and $m_2' = n_1 + n_2 - n_1' - n_2'$, construct the above tests which are UMP tests. #

<u>Problem 11.11.18</u>. Continue Problems 11.11.16 and 11.11.17. Show that under the hypotheses of these two problems, if $\emptyset \in R_k$ then

(11.11.20) $$\beta_1(\|A_1^t B\emptyset\|^2/2\sigma^2,\alpha) \geq \beta_2(\|A_2^t B\emptyset\|^2/2\sigma^2,\alpha) .$$

<u>Problem 11.11.19</u>. Let $\underline{X}$ be a $n\times 1$ random vector with a joint normal density function such that

(11.11.21) $$E\,\underline{X} = M \quad\text{and}\quad E(\underline{X} - M)(\underline{X} - M)^t = \Sigma .$$

Suppose $A$ is a $n \times n$ symmetric matrix such that $\underline{X}^t A\underline{X}$ has a non-central Chi-square probability density function. This can hold if and only if $(A\Sigma)^2 = A\Sigma$.

<u>Problem 11.11.20</u>. Suppose $A$ and $B$ are $n \times n$ symmetric matrices and $AB = 0$. Show that $BA = 0$ and that there exists $U \in \underline{O}(n)$ such that $UAU^t$ and $UBU^t$ are both diagonal matrices.

<u>Problem 11.11.21</u>. If $\underline{X}$ is as in Problem 11.11.19 and if $A$ and $B$ are $n \times n$ symmetric matrices such that $A\Sigma B = 0$ then $\underline{X}^t A\underline{X}$ and $\underline{X}^t B\underline{X}$ are stochastically independent.

<u>Hint</u>. Show $B\Sigma A = 0$. Thus, use Problem 11.11.20 and simultaneously diagonalize $\Sigma^{1/2}A\Sigma^{1/2}$ and $\Sigma^{1/2}B\Sigma^{1/2}$. #

<u>Problem 11.11.22</u>. (Graybill and Milliken (1969)) Let $\underline{X}$ be a $n \times 1$ random vector having a joint normal probability density function such that $E\,\underline{X} = M$ and $E(\underline{X} - M)(\underline{X} - M)^t = I_n$. Let $K$ be a $r \times n$ matrix and $L$ a $n \times n$ matrix such that $KL^t = 0$. Let $\underline{A}$ be a random symmetric $n \times n$ matrix whose entries are measurable functions of $K\underline{X}$. Suppose

(i) $\underline{A}^2 = \underline{A}$

(ii) $\underline{A} = L^t\underline{A}L$

(iii) $\mathrm{tr}\ \underline{A} = m$, a nonrandom constant

(iv) $M^t AM = 2\lambda$, a nonrandom constant.

Then $\underline{X}^t \underline{AX}$ has a noncentral $\chi^2_m$ probability density function with noncentrality parameter $\lambda$. If $m$ and $\lambda$ are random, what mixture of probability distributions results?

Hint. The random variables $L\underline{X}$ and $K\underline{X}$ are independent and $\underline{X}^t \underline{AX} = \underline{X}^t \underline{L}^t \underline{ALX}$. Use 11.1. #

Problem 11.11.23. (Graybill and Milliken (1969)) Continue Problem 11.11.22. Suppose $\underline{A}$ and $\underline{B}$ are $n \times n$ random symmetric matrices such that the entries of $\underline{A}$ and $\underline{B}$ are functions of $K\underline{X}$. Assume

(i) $\underline{A}^2 = \underline{A}$ and $\underline{B}^2 = \underline{B}$

(ii) $\underline{A} = L^t \underline{A}L$ and $\underline{B} = L^t \underline{B}L$

(iii) $\underline{AB} = 0$

(iv) $\mathrm{tr}\ \underline{A}$, $\mathrm{tr}\ \underline{B}$, $M^t \underline{A}M$ and $M^t \underline{B}M$ are nonrandom constants.

Then the random variables $\underline{X}^t \underline{AX}$ and $\underline{X}^t \underline{BX}$ are stochastically independent.

Hint. By Section 11.1, since conditional on $K\underline{X}$ the random variables $\underline{A}L\underline{X}$ and $\underline{B}L\underline{X}$ have a joint normal probability density function with zero covariances, conditional independence follows. Since the joint distribution is independent of $K\underline{X}$, unconditional independence follows. #

Problem 11.11.24. (Graybill and Milliken (1969)) Continue Problem 11.11.23.

In the analysis of variance model, $E\ \underline{X} = B\emptyset$. In the notations of Problem 11.11.23 let

$$L = I_n - B(B^tB)^+B^t; \tag{11.11.22}$$

$$K = B^t\ ;$$

$$M = B\emptyset\ ;$$

and "+" means generalized inverse. Graybill and Milliken, op. cit., use conditional inverses instead of generalized inverses. See Sections 8.5, 11.10 and the paper by Graybill and Milliken.

Let $\underline{Q}$ be a random $n \times r$ matrix whose entries are functions of $K\underline{X} = B^t\underline{X}$ such that

(11.11.23) $$\text{tr } L\underline{Q} = m, \text{ a nonrandom constant.}$$

Define

(11.11.24) $$\underline{A} = (L\underline{Q})((L\underline{Q})^t(L\underline{Q}))^+(L\underline{Q})^t, \text{ and}$$

$$\underline{C} = L - \underline{A}.$$

Show in sequence that

(i) $\underline{A}^2 = \underline{A}$ ;

(ii) $\underline{A} < \underline{C}$ ;

(iii) $L\underline{A} = \underline{A}L = \underline{A}$ and $L^2 = L$ ;

(iv) $\underline{C}^2 = \underline{C}$ and $\underline{C}L = L\underline{C} = \underline{C}$ ;

(v) $\underline{AC} = 0$ ;

(vi) tr $\underline{A} = m$, a nonrandom constant ;

(vii) tr $\underline{C} = n - \text{rank } B - M$, a nonrandom constant;

(viii) $M^t\underline{A}M = M^t\underline{C}M = 0$;

(ix) $\underline{X}^t\underline{AX}$ and $\underline{X}^t\underline{CX}$ are independently distributed noncentral Chi-square random variables.

Problem 11.11.25. (Graybill and Milliken (1969)) Continue Problem 11.11.24.

In the two-way classification we let $n = IJ$ and speak of the ij component of $\underline{X}$. Our model is

(11.11.25) $$E\,\underline{X} = \{\phi_{ij}\}, \quad 1 \leq i \leq I, \quad 1 \leq j \leq J,$$

subject to the side conditions

(11.11.26) $$I^{-1} \sum_{i=1}^{I} \phi_{ij} = J^{-1} \sum_{j=1}^{J} \phi_{ij}, \quad \text{i.e.,} \quad \phi_{i\cdot} = \phi_{\cdot j}.$$

A test due to Tukey is a test of whether the interactions $\phi_{ij} - \phi_{i.} - \phi_{.j} + \phi_{..}$ all vanish. We assume that they do all vanish so that $E\,\underline{X} = \phi_{i.} + \phi_{.j} - \phi_{..}$. The statistic is described as follows.

$$(11.11.27)\qquad s_1^2 = \frac{\left(\sum_{i=1}^{I}\sum_{j=1}^{J}(x_{ij}-x_{i.}-x_{.j}+x_{..})(x_{i.}-x_{..})(x_{.j}-x_{..})\right)^2}{\sum_{i=1}^{I}(x_{i.}-x_{..})^2\ \sum_{j=1}^{J}(x_{j.}-x_{..})^2},$$

and

$$s_2^2 = \sum_{i=1}^{I}\sum_{j=1}^{J}(x_{ij}-x_{i.}-x_{.j}+x_{..})^2 .$$

The numerator of the F-test is $s_1^2 = \underline{X}^t\underline{AX}$ and the denominator is $s_2^2 - s_1^2 = \underline{X}^t\underline{CX}$. The problem is to obtain descriptions of these statistics in the notations of Problem 11.11.24. Let $P_1$ and $P_2$ be the $IJ \times IJ$ orthogonal projection matrices such that

$$(11.11.28)\qquad (P_1X)^t = (x_{.1}, x_{.2},\ldots,x_{.J},x_{.1}, x_{.2},\ldots,x_{.J},\ldots,x_{.1}, x_{.2},\ldots,x_{.J})$$

and

$$(P_2X)^t = (x_{1.},\ldots,x_{1.},x_{2.},\ldots,x_{2.},\ldots,x_{I.},\ldots,x_{I.}).$$

Then $P = P_1P_2 = P_2P_1$ is a rank one projection matrix. Then

$$(11.11.29)\qquad I_n-B(B^tB)^+B^t = L = I_n-P_1-P_2+P = (I_n-P_1)(I_n-P_2).$$

The matrix $Q$ is a $n \times 1$ matrix of rank 1 defined to have $ij$ entry equal $x_{i.}x_{.j}$. This is a function of $B^tX$. Then

(11.11.30) $LQ$ has $ij$ entry $(x_{i.} - x_{..})(x_{.j} - x_{..})$.

Find $((LQ)^t(LQ))^+$ and the matrix $A = LAL = L^tAL$. Since

(11.11.31) the $ij$ entry of $LX$ is $x_{ij} - x_{i.} - x_{.j} + x_{..}$ it follows that $\underline{s}_1^2 = \underline{X}^t\underline{AX}$ and $\underline{s}_2^2 - \underline{s}_1^2 = \underline{X}^t\underline{BX}$. What is the rank of $A$?

By Problem 11.11.24 the statistic

$$s_1^2/(s_2^2 - s_1^2) = \chi_1^2/\chi^2_{(I-1)(J-1)-1} \tag{11.11.32}$$

is a unnormalized central F-statistic. In case the second order interactions do not vanish then $B = I_{IJ}$ and the above analysis no longer applies.

## Chapter 12. The construction of zonal polynomials.

### 12.0. Introduction.

The discussion of previous chapters has shown that in many of the noncentral problems the answer involves an integral that cannot be evaluated in closed form, typically an integral over a locally compact topological group with respect to a Haar measure. In this book the groups have been matrix groups. For example the discussion of James (1955a) given in Chapter 5 obtains the probability density function of the noncentral Wishart distribution in terms of an integral $\int_{\underline{O}(n)} \exp(\operatorname{tr} X^t HM)\,dH$, where $dH$ means the Haar measure of unit mass on $\underline{O}(n)$. And in the examples of Chapters 9 and 10 examples of integrals over $\underline{T}(n)$ and $GL(n)$ have been given.

The theory of zonal polynomials allows the evaluation of some of these integrals in terms of infinite series whose summands are a coefficient multiplying a zonal polynomial. The general theory was presented in a series of papers in the years 1960, 1961, 1962, 1963, 1964, and 1968 by James and the papers by Constantine (1963, 1966). The second paper by Constantine is really about a related topic of Laguerre polynomials of a matrix argument. In the 1963 paper Constantine produced some very basic computational formulas without which the series representation in zonal polynomials would not have succeeded. Constantine's paper showed that the series being obtained in the multivariate problems were in fact series representations of hypergeometric functions as defined by Herz (1955). Herz also defined Laguerre polynomials and the 1966 paper by Constantine is a further development of this theory.

Roughly speaking the substance of this chapter is the necessary algebra needed for the existence and uniqueness theory of the polynomials. The chapter provides sufficient background for the reading of Constantine (1963) and James (1964) but the reader should refer to these papers for the actual calculation of multivariate examples. The original definition of a zonal polynomial in James (1960) defined the polynomials implicitly using group representations. The main theoretical paper by James was James (1961b) in which enough combinatorial analysis was done to explicitly calculate the polynomials of low degree and to give an algorithm for the calculation of polynomials of higher degree. James (1964) is a survey paper giving a complete summary of all results known to Constantine and James at that time. This remarkable survey paper is apparently completely without error but contains a number of unproven assertions. Proofs of some of the unproven results of James (1964) may be found in this chapter. The calculation of the polynomials is discussed in Section 12.12 and again in Section 13.5. Over the years James (1961b, 1964, 1968) has developed three essentially different computational algorithms which have produced the same polynomials, which is the main evidence for the correctness of the existing tables. More recently James and Parkhurst (1974) have published tables of the polynomials of degree less than or equal twelve. In the survey paper, James (1964), he makes a reference to Helgason (1962) which if followed through implies that the zonal polynomials are spherical functions in the meaning of Helgason. This fact is used implicitly in James (1968) where the Laplace Beltrami operator is used to derive a differential equation for the zonal polynomials. The first two algorithms of James in earlier papers were combinatorial in character. To this author it has not seemed possible to pull these results into a coherent readable form based on group representations. However if one takes the algebra as developed by Weyl (1946) and makes a direct algebraic attack on the subject a unified self contained presentation does result. The subject is very deep and not rewarding to any but the brightest students. However it should be noted that parts of the tensor algebra discussed in the sequel can be taken out of the present context and used in subjects like the analysis of variance.

James in his development of the subject depended heavily on Littlewood (1940, 1950) for the theory of algebras which decompose into a direct sum of closed minimal ideals. A more modern presentation using normed algebras may be found in Loomis (1953) which helps in getting more directly to the essential things.

An integral

$$\int_{\underline{O}(n)} \exp(\mathrm{tr}\, HX)\,dH = \sum_{m=0}^{\infty} (m!)^{-1} \int_{\underline{O}(n)} (\mathrm{tr}\, HX)^m\, dH \tag{12.0.1}$$

is an infinite series of homogeneous polynomials of even degree. Our theory is therefore a theory of homogeneous polynomials of degree $2m$ in the variables of a matrix $X$. Note that if $m$ is odd

$$\int_{\underline{O}(n)} (\mathrm{tr}\, HX)^m dH = \int_{\underline{O}(n)} (\mathrm{tr}{-HX})^m dH = -\int_{\underline{O}(n)} (\mathrm{tr}\, HX)^m dH, \tag{12.0.2}$$

so that the odd degree terms vanish. The theory which follows establishes an isomorphism between bi-symmetric linear transformations and the homogeneous polynomials. The bi-symmetric linear transformations are an $H^*$ algebra in the meaning of Loomis (1953) and the space of bi-symmetric transformations is a direct sum of its minimal closed ideals.

The bi-symmetric linear transformations act on $M(E^m, C)$, the space of bi-linear m-forms over the complex numbers $C$. Within the algebra of all endomorphisms of $M(E^m, C)$ the commutator algebra with the bi-symmetric transformations has a very special form which plays a key role in the theory. This algebra is a representation of the group algebra of the symmetric group. A complete description may be obtained from Weyl (1946). We use Weyl as our source.

The idempotents in the center of the commutator algebra are directly identifiable with the zonal polynomials of complex transpose symmetric matrices. This fills in a part of the theory mentioned without proof by James (1964). When $X$ is an $n \times n$ matrix and we are considering polynomials of degree $m$ the idempotents in the center of the commutator algebra are in one to one correspondence with partitions of the integer $m$ into $n$ parts. This correspondence is established using Young's diagrams, discussed below.

Each polynomial $\int_{\underline{O}(n)} (\operatorname{tr} HX)^m dH = f(X)$ satisfies, if $G_1, G_2 \in \underline{O}(n)$ then

$$(12.0.3) \qquad f(X) = f(G_1 X G_2) = f(G_2^t (X^t X)^{1/2} G_2) = f(\begin{pmatrix} x_1 & & 0 \\ & \ddots & \\ 0 & & x_n \end{pmatrix}),$$

where we write $x_1, \ldots, x_n$ for the eigenvalues of $(X^t X)^{1/2}$. From this it is easily seen that $f$ is a homogeneous symmetric polynomial of degree $m$ in $x_1, \ldots, x_n$, which vanishes if $m$ is odd. In fact (see the problems of Chapter 5) for even exponents $2m$,

$$(12.0.4) \qquad f(X) = \int_{\underline{O}(n)} (H_{11} x_1 + \ldots + H_{nn} x_n)^{2m} dH$$

$$= \sum_{m_1 + \ldots + m_n = m} \int x_1^{2m_1} \ldots x_n^{2m_n} H_{11}^{2m_1} \ldots H_{nn}^{2m_n} dH$$

so that $f$ is a polynomial in the symmetric functions

$$(12.0.5) \qquad \operatorname{tr} X^t X, \ (\operatorname{tr} X^t X)^2, \ldots, (\operatorname{tr} X^t X)^m.$$

Therefore part of the algebraic theory is a theory of polynomials which are symmetric functions. Our sources are Littlewood (1940) and Weyl (1946).

Zonal polynomials as defined by James have the invariance property (12.0.3) and form a basis for the symmetric functions.

Therefore, given the degree $2m$, the number of zonal polynomials of that degree is the number of linearly independent symmetric polynomials of degree $m$ in $n$ variables.

Zonal polynomials of a real symmetric matrix are harder to define. Section 12.10 gives an algebraic definition along the lines of James (1961b). Section 12.11 gives a definition in terms of spherical functions and integrals of group characters as suggested by James (1964). Sections 12.12 and 13.6 explore the consequences of an idea of Saw (1975) resulting in a third definition and version of the theory that is more easily taught to the uninitiated.

## 12.1. Kronecker products and homogeneous polynomials.

We assume $E$ is a n-dimensional vector space over the complex numbers $C$ with fixed basis $e_1,\dots,e_n$ and let $u_1,\dots,u_n$ be the canonical basis of the dual space. As in Chapter 6 we let $M(E^m,C)$ be the space of multilinear m-forms with coefficients in the complex numbers. Linear transformations $X_1,\dots,X_m$ of $E$ induce a linear transformation $X_1 \otimes X_2 \otimes \dots \otimes X_m$ of $M(E^m,C)$. Here, relative to the fixed basis, linear transformations will be represented by the transformation's matrix which allows us to make the definition

Definition 12.1.1. If $f \in M(E^m,C)$ then

$$(12.1.1)\qquad (X_1 \otimes \dots \otimes X_m)(f)(e_{i_1},\dots,e_{i_m})$$

$$= f(X_1^t e_{i_1},\dots,X_m^t e_{i_m}),\ 1 \le i_1 \le n,\dots,1 \le i_m \le n.$$

We will think of this definition as defining a matrix $X_1 \otimes \dots \otimes X_m$, and below we compute the entries of this matrix. In (12.1.1) the use of the transpose enters because of the bracket product

$$(12.1.2)\qquad [Xu_i, e_j] = [u_i, X^t e_j].$$

If

(12.1.3) $$X_p^t e_i = \sum_{j=1}^{n} (X_p^t)_{ij} e_j ,$$

then one obtains

(12.1.4) $$(X_1 \otimes \dots \otimes X_m)(u_{j_1} \dots u_{j_m})(e_{i_1}, \dots, e_{i_m})$$

$$= (u_{j_1} \dots u_{j_m})\Big(\sum_{k=1}^{n} (X_1^t)_{i_1 k} e_k, \dots, \sum_{k=1}^{n} (X_m^t)_{i_m k} e_k\Big)$$

$$= (X_1^t)_{i_1 j_1} (X_2^t)_{i_2 j_2} \cdots (X_m^t)_{i_m j_m} ,$$

$$1 \leq j_1, \dots, j_m, \ i_1, \dots, i_m \leq n .$$

The identity (12.1.4) gives the $((i_1, \dots, i_m), (j_1, \dots, j_m))$ entry of the Kronecker product matrix $X_1 \otimes \dots \otimes X_m$. Following are some useful lemmas.

<u>Lemma 12.1.2</u>. $\operatorname{tr} X_1 \otimes \dots \otimes X_m = \prod_{i=1}^{m} (\operatorname{tr} X_i)$.

<u>Lemma 12.1.3</u>. $X_1^t \otimes \dots \otimes X_m^t = (X_1 \otimes \dots \otimes X_m)^t$.

<u>Lemma 12.1.4</u>. $(X_1 \otimes \dots \otimes X_m)(Y_1 \otimes \dots \otimes Y_m) = (X_1 Y_1) \otimes \dots \otimes (X_m Y_m)$.

Note that the use of transposition is needed in Definition 12.1.1 in order that Lemma 12.1.4 be correct. In the sequel, if $c \in C$ then $\overline{c}$ is the complex conjugate of $c$.

<u>Lemma 12.1.5</u>. $\overline{X}_1 \otimes \dots \otimes \overline{X}_m = \overline{X_1 \otimes \dots \otimes X_m}$ .

<u>Lemma 12.1.6</u>. If $A$ is a $n_1 \times n_1$ matrix and $B$ is a $n_2 \times n_2$ matrix then

(12.1.5) $$\det A \otimes B = (\det A)^{n_2} (\det B)^{n_1} .$$

Proof. It is easily verified that $\det I_{n_1} \otimes B = (\det B)^{n_1}$, and that, using Lemma 12.1.4, $A \otimes B = (I_{n_1} \otimes B)(A \otimes I_{n_2})$. #

We will be interested mostly in the case that $X_1 = X_2 = \ldots = X_n = X$. The $(i_1,\ldots,i_m)$, $(j_1,\ldots,j_m)$ entry of $X \otimes \ldots \otimes X$ is $(X)_{i_1 j_1} \cdots (X)_{i_m j_m}$ which is a homogeneous polynomial of degree $m$ in the entries of $X$. The arbitrary homogeneous polynomial of degree $m$ is

$$(12.1.6) \qquad \Sigma \ldots \Sigma\, a_{i_1 j_1 \ldots i_m j_m} (X)_{i_1 j_1} \cdots (X)_{i_m j_m}$$

$$= \operatorname{tr} A^t (X \otimes \ldots \otimes X),$$

where we use the observation that if $B = (b_{ij})$ and $C = (c_{ij})$ then

$$(12.1.7) \qquad \operatorname{tr} B^t C = \Sigma_i \Sigma_j b_{ij} c_{ij}.$$

Usually there will be several different coefficient matrices $A$ resulting in the same homogeneous polynomial. Uniqueness is introduced by requiring $A$ to be a bi-symmetric matrix. To define this property we consider the action of permutations $\sigma$ of $1,2,\ldots,m$. We let $\sigma$ act on $M(E^m,C)$ by means of the definition of $P_\sigma$, to be extended by linearity, that

$$(12.1.8) \qquad (P_\sigma f)(e_{i_1},\ldots,e_{i_m}) = f(e_{i_{\sigma(1)}},\ldots,e_{i_{\sigma(m)}}),$$

$$1 \le i_1,\ldots,i_m \le n .$$

Note that $P_\sigma(u_{j_1}\cdots u_{j_m})(e_{i_1},\ldots,e_{i_m}) = u_{j_1}(e_{i_{\sigma(1)}})\cdots u_{j_m}(e_{i_{\sigma(m)}})$, and this is zero unless

$$(12.1.9) \qquad j_1 = i_{\sigma(1)},\ldots,j_m = i_{\sigma(m)} .$$

Therefore $P_\sigma$ is a permutation of the canonical basis of $M(E^m, C)$ and $P_\sigma$ is therefore an orthogonal matrix. In particular $P_\sigma P_\sigma^t$ = the identity matrix.

Lemma 12.1.7. $P_\sigma(X \otimes \ldots \otimes X) = (X \otimes \ldots \otimes X)P_\sigma$.

Proof. If $f \in M(E^m, C)$ then, with $g = (X \otimes \ldots \otimes X)(f)$,

(12.1.10) $$P_\sigma(X \otimes \ldots \otimes X)(f)(e_{i_1}, \ldots, e_{i_m})$$

$$= P_\sigma(g)(e_{i_1}, \ldots, e_{i_m}) = g(e_{i_{\sigma(1)}}, \ldots, e_{i_{\sigma(m)}})$$

$$= f(X^t e_{i_{\sigma(1)}}, \ldots, X^t e_{i_{\sigma(m)}}) = (P_\sigma f)(X^t e_{i_1}, \ldots, X^t e_{i_m})$$

$$= ((X \otimes \ldots \otimes X)(P_\sigma f)(e_{i_1}, \ldots, e_{i_m}),$$

and this holds if $1 \leq i_1, \ldots, i_m \leq n$. #

Definition 12.1.8. A matrix $A$ representing an element of End $M(E^m, C)$, in the basis $u_{j_1} \ldots u_{j_m}$, $1 \leq j_1, \ldots, j_m \leq n$, is said to be a bi-symmetric matrix if and only if for all permutations $\sigma$ of $1, \ldots, m$,

(12.1.11) $$P_\sigma A = AP_\sigma .$$

Lemma 12.1.9. Let $B = (m!)^{-1} \sum_\sigma P_\sigma A P_\sigma^t$. Then as polynomials in the

(12.1.12) $$\text{tr } A\, X \otimes \ldots \otimes X \quad \text{and} \quad \text{tr } B\, X \otimes \ldots \otimes X$$

are the same polynomial. The matrix $B$ is bi-symmetric.

Lemma 12.1.10. If a homogeneous polynomial of degree $m$ in $p$ variables is identically zero then all coefficients are zero (coefficients in $C$).

Lemma 12.1.11. The set of bi-symmetric matrices is the linear span of the matrices of the form $X \otimes \ldots \otimes X$.

Proof. If not, there is a linear functional on End $M(E^m, C)$, which we represent by a matrix B, and a bi-symmetric matrix A, such that if X is $n \times n$ then

$$\text{(12.1.13)} \qquad \operatorname{tr} B^t X \otimes \ldots \otimes X = 0 \quad \text{and} \quad \operatorname{tr} B^t A \neq 0.$$

Since

$$\text{(12.1.14)} \qquad 0 \neq \operatorname{tr} B^t A = \operatorname{tr} P_\sigma B^t A\, P_\sigma^t = \operatorname{tr} P_\sigma B^t P_\sigma^t A,$$

it follows that

$$\text{(12.1.15)} \qquad 0 \neq \operatorname{tr}((m!)^{-1}\Sigma_\sigma P_\sigma B P_\sigma^t)^t A,$$

and therefore we may assume the matrix B is bi-symmetric. By hypothesis and Lemma 12.1.10, all coefficients of the polynomial $\operatorname{tr} B^t X \otimes \ldots \otimes X$ are zero. The term $x_{i_1 j_1} \ldots x_{i_m j_m}$ occurs also as $x_{i_{\sigma(1)} j_{\sigma(1)}} \ldots x_{i_{\sigma(m)} j_{\sigma(m)}}$ and these are the only occurrences, so the coefficient involved is

$$\text{(12.1.16)} \qquad 0 = \Sigma_\sigma b_{i_{\sigma(1)} i_{\sigma(2)} \ldots j_{\sigma(1)} \ldots j_{\sigma(m)}}$$

$$= (m!) b_{i_1 \ldots i_m j_1 \ldots j_m},$$

since B is bi-symmetric. Therefore B = 0. This contradiction shows that the conclusion of the lemma must hold. #

We summarize the results above in a Theorem.

Theorem 12.1.12. The vector space of homogeneous polynomials of degree m in the variables of X is isomorphic to the vector space of bi-symmetric matrices under the representation (12.1.6).

Lemma 12.1.13. The product of two bi-symmetric matrices is a bi-symmetric matrix. Hence, under the matrix operations of addition, scalar multiplication and matrix multiplication, the bi-symmetric matrices form an algebra over the complex numbers which is closed under conjugation and transposition.

We now show that the algebra of bi-symmetric matrices is an $H^*$ algebra in the sense of Loomis (1953). The required inner product is defined by

Definition 12.1.14. $(A,B) = \mathrm{tr}(\overline{B})^t A = \mathrm{tr}\, A(\overline{B})^t$.

Theorem 12.1.15. The bilinear functional $(\ ,\ )$ is an inner product under which the bi-symmetric matrices are an $H^*$ algebra.

Proof. The involution $A^* = (\overline{A})^t$, i.e., the conjugate transpose, clearly satisfies

(12.1.17a) $A^{**} = A$;

(12.1.17b) $(A+B)^* = A^* + B^*$;

(12.1.17c) if $c \in C$ then $(cA)^* = \overline{c}A^*$;

(12.1.17d) $(AB)^* = B^*A^*$.

Further,

(12.1.18) $(AB,C) = \mathrm{tr}(AB)C^* = \mathrm{tr}\, B(C^*A) = \mathrm{tr}\, B(A^*C)^* = (B,A^*C)$,

and

(12.1.19) $\|A\|^2 = \mathrm{tr}\, AA^* = \mathrm{tr}\, A^*A^{**} = \|A^*\|^2$.

Also, $A^*A = 0$ implies $\mathrm{tr}\, AA^* = 0$, that is, $\|A\| = 0$ and $A = 0$. Therefore the defining conditions given by Loomis are satisfied.

By Lemma 12.1.13, the inner product $(\ ,\ )$ is defined on the algebra of bi-symmetric matrices, so this algebra is also an $H^*$ algebra. #

Remark 12.1.16. The algebra of bi-symmetric matrices is finite dimensional. By the results on $H^*$ algebras in Loomis, op. cit., it follows that the algebra of bi-symmetric matrices is a direct sum of its minimal closed ideals. Each minimal closed ideal is isomorphic to a full matrix algebra with unit. (Note that each ideal, being a finite dimensional subspace, is closed topologically.)

## 12.2. Symmetric polynomials in n variables.

The material of the section is taken from Littlewood (1950) and Weyl (1946). Following the notations of Littlewood, given $c_1,\dots,c_n \in C$ we define a polynomial

$$x^n f(x^{-1}) = \prod_{i=1}^{n} (x - c_i) = x^n - a_1 x^{n-1} + \dots + (-1)^n a_n. \tag{12.2.1}$$

As is well known the coefficients $a_1,\dots,a_n$ are the elementary symmetric functions of $c_1,\dots,c_n$. We note that

$$\begin{aligned} F(x) = 1/f(x) &= \prod_{i=1}^{n} (1/(1-c_i x)) \\ &= \prod_{i=1}^{n} (1 + c_i x + c_i x^2 + \dots) \\ &= 1 + h_1 x + h_2 x^2 + \dots + h_m x^m + \dots . \end{aligned} \tag{12.2.2}$$

The coefficient $h_m$ is a symmetric polynomial of degree $m$ which is the sum of the homogeneous products of degree $m$ of $c_1,\dots,c_n$. Also,

$$\log f(x) = \sum_{i=1}^{n} \log (1 - c_i x), \tag{12.2.3}$$

so that by taking derivatives

$$\begin{aligned} f'(x)/f(x) &= \sum_{i=1}^{n} -c_i/(1-c_i x) \\ &= - \sum_{i=1}^{n} (c_i + c_i^2 x + \dots + c_i^{m+1} x^m + \dots) \\ &= - (S_1 + S_2 x + \dots + S_{m+1} x^m + \dots) . \end{aligned} \tag{12.2.4}$$

In (12.2.4) the term $S_m = \sum_{i=1}^{n} c_i^m$. From the relation (12.2.4) upon substitution of (12.2.1) for $f$ and $f'$, we obtain

(12.2.5) $$(a_1 - 2a_2x + 3a_3x^2 - \ldots) = (1 - a_1x + a_2x^2 - \ldots)(S_1 + S_2x + \ldots) .$$

By matching coefficients we obtain the Newton identities, which are stated in the next Lemma.

Lemma 13.2.1.

(12.2.6) $$ma_m = S_1a_{m-1} - S_2a_{m-2} + S_3a_{m-3} - \ldots + (-1)^{m+1}S_m .$$

It follows from the identities (12.2.6) that $a_1,\ldots,a_m$ are polynomials of the variables $S_1,\ldots,S_m$ and that by solving the necessary equations, $S_1,\ldots,S_m$ are polynomials of the variables $a_1,\ldots,a_m$. In particular every term $S_{i_1}\ldots S_{i_p}$ entering into $a_m$ is homogeneous of degree $m$, i.e., $i_1 + \ldots + i_p = m$, and similarly when expressing $S_m$ as a polynomial of the variables $a_1,\ldots,a_m$.

Lemma 12.2.2. The symmetric functions $a_1,\ldots,a_n$ of the variables $c_1,\ldots,c_n$ are functionally independent in the sense that, if $F$ is a polynomial of $n$ variables over $C$ such that $F(a_1,\ldots,a_n)$ vanishes identically in the variables $c_1,\ldots,c_n$, then $F = 0$.

Proof. Given complex numbers $a_1,\ldots,a_n$ the polynomial $x^n - a_1x^{n-1} + \ldots + (-1)^n a_n$ has $n$ complex roots $c_1,\ldots,c_n$. Therefore the mapping $(c_1,\ldots,c_n) \to (a_1,\ldots,a_n)$ is onto $C^n$. If $F$ is a polynomial of $n$ variables such that

(12.2.7) $$F(a_1(c_1,\ldots,c_n),\ldots,a_n(c_1,\ldots,c_n)) \equiv 0,$$

then $F(a_1,\ldots,a_n) = 0$ for every n-tuple of complex numbers. Hence $F$ is the zero function, i.e., $F = 0$. #

Lemma 12.2.3. The symmetric functions $S_1,\dots,S_n$ as functions of $c_1,\dots,c_n$, are functionally independent. As functions of $a_1,\dots,a_n$, they are functionally independent.

Proof. The same as above after noting from the Newton identities (12.2.6) that the mapping $(a_1,\dots,a_n) \to (S_1,\dots,S_n)$ is onto $C^n$. #

To an integer $f$ we express (not uniquely)

(12.2.8) $$f = 1.m_1 + 2m_2 + \dots + nm_n, \text{ and if } 1 \leq i \leq n,$$

$$f_i = m_i + \dots + m_n .$$

Then clearly

(12.2.9) $$f_1 \geq f_2 \geq \dots \geq f_n, \text{ and } f = f_1 + \dots + f_n .$$

Conversely, if numbers $f_1,\dots,f_n$ satisfy (12.2.9) then we can define

(12.2.10) $$f_i = m_i + \dots + m_n, \ 1 \leq i \leq n ,$$

and solve this system of equations for nonnegative integers $m_1,\dots,m_n$.

Definition 12.2.4. A partition of the integer $f > 0$ into $n \geq 1$ parts is a sequence $f_1 \geq f_2 \geq \dots \geq f_n$ of $n$ integers such that $f = f_1 + \dots + f_n$. We use the notation $(f)$ as the generic notation for a partition of $f$ into $n$ parts.

To each partition $(f)$ we associate a homogeneous polynomial via the equations (12.2.10) by the definition

(12.2.11) $$S_1^{m_1} S_2^{m_2} \dots S_n^{m_n} = (c_1+\dots+c_n)^{m_1}(c_1^2+\dots+c_n^2)^{m_2}\dots(c_1^n+\dots+c_n^n)^{m_n}.$$

We now prove

Theorem 12.2.5. The symmetric functions (12.2.11) corresponding to the possible partitions $(f)$ of $f$ into $n$ parts are a basis of the space of symmetric homogeneous polynomials of degree $f$.

Proof. A linear combination

$$\sum_{(f)} a_{(f)} S_1^{m_1} S_2^{m_2} \cdots S_n^{m_n} = 0 \tag{12.2.12}$$

is a polynomial of $n$ variables vanishing identically in $c_1, \ldots, c_n$. By Lemma 12.2.3, the coefficients all vanish. That is, the indicated symmetric functions are linearly independent over the complex numbers.

To show that the functions (12.2.11) span is harder. We order the partitions of $f$ and use this ordering for a mathematical induction. If $(f)_i$ is given by $f_{i1} \geq f_{i2} \geq \cdots \geq f_{in}$, $i = 1, 2$, then we say that $(f)_1 > (f)_2$ if there exists a least integer $j$ such that $f_{1j} \neq f_{2j}$ and $f_{1j} > f_{2j}$.

For the argument which follows we suppose $g$ is a homogeneous symmetric polynomial of degree $f$ and that among all terms $c_1^{p_1} c_2^{p_2} \cdots c_n^{p_n}$ which occur in $g$, $p_1 \geq p_2 \geq \cdots \geq p_n$, there is a term $c_1^{p_1} c_2^{p_2} \cdots c_j^{p_j}$ with $p_j \neq 0$ and $p_1 \geq p_2 \geq \cdots \geq p_j$. This term can be written

$$c_1^{p_1} c_2^{p_2} \cdots c_j^{p_j} = (c_1 \cdots c_j)^{p_j} (c_1 \cdots c_{j-1})^{p_{j-1}-p_j} \cdots c_1^{p_1-p_2} \tag{12.2.13}$$

which is a term of

$$a_1^{p_1-p_2} a_2^{p_2-p_3} \cdots a_j^{p_j} . \tag{12.2.14}$$

We prove first that terms of the form $a_1^{m_1} a_2^{m_2} \cdots a_n^{m_n}$ with $f = m_1 + 2m_2 + \cdots + nm_n$ span. In reference to (12.2.13) we suppose $j$ is the least integer such that a term of $j$ factors occurs and further suppose that among all partitions of $f$ into $j$ parts the partition $p_1 \geq p_2 \geq \cdots \geq p_j$ is maximal. Let a term

$$c_1^{q_1} c_2^{q_2} \cdots c_n^{q_n} , \quad q_1 \geq q_2 \geq \cdots \geq q_n , \tag{12.2.15}$$

occur in the expansion of (12.2.14). Then clearly $q_1 \leq p_1$. Suppose

$q_1 = p_1, \ldots, q_i = p_i$, and $q_{i+1} \neq p_{i+1}$. Then the term (12.2.15) contains the factor

$$(12.2.16) \qquad c_1^{p_1-p_2}(c_1c_2)^{p_2-p_3}\ldots(c_1\ldots c_i)^{p_i} = c_1^{p_1}c_2^{p_2}\ldots c_i^{p_i}$$

$$= c_1^{q_1}c_2^{q_2}\ldots c_i^{q_i} .$$

Then the factor $c_{i+1}^{q_{i+1}}$ of (12.2.15) can arise only as a factor from terms $a_{i+1}, \ldots, a_n$. Therefore the exponent $q_{i+1} \leq p_{i+1}$ and since $q_{i+1} \neq p_{i+1}$, we have $q_{i+1} < p_{i+1}$. In particular any term in (12.2.14) with exactly $j$ factors arises from a partition of $f$ into $j$ parts which is less, in the ordering of partitions, than the partition $p_1 \geq p_2 \geq \cdots \geq p_j$. And clearly any term from (12.2.14) must contain $j$ or more of the variables $c_1, \ldots, c_n$. Therefore there exists a complex number $c$ such that

$$(12.2.17) \qquad g - ca_1^{p_1-p_2} \ldots a_j^{p_j}$$

contains only terms (i) of more than $j$ factors, or (ii) terms corresponding to partitions of $f$ into $j$ factors with said partition being less than the partition $p_1 \geq p_2 \geq \cdots \geq p_j$.

The induction is now clear. If the degree $f \leq n$, the number of variables being $n$, the process terminates with the polynomial $g$ having been reduced to the single term $a_f$. If $f > n$ the process produces a polynomial every term of which involves a product $c_1 \ldots c_n$. The degree is then reduced by dividing out by $(c_1 \ldots c_n)^k$, where $k$ is the largest integer such that $(c_1 \ldots c_n)^k$ divides. This reduces the degree by $nk$. The preceeding process may then be applied to the resulting polynomial of degree $f - nk$, etc., thereby reducing the degree, ultimately, to zero.

Each step of reduction (12.2.17) subtracts off a symmetric homogeneous polynomial of degree $f$ in the variables $a_1,\ldots,a_n$. Therefore the terminal result is a polynomial in the variables $a_1,\ldots,a_n$ such that if the term $a_1^{m_1}\ldots a_n^{m_n}$ occurs then $f = m + 2m_2 + \ldots + nm_n$. Since every such term is a polynomial in the variables $S_1,\ldots,S_n$, it follows that $g$ is expressible as a polynomial in the variables $S_1,\ldots,S_n$ such that if the term $S_1^{m_1}\ldots S_n^{m_n}$ occurs then $f = n + 2m_2 + \ldots + nm_n$. # (See also Section 12.12, particularly (12.12.16).)

## 12.3. The symmetric group algebra.

We begin this section with definitions and lemmas that apply to the group algebras of finite groups generally. Then towards the end of the section results are specialized to the needed results about the symmetric group on $m$ letters.

We let $\mathfrak{G}$ be a finite group with $m$ elements $g_1,\ldots,g_m$. By considering $g_1,\ldots,g_m$ as linearly independent elements over the complex numbers $C$ we may form the $m$-dimensional vector space with elements

$$(12.3.1) \qquad \sum_{g\in\mathfrak{G}} a(g)\, g\ ,$$

where $a: \mathfrak{G} \to C$ is a coefficient function. As elements of a vector space addition and scalar multiplication are done coordinate wise. As elements of an algebra products are defined by

$$(12.3.2) \qquad \Big(\sum_{g\in\mathfrak{G}} a(g)g\Big)\Big(\sum_{g\in\mathfrak{G}} b(g)g\Big) = \sum_{k\in\mathfrak{G}} \Big(\sum_{gh=k} a(g)b(h)\Big)k\ .$$

Since each element of the algebra has a unique representation this bilinear functional is well defined. It is easy to show that this product is associative and distributive.

An involution $*$ is defined by

$$(12.3.3) \qquad (\sum_{g\in\mathcal{G}} a(g)g)^* = \sum_{g\in\mathcal{G}} \overline{a(g^{-1})}g = \sum_{g\in\mathcal{G}} \overline{a(g)}\, g^{-1}.$$

An inner product is defined by

$$(12.3.4) \qquad (\sum_g a(g)g, \sum_g b(g)g) = \sum_g a(g)\overline{b(g)} .$$

It is clear that under this inner product $g_1,..,g_m$ form an orthonormal basis of the additive group of the algebra. Also, if $e$ is the multiplicative unit of $\mathcal{G}$ then $e$ is the unit of the ring of the algebra.

<u>Lemma 12.3.1</u>. The group algebra of a finite group $\mathcal{G}$ is a $H^*$ algebra.

<u>Proof</u>. The involution $*$ clearly satisfies (i) $a^{**} = a$, (ii) $(a+b)^* = a^*+b^*$, and (iii) $(ca)^* = \overline{c}a^*$. We now verify (iv) $(ab)^* = b^*a^*$. We find that

$$(12.3.5) \qquad (ab)^* = \sum_k(\sum_{gh=k} \overline{a(g)b(h)})k^{-1}, \text{ and}$$

$$(12.3.6) \qquad b^*a^* = (\sum_h \overline{b(h)}\, h^{-1})(\sum_g \overline{a(g)}\, g^{-1})$$

$$= \sum_k(\sum_{h^{-1}g^{-1}=k^{-1}} \overline{b(h)a(g)})k^{-1}$$

$$= \sum_k(\sum_{gh=k} \overline{a(g)b(h)})k^{-1}. \; \#$$

The property $\| a \| = \| a^* \|$ is obvious. We compute

$$(12.3.7) \qquad a^*a = (\sum_g \overline{a(g)}g^{-1})(\sum_h a(h)h)$$

$$= \sum_k(\sum_{g^{-1}h=k} \overline{a(g)}a(h))k .$$

The coefficient of $e$ in the expression for $a^*a$ is $\sum_g a(g)\overline{a(g)} = \|a\|^2$.

Thus $a^*a = 0$ implies $\|a\| = 0$ or $a = 0$. Last, we establish that $(ab,c) = (b,a^*c)$. Note that

$$(12.3.8) \qquad (ab,c) = \sum_k \Big( \sum_{gh=k} a(g)b(h) \Big) \overline{c(k)}, \text{ and}$$

$$(b,a^*c) = \sum_h \overline{\Big( \sum_{gk=h} \overline{a(g^{-1})} c(k) \Big)} b(h)$$

$$= \sum_k \Big( \sum_{g^{-1}h=k} a(g^{-1})b(h) \Big) \overline{c(k)} = (ab,c). \ \#$$

Corollary 12.3.2. The group algebra of a finite group $\mathfrak{G}$ is a direct sum of its minimal closed ideals. Each minimal closed ideal is isomorphic to a full matrix algebra and thus each minimal closed ideal contains an idempotent that acts as multiplicative identity in the ideal.

Proof. Directly from Loomis (1953) on $H^*$ algebras. #

Definition 12.3.3. The class of an element $h \in \mathfrak{G}$ is the set $[h]$ of all elements conjugate to $h$.

Lemma 12.3.4. The set $\mathfrak{G}_h$ of all elements of $\mathfrak{G}$ that commute with $h$ is a subgroup of $\mathfrak{G}$.

Lemma 12.3.5. If $g_1$, $g_2$, $h \in \mathfrak{G}$ and $g_1hg_1^{-1} = g_2hg_2^{-1}$ then $g_1^{-1}g_2 \in \mathfrak{G}_h$ and conversely, so that $g_1$ and $g_2$ are in the same coset of $\mathfrak{G}_h$. Hence the number of elements in the class $[h]$ divides the order of $\mathfrak{G}$.

Lemma 12.3.6. If $\mathfrak{G}$ has $m$ elements and $[h]$ has $p$ elements then the sequence $g_1hg_1^{-1},\ldots,g_mhg_m^{-1}$ repeats each element of $[h]$ exactly $m/p$ times.

Theorem 12.3.7. Let the classes of $\mathfrak{G}$ be $C_0 = [e], C_1,\ldots,C_{p-1}$ containing $1, m_1,\ldots,m_{p-1}$ elements respectively. An element $a$ of the group algebra is in the center of the algebra if and only if

(12.3.9) $$a = \sum_{i=0}^{p-1} \phi_i \left( \sum_{g \in C_i} g \right) .$$

Proof. If $a$ is in the center and $h \in \mathcal{G}$ then $a = hah^{-1}$ so

(12.3.10) $$\begin{aligned} ma &= \sum_h hah^{-1} = \sum_g a(g) \sum_h hgh^{-1} \\ &= \sum_{i=0}^{p-1} \left( \sum_{g \in C_i} a(g) \right) \left( (m/m_i) \sum_{k \in C_i} k \right) \end{aligned}$$

and the coefficient $\phi_i$ of (12.3.9) is thus $\phi_i = \sum_{g \in C_i} a(g)/m_i$.

The converse is obvious after observing that $g\left( \sum_{k \in C_i} k \right) g^{-1} = \sum_{k \in C_i} k$. #

Theorem 12.3.8. If the group algebra of $\mathcal{G}$ has $p$ classes then the group algebra is the direct sum of $p$ minimal closed ideals, hence, of $p$ simple matrix algebras.

Proof. Let the minimal closed ideals be $N_1, \ldots, N_q$ so that $N_1 \oplus \ldots \oplus N_q$ is the group algebra. $N_i$ has an idempotent $e_i$ which is the multiplicative unit in $N_i$, $1 \leq i \leq q$. If $a = a_1 + \ldots + a_q$, $a_i \in N_i$, $1 \leq i \leq q$, then $a$ being in the center requires that $e_i a = a e_i = a_i e_i$, $1 \leq i \leq q$. Since $N_i$ is a full matrix algebra, and since $a_i$ must commute with all elements of $N_i$, it follows that $a_i = \phi_i e_i$ for some $\phi_i \in C$, $1 \leq i \leq q$. Conversely, every element $\phi_i e_i$ is in the center of the group algebra. By Theorem 12.3.7, the center is p-dimensional, so that $p = q$ follows. #

Lemma 12.3.9. In the symmetric group on $m$ letters two permutations are conjugate if and only if

(a) each is the product of the same number of cycles, and,

(b) after ordering in decreasing order the sequence of cycle lengths are the same partition of $m$.

Proof. A permutation $f$ of $1,\ldots,r$ is said to be a cycle if no proper nonempty subset of $1,\ldots,r$ is invariant under $f$. In this case $f(1)$, $f^2(1) = f(f(1)),\ldots,f^r(1) = f(f^{r-1}(1))$ are pairwise distinct. Conversely, if $f(1)$, $f^2(1),\ldots,f^r(1)$ are pairwise distinct then $f$ is a cycle.

Clearly given a permutation $f$ of $1,\ldots,m$ the minimal invariant subsets are uniquely determined and $f$ is a cycle on each of the minimal invariant subsets, giving the cyclic decomposition of $f$. Also, if $g$ is a permutation of $1,\ldots,m$ then $g^{-1}fg$ has the same number of invariant sets each of the same ordinality as $f$.

Conversely, given the invariant sets $\{a_{11},\ldots,a_{1n_n}\}$, $\{a_{21},\ldots,a_{2n_2}\}$, etc., of $f$ and invariant sets $\{b_{11},\ldots,b_{1n_1}\}$, $\{b_{21},\ldots,b_{2n_2}\},\ldots$ of $g$ such that

(12.3.11) $$f = (a_{11},\ldots,a_{1n_1})(a_{21},\ldots,a_{2n_2})\ldots$$

$$g = (b_{11},\ldots,b_{1n_1})(b_{21},\ldots,b_{2n_2})\ldots$$

then the permutation $h$ such that

(12.3.12) $$h(b_{ij}) = a_{ij} \quad \text{satisfies}$$

(12.3.13) $$h^{-1}fh = g. \ \#$$

Remark 12.3.10. The group algebra of the symmetric group on $1,\ldots,m$ becomes an algebra of linear transformations of multilinear m-forms by means of the mapping

(12.3.14) $$\sum_{g\in\mathfrak{Z}} a(g)g \to \sum_{g\in\mathfrak{Z}} a(g)P_g .$$

The permutation matrices $P_g$ were defined in (12.1.8). As we will see below, this algebra homomorphism is in general not one to one if $m > n$, corresponding to the fact that some m-forms in $n < m$ variables vanish. In particular for the alternating operator $A$ defined in Chapter 6,

$$(12.3.15) \qquad A = \sum_g \epsilon(g)(m!)^{-1} g,$$

where $\epsilon(g)$ is the sign of the permutation. In case $m > n$ and $f \in M(E^m, C)$, then

$$(12.3.16) \qquad Af = 0 .$$

In the discussion to follow the relationship between alternation and being a zero transformation will become more clear.

12.4. Young's symmetrizers. The terminology and results of this section are taken from Weyl (1946). In our statement and proofs we have found it necessary to interchange "p" and "q" throughout and we are thus in partial disagreement with the results in Weyl, op.cit.

Associated with a partition $m_1 \geq m_2 \geq \cdots \geq m_p$ of $m$, call it $(m)$, is a diagram $T(m)$ consisting of $p$ rows and $m_1$ columns. Each row is divided into $1 \times 1$ cells, the ith row has length $m_i$ cells which are numbered from left to right,

$$(12.4.1) \qquad m_1 + \cdots + m_{i-1} + 1, \ldots, m_1 + \cdots + m_{i-1} + m_i .$$

The ith and (i+1)st rows, $1 \leq i \leq p-1$, are in the relation shown in

(12.4.2)

| | | | | | |
|---|---|---|---|---|---|
| $m_1+\cdots+m_{i-1}+1$ | $m_1+\cdots+m_{i-1}+2$ | $,\ldots,$ | $m_1+\cdots+m_{i-1}+m_{i+1}$ | $,\ldots,$ | $m_1+\cdots+m_i$ |
| $m_1+\cdots+m_i+1$ | $m_1+\cdots m_i+2$ | $,\ldots,$ | $m_1+\cdots+m_{i+1}$ | | |

A permutation of $1,\ldots,m$ is said to be of type $p$ if the permutation leaves the row sets of $T(m)$ fixed; it is of type $q$ if the permutation leaves the column sets of $T(m)$ fixed.

Definition 12.4.1. The Young's symmetrizer of the diagram $T(m)$ is

$$(12.4.3) \qquad \sum_p \sum_q \epsilon(q) pg ,$$

the sum being taken over all permutations of type $p$ and $q$ relative to $T(m)$.

Lemma 12.4.2. Relative to a fixed diagram the permutations of type $p$ form a group and those of type $q$ form a group. A permutation that is both of type $p$ and type $q$ is the identity permutation. Therefore, $p_1q_1 = p_2q_2$ implies $p_1 = p_2$ and $q_1 = q_2$.

Proof. The first two statements are obvious. Suppose a permutation $\sigma$ is of type $p$ and of type $q$. If $i$ is in row $j_1$ and column $j_2$, then it follows that $\sigma(i)$ is also in row $j_1$ and column $j_2$. Since the row number and column number as a pair uniquely determine the cell of the diagram, $\sigma(i) = i$. #

Young's diagrams are partially ordered using the partial ordering of permutations described after (12.2.12). Thus $T(m) \geq T(m)'$ if and only if $(m) \geq (m)'$. Given an Young's diagram $T(m)$ we may define configurations $\sigma T(m)$ obtained from $T$ by replacing the entry $i$ by $\sigma(i)$ throughout, $1 \leq i \leq m$.

Lemma 12.4.3. Let $T(m) \geq T(m)'$ and let $\sigma$ be a permutation of $1,\ldots,m$. Then either

(a) there are two numbers $i_1$ and $i_2$ occurring in the same row of $T(m)$ and the same column of $\sigma T(m)'$, or,

(b) $T(m) = T(m)'$ and the permutation $\sigma$ is a product $pq$.

Proof. We have $(m) \geq (m)'$. The $m_1$ numbers of the first row of $T(m)$ must be spread among the $m_1'$ columns of $T(m)'$. If $m_1 > m_1'$ then two numbers in the first row of $T(m)$ must occur in some column of $\sigma T(m)'$.

For the remainder of the argument we assume $m_1 = m_1'$ and that conclusion (a) is false. Then there exists a permutation $\tau_1$ of the column sets of $\sigma T(m)'$ such that the first row of $(\tau_1\sigma)T(m)'$ is the same as the first row of $T(m)$. Delete the first row of $T(m)$

and of $(\tau_1\sigma)T(m)'$ and make an induction on the reduced diagrams, with the induction being on the number of rows. By inductive hypotheses the partitions determined by the reduced diagrams are the same, so that $(m) = (m)'$ and $T(m) = T(m)'$ and there exists a permutation $\tau_2$ of the column sets of $(\tau_1\sigma)T(m)'$ such that $(\tau_2\tau_1\sigma)T(m)'$ has the same row sets as does $T(m) = T(m)'$. Therefore there is a permutation $\pi$ of the row sets of $(\tau_2\tau_1\sigma)T(m)$ such that $(\pi\tau_2\tau_1\sigma)T(m) = T(m)$. Therefore with $\tau_3 = \tau_2\tau_1$,

$$\pi\tau_3\sigma = \text{identity and } \sigma = \tau_3^{-1}\pi^{-1}. \tag{12.4.4}$$

Recall that $\tau_3$ is a permutation of the column sets of $\sigma T(m)$. It is clear that there exists a permutation $\tau_4$ of the column sets of $T(m)$ such that

$$\tau_3\sigma = \sigma\tau_4. \tag{12.4.5}$$

Therefore

$$\pi\tau_3\sigma = \pi\sigma\tau_4 = \text{identity, and } \sigma = \pi^{-1}\tau_4^{-1}. \tag{12.4.6}$$

This is permutation of type pq.

In the converse case $m_1 = m_1'$ and conclusion (a) holds on the reduced diagrams. Then (a) holds for the original diagrams. #

<u>Lemma 12.4.4</u>. Let $T(m)$ be a Young's diagram and let $s$ be a permutation of $1,\ldots,m$ which is not of type pq. Then there exists a transposition $u$ of type $p$ and a transposition $v$ of type $q$ such that

$$us = sv. \tag{12.4.7}$$

<u>Proof</u>. By Lemma 12.4.3 with $(m) = (m)'$, $\sigma = s$, and configuration $sT(m)$, by conclusion (a), there exists a row and entries in the row $i_1$, $i_2$ which occur in the same column of $sT(m)$. We let $v$ be the

transposition interchanging $s^{-1}(i_1)$ and $s^{-1}(i_2)$, and $u$ be the transposition interchanging $i_1$ and $i_2$. Then $u$ is of type $p$ and $v$ is of type $q$ since $s^{-1}(i_1)$ and $s^{-1}(i_2)$ are in the same column of $T(m)$. Further

$$(sv)T(m) = (us)T(m) \quad \text{and thus} \quad sv = us. \ \#$$

Lemma 12.4.5. If $(m) \geq (m)'$ and $(m) \neq (m')$ and $s$ is a permutation of $1,\ldots,m$ then there exist transpositions $u$ of type $p$ and $v$ of type $q'$ such that

(12.4.8) $$us = sv .$$

Proof. In Lemma 12.4.3 with $T(m) \geq T(m)'$ conclusion (a) holds and in some row of $T(m)$ exist $i_1$, $i_2$ which occur in the same column of $sT(m)'$. As above, let $u$ transpose $i_1$ and $i_2$ and $v$ transpose $s^{-1}(i_1)$ and $s^{-1}(i_2)$. Then clearly $us = sv$, $u$ is of type $p$ and $v$ is of type $q'$ since $s^{-1}(i_1)$ and $s^{-1}(i_2)$ occur in the same column of $T(m)'$. #

Lemmas 12.4.3 to 12.4.5 give the combinatorial basis for the study of Young's symmetrizers defined in (12.4.3). For the purpose of studying the symmetrizers we introduce coefficient functions $c(g)$, $g \in \mathfrak{S}$, defined by (relative to a partition $(m)$)

(12.4.9) if $g$ is of type $pq$ and $g = pq$ then $c(g) = \varepsilon(q)$; otherwise, $c(g) = 0$.

Then relative to a partition $(m)$ the Young's symmetrizer is

(12.4.10) $$c = \sum_g c(g)g .$$

Lemma 12.4.6. The coefficient function defined in (12.4.9) satisfies,

(12.4.11) $$c(pg) = c(g),\ g \in \mathfrak{S},\ \text{and } p \text{ of type } p;$$
$$c(gq) = \varepsilon(q)c(g),\ g \in \mathfrak{S},\ \text{and } q \text{ of type } q.$$

Proof. $c(g) = 0$ unless $g = p_1q_1$. Then $pg = (pp_1)q_1$ and $c(pg) = 0$ unless $g = p_1q_1$ in which case $c(pg) = c(q_1) = c(g)$ by definition. Likewise if $g$ is not of type $pq$ then $gq$ is not of type $pq$ and $c(g) = c(gq) = 0$. Otherwise $c(gq) = c(p_1(q_1q))$ $= \epsilon(q_1q) = \epsilon(q_1)\epsilon(q) = \epsilon(q)c(g)$. #

Lemma 12.4.7. Let $d = \sum_g d(g)g$ and assume the coefficient function of $d$ relative to the partition $(m)$ satisfies

(12.4.12) $$d(pg) = d(g) \text{ and } d(gp) = \epsilon(q)d(g) \text{ for } g \in \mathcal{S},$$
$p$ of type $p$ and $q$ of type $q$.

Then the complex number $d(e)$ satisfies

(12.4.13) $$\text{if } g \in \mathcal{S} \text{ then } d(e)c(g) = d(g),$$

where $c$ is the coefficient function of the Young's symmetrizer.

Proof. If $e$ is the unit of $\mathcal{S}$ then $e$ is both a $p$ and a $q$. Therefore

(12.4.14) $$d(e) = d(pe) = d(p), \text{ and}$$
$$d(e) = \epsilon(q)d(eq) = \epsilon(q)d(q).$$

If the permutation $g$ is not of type $pq$ then there exist transpositions $u$ of type $p$ and $v$ of type $q$ such that $ug = gv$. Then

(12.4.15) $$d(g) = d(ug) = d(gv) = \epsilon(v)d(g) = -d(g).$$

Therefore $d(g) = 0$. Together (12.4.14) and (12.4.15) say

(12.4.16) $$d(g) = d(e)c(g), \; g \in \mathcal{S}. \; \#$$

Lemma 12.4.8. Let $(m) \geq (m)'$ and $(m) \neq (m)'$. Let the coefficient function $d$ satisfy

(12.4.17) $$d(pg) = d(g) \quad \text{and} \quad d(gq') = \epsilon(q')d(g)$$

for all $g \in \mathcal{G}$, and permutations p, q' of types p,q'.

Then $d(g) = 0$ for all $g \in \mathcal{G}$.

Proof. By Lemma 12.4.5 there exist transpositions u of type p and v of type q' such that $ug = gv$. Then

(12.4.18) $$d(g) = d(ug) = d(gv) = -d(g). \; \#$$

Lemma 12.4.9. Let (m) be a partition of m with Young's diagram T(m) and Young's symmetrizer $c = \Sigma_{g \in \mathcal{G}} c(g)g$. If $g_0 \in \mathcal{G}$ there exists a complex number $\emptyset(g_0)$ such that

(12.4.19) $$cg_0c = \emptyset(g_0)c \; .$$

Proof. Let $d = cg_0c$ and compute the coefficient function of d. It is

(12.4.20) $$d(g) = \sum_{rst=g} c(r)g_0(s)c(t) \; .$$

For a permutation p of type p,

(12.4.21) $$d(pg) = \sum_{rst=pg} c(r)g_0(s)c(t) = \sum_{rst=g} c(pr)g_0(s)c(t) = d(g).$$

Similarly for a permutation q of type q, $d(gq) = \epsilon(q)d(g)$. By Lemma 12.4.7, the number $\emptyset(g_0)$ exists and its value is

(12.4.22) $$\emptyset(g_0) = \sum_{rst=e} c(r)g_0(s)c(t). \; \#$$

Lemma 12.4.10. Let $(m) \geq (m)'$ and $(m) \neq (m)'$, with respective Young's symmetrizers c and c'. Then

(12.4.23) $$\text{if } g \in \mathcal{G}; \; cgc' = 0.$$

Proof. We let $d = cgc'$ and compute the coefficient function of $d$. This function satisfies (12.4.17) and hence $d = 0$. See Lemma 12.4.8 and the proof of Lemma 12.4.9. #

In the group algebra of the symmetric group $\mathfrak{H}$ on $m$ letters the partition $(m)$ and diagram $T(m)$ generate a left ideal $I_{(m)} = \{x | x = yc, y \in \text{group algebra}\}$, and $c$ the Young's symmetrizer for $(m)$. We let the dimension of $I_{(m)}$ be $n_{(m)}$.

Lemma 12.4.11. If $c$ is the Young's symmetrizer for $(m)$ then

$$c^2 = \phi c, \text{ and } \phi = (m!)/n_{(m)} \neq 0 . \tag{12.4.24}$$

Proof. Define a linear transformation of the group algebra by

$$L(x) = xc. \tag{12.4.25}$$

The range of $L$ is $I_{(m)}$ so the matrix of $L$ has rank $n_{(m)}$. Using as basis the elements $h \in \mathfrak{H}$, we compute

$$L(h) = hc = \sum_g c(g)hg = \sum_g c(h^{-1}g)g . \tag{12.4.26}$$

The trace of the matrix is

$$c(e) + c(e) + \dots + c(e) = (m!)c(e). \tag{12.4.27}$$

On the other hand the action of $L$ on $I_{(m)}$ is

$$L(xc) = (xc)c = \phi xc, \tag{12.4.28}$$

where we have used (12.4.24). If we take a new basis, consisting of a basis $h_1, \dots, h_{n_{(m)}}$ of $I_{(m)}$ together with other elements outside the ideal then the matrix of $L$ is

$$\begin{pmatrix} \phi I_{n_{(m)}} & 0 \\ A & 0 \end{pmatrix}, \text{ which has trace} \tag{12.4.29}$$

(12.4.30) $$\phi n_{(m)} = (m!)c(e) = m! \ ,$$

since $c(e) = 1$. Therefore $\phi$ has the value (12.4.24) and $c^2 \neq 0$.#

Lemma 12.4.12. If $e = c/\phi$ with $c$ the Young's symmetrizer relative to $(m)$ and $\phi$ as in (12.4.24), then $e$ is a primitive idempotent.

Proof. $e^2 = c^2/\phi^2 = c/\phi = e$, by (12.4.24). Suppose $e = e_1 + e_2$ such that

(12.4.31) $$e_1^2 = e_1,\ e_2^2 = e_2, \text{ and } e_1e_2 = e_2e_1 = 0.$$

Then $ee_1 = e_1$ and $ee_1e = e_1e = e_1$. By Lemma 12.4.9,

(12.4.32) $$e_1 = ce_1c/\phi^2 = \phi(e_1)c/\phi^2 = (\phi(e_1)/\phi)e$$

and $\phi(e_1)/\phi = 1$ or $0$ since $e_1^2 = e_1$ and $e = e^2$. Hence $e_1 = e$ or $e_2 = e$. #

Lemma 12.4.13. If $(m)$ and $(m)'$ are distinct partitions of $m$ with Young's symmetrizers $c$ and $c'$ respectively then the subspaces $I_{(m)}$ and $I_{(m)'}$ are invariant irreducible subspaces and they are inequivalent.

Proof. Assume $(m) > (m)'$. The left ideals $I_{(m)}$ and $I_{(m)'}$ are irreducible since the corresponding idempotents $c/\phi$ and $c'/\phi'$ are irreducible. We are to show there cannot exist $x$ in the group ring such that

(12.4.33) $$x(yc)x^{-1} = y'c'$$

maps $I_{(m)}$ onto $I_{(m)'}$ in a one to one fashion. To show this, if (12.4.33) holds, then

(12.4.34) $$ycx^{-1} = x^{-1}y'c' \text{ and } cycx^{-1} = cx^{-1}y'c' = 0$$

by (12.4.23). By (12.4.19) there exists a complex number $\phi(y)$ such that $\phi(y)c = cyc$ so that

(12.4.35) $$0 = \phi(y)cx^{-1}.$$

Since $y$ is arbitrary, the choice $y = c$ gives $\phi(y) \neq 0$ and therefore $cx^{-1} = 0$. Therefore, if $y'$ is in the group ring, $y'c' = 0$, so that in particular $c'c' = 0$. Contradiction. Therefore an invertible $x$ satisfying (12.4.33) cannot exist. #

From the Young's symmetrizer $c$ for the partition $(m)$ we construct the element in the center of the group ring

(12.4.36) $$(\phi)^{-2} \sum_g gcg^{-1} = \epsilon_1,$$

where $\phi c = c^2$. We show $\epsilon_1^2 = \epsilon_1$ and that the principal ideal determined by $\epsilon_1$ is a minimal closed ideal. To do this we consider something that seems different. Take the coefficient function

(12.4.37) $$\epsilon(g) = (\phi)^{-2} \sum_h c(h^{-1}gh)$$

and define

(12.4.38) $$\epsilon = \sum_g \epsilon(g)g .$$

Relative to a second partition $(m)'$ we similarly define $\epsilon'$.

<u>Lemma 12.4.14</u>. $\epsilon^2 = \epsilon$. If $(m) \neq (m)'$ then $\epsilon\epsilon' = 0$. Both $\epsilon$ and $\epsilon'$ are in the center and $\epsilon = \epsilon_1$ (see (12.4.36)).

<u>Proof</u>. Refer to (12.4.14). $\Sigma_1 tct^{-1} = \Sigma_t\Sigma_g\, c(g)tgt^{-1}$ $= \Sigma_t\Sigma_g c(t^{-1}(tgt^{-1})t)(tgt^{-1}) = \Sigma_t\Sigma_g c(t^{-1}gt)g = (\phi)^2\epsilon$. Therefore $\epsilon_1 = \epsilon$ and $\epsilon$ is in the center. Given $(m) > (m)'$ we have by (12.4.23) that $(\phi)^2(\phi')^2\epsilon\epsilon' = \Sigma_t\Sigma_s tct^{-1}sc's^{-1} = 0$. Since both $\epsilon$ and $\epsilon'$ are in the center it follows that $\epsilon'\epsilon = 0$.

It remains to show that $\epsilon^2 = \epsilon$. We first compute $\epsilon c = c$ (since $\epsilon$ is in the center). By the proof of the first part of the proof,

(12.4.39) $$\varepsilon c = (\phi)^{-2}\Sigma_t\ tct^{-1}c = (\phi)^{-2}\Sigma_t\ \phi(t^{-1})\,tc\ ,$$

where $ct^{-1}c = \phi(t^{-1})c$. We compute the value of $\phi(t^{-1})$.

(12.4.40) $$ct^{-1}c = \Sigma_g\Sigma_h\ c(g)c(h)gt^{-1}h = \phi(t^{-1})\sum_h c(h)h.$$

Since $c(e) = 1$ we have on the coefficient of $e$ that

(12.4.41) $$\phi(t^{-1}) = \sum_{gt^{-1}h=e} c(g)c(h) = \sum_h c(h^{-1}t)c(h).$$

Also

(12.4.42) $$c^2 = \Sigma_g\Sigma_h\ c(g)c(h)gh = \Sigma_g\Sigma_h\ c(g)c(g^{-1}gh)gh$$
$$= \Sigma_k\Sigma_g\ c(g)c(g^{-1}k)k\ .$$

Therefore

(12.4.43) $$\phi(t^{-1}) = c^2(t) = \phi c(t).$$

Resume the calculation of (12.4.39).

(12.4.44) $$\varepsilon c = (\phi)^{-2}\ \Sigma_t\ \phi c(t)tc = \phi^{-1}c^2 = c.$$

Since $\varepsilon$ is in the center,

(12.4.45) $$tct^{-1} = t\varepsilon ct^{-1} = \varepsilon tct^{-1}\ .$$

The sum of (12.4.45) over $t$ gives

(12.4.46) $$(\phi)^2\varepsilon = (\phi)^2\varepsilon^2 \quad \text{or} \quad \varepsilon = \varepsilon^2.\ \#$$

Theorem 12.4.15. To each partition $(m)$ of $m$ and diagram $T(m)$ let $\varepsilon_{(m)}$ be the corresponding idempotent in the center of the group ring. If $(m) \neq (m)'$ then $\varepsilon_{(m)}\varepsilon_{(m)'} = 0$. The sum $\Sigma_{(m)}\varepsilon_{(m)}$ gives a complete decomposition of the identity of the group algebra. The principle ideals are minimal closed ideals.

Proof. The idempotents $\epsilon_{(m)}$ are linearly independent and the dimension of their linear span is the number of partitions of $(m)$. By Lemma 12.3.9 the number of partitions of $m$ is the number of conjugate classes, and by Theorem 12.3.8 this is the dimension of the center of the group algebra. Therefore each $\epsilon_{(m)}$ is irreducible in the center. And the identity being an element of the center clearly is $\Sigma_{(m)}\epsilon_{(m)}$. It follows that the principal ideals determined by the $\epsilon_{(m)}$ are the minimal closed ideals. #

## 12.5. Realization of the group algebra as linear transformations.

We have defined permutation matrices $P_\sigma$ by, if $f \in M(E^m, C)$ then $(P_\sigma f)(x_1,\ldots,x_m) = f(x_{\sigma(1)},\ldots,x_{\sigma(m)})$. Then $(P_\tau(P_\sigma f))(x_1,\ldots,x_m) = (P_\sigma f)(y_1,\ldots,y_m) = f(y_{\sigma(1)},\ldots,y_{\sigma(m)})$ where $y_j = x_{\tau(j)}$. Thus $(P_\tau(P_\sigma f))(x_1,\ldots,x_m) = f(x_{\tau\sigma(1)},\ldots,x_{\tau\sigma(m)}) = (P_{\tau\sigma}f)(x_1,\ldots,x_m)$. That is,

$$P_\tau P_\sigma = P_{\tau\sigma} \; . \tag{12.5.1}$$

Consequently corresponding to a Young's diagram and symmetrizer $c = \Sigma_p\Sigma_q\epsilon(q)pq$ is the linear transformation

$$\Sigma_p\Sigma_q\epsilon(q)P_{pq} = \Sigma_p\Sigma_q\epsilon(q)P_pP_q \; . \tag{12.5.2}$$

The action on a multilinear form is

$$\Sigma_pP_p(\Sigma_q\epsilon(q)P_qf). \tag{12.5.3}$$

The subgroup $\mathfrak{S}_q$ of permutations leaving the column sets of $T(m)$ invariant is a direct sum $\mathfrak{S}_q = \mathfrak{S}_{q1} \oplus \ldots \oplus \mathfrak{S}_{qm_1}$, where $\mathfrak{S}_{qi}$ is the group of permutations of the letters in the i-th column of $T(m)$. The first column of $T(m)$ is the longest containing some number $r \geq 1$ cells corresponding to a partition of $m$ into $r$ nonzero parts. We write

(12.5.4) $$q = q_r q_{r-1} \cdots q_1,\ q_i \in \mathcal{S}_{qi},\ 1 \leq i \leq r.$$

The factors $q_1, \ldots, q_r$ commute, of course. Then

(12.5.5) $$\Sigma_q \epsilon(q) P_q = \Sigma_{q_1} \cdots \Sigma_{q_r} \epsilon(q_1) \cdots \epsilon(q_r) P_{q_1} \cdots P_{q_r}$$

and if $r > n$ then

(12.5.6) $$\Sigma_{q_1} \epsilon(q_1) P_{q_1} f = 0,$$

as was shown in Chapter 6, since the maximal degree of a nonzero alternating m-form in $n$ variables is $n$.

Lemma 12.5.1. Let $\epsilon$ be the idempotent in the center of the group algebra corresponding to the partition $m_1 \geq m_2 \geq \cdots \geq m_r \geq 1$ of $m$. Let $\epsilon = \Sigma_q \epsilon(q) q$ and assume $r > n$. Then the linear operator $\Sigma_g \epsilon(g) P_g$ is the zero operator on $M(E^m, C)$.

Proof. By (12.4.36) and the proof of Lemma 12.4.14, $\emptyset^2 = \Sigma_t tct^{-1}$ so that the corresponding linear operator is

(12.5.7) $$(\Sigma_t P_t (\Sigma_g c(g) P_g) P_{t^{-1}}) f = \Sigma_t P_t (\Sigma_g c(g) P_g)(P_{t^{-1}} f) = 0. \ \#$$

Lemma 12.5.2. The map $d \to P_d$ of the group algebra into the endomorphisms of $M(E^m, C)$ defined by

(12.5.8) $$\Sigma d(g) g \to \Sigma d(g) P_g = \text{def } P_d$$

is an algebra homomorphism. The kernel is an ideal which is a direct sum of the minimal ideals contained in the kernel. If $\epsilon$ is the idempotent in the center corresponding to the partition $(m)$ and if $P_\epsilon \neq 0$, then the map $d\epsilon \to P_{d\epsilon}$ is one to one from the principal ideal $\{d\epsilon\}$ to the endomorphisms of $M(E^m, C)$.

Proof. We use the results of Loomis (1953) as applied to the group algebra which is an $H^*$ algebra, as stated in Lemma 12.3.1. It follows from Loomis that every ideal in the group algebra is a direct sum of the minimal closed ideals contained in it.

The mapping $d\epsilon \to P_{d\epsilon}$ is an algebra homomorphism of the ideal $\{d\epsilon\}$ into End $M(E^m, C)$ and is either one to one or has as kernal a proper ideal in $\{d\epsilon\}$. However $\{d\epsilon\}$ being a full matrix algebra has no proper nonzero ideals. The Lemma now follows. #

Lemma 12.5.3. If $(m)$ is a partition of $m$ into $\leq n$ parts and $\epsilon_{(m)}$ is the idempotent in the center corresponding to $(m)$, then $P_{\epsilon_{(m)}} \neq 0$.

Proof. For brevity we write $\epsilon = \epsilon_{(m)}$. If $P_\epsilon = 0$ then $P_{d\epsilon} = 0$ for all $d$ in the group algebra. From (12.4.44) we have $c\epsilon = \epsilon c = c$ so that it is sufficient to show $P_c \neq 0$. In turn, $c = \Sigma_p \Sigma_q \epsilon(q) pq$ $= (\Sigma_p p)(\Sigma_q \epsilon(q) q)$ so that it is necessary to show that $\Sigma_q \epsilon(q) P_q \neq 0$.

Choose $\sigma$ so that the configuration $\sigma T(m)$ has as columns $1, \ldots, s_1;\ s_1+1, \ldots, s_2; \ldots ; s_{r-1}+1, \ldots, s_r = m$. We show $\Sigma_q \epsilon(q) P_{\sigma q \sigma^{-1}} \neq 0$ and the permutations $\sigma q \sigma^{-1}$ leave the column sets of $\sigma T(m)$ fixed. If we factor $\Sigma_q \epsilon(q) P_{\sigma q \sigma^{-1}} = \prod_{i=1}^{r} (\Sigma_{q_i} \epsilon(q_i) P_{\sigma q_i \sigma^{-1}})$ as in (12.5.5) and apply this operator to $f = u_1 u_2 \ldots u_m$ then

(12.5.9) $$g = \Sigma_q \epsilon(q) P_{\sigma q \sigma^{-1}} f$$
$$= (u_1 \wedge \ldots \wedge u_{s_1})(u_{s_1+1} \wedge \ldots \wedge u_{s_2}) \ldots (u_{s_{r-1}+1} \wedge \ldots \wedge u_m).$$

If we evaluate $P_{\sigma P \sigma^{-1}} g$ at $e_{i(1)}, \ldots, e_{i(m)}$ then

(12.5.10) $$(P_{\sigma p \sigma^{-1}} g)(e_{i(1)}, \ldots, e_{i(m)})$$
$$= \prod_{j=0}^{r-1} (u_{s_j+1} \wedge \ldots \wedge u_{s_{j+1}})(e_{i(\sigma p \sigma^{-1}(s_j+1))}, \ldots, e_{i(\sigma p \sigma^{-1}(s_{j+1}))}).$$

We choose the subscript function $i$ of (12.5.10) to be the permutation $\sigma p_0 \sigma^{-1}$ of $1, \ldots, m$. Then $i(\sigma p \sigma^{-1}(j)) = \sigma p_0 \sigma^{-1} \sigma p \sigma^{-1}(j) = \sigma p_0 p \sigma^{-1}(j)$.

If $p$ and $p_0$ are permutations of type $p$ then $\sigma p_0 p \sigma^{-1}$ permutes the row sets of the configuration $\sigma T(m)$. However (12.5.10) vanishes unless $\sigma p_0 p \sigma^{-1}(s_j+1), \ldots, \sigma p_0 p \sigma^{-1}(s_{j+1})$ is a permutation of the column set $s_j+1, \ldots, s_{j+1}$. This requires $p_0 p$ to be the identity permutation and $p = p_0^{-1}$. Therefore if $i = \sigma p_0 \sigma^{-1}$ then (12.5.10) vanishes for all permutations $p$ of type $p$ except for $p = p_0^{-1}$. This clearly means $\Sigma_p P_{\sigma p \sigma^{-1}} \Sigma_q \epsilon(q) P_{\sigma q \sigma^{-1}} \neq 0$, as was to be shown.

Lemma 12.5.4. The idempotents $\epsilon_{(m)}$ of the center of the group algebra have matrices $P_{\epsilon_{(m)}}$ which are bi-symmetric. The matrices $P_{\epsilon_{(m)}}$ commute with all bi-symmetric matrices.

Proof. $P_\sigma P_{\epsilon_{(m)}} = P_{\sigma\epsilon_{(m)}} = P_{\epsilon_{(m)}\sigma} = P_{\epsilon_{(m)}} P_\sigma$, where $\sigma$ is a permutation of $1, \ldots, m$. If $A$ is any bi-symmetric matrix then

$$AP_{\epsilon_{(m)}} = A\, \Sigma_p \Sigma_q \epsilon(q) P_p P_q = \Sigma_p \Sigma_q \epsilon(q) P_p P_q A = P_{\epsilon_{(m)}} A. \ \#$$

## 12.6. The center of the bi-symmetric matrices, as an algebra.

It is the purpose of this section to show that the matrices $P_{\epsilon_{(m)}}$ form the center of the algebra of bi-symmetric matrices. In order to obtain this conclusion some abstract algebraic argument is needed. We introduce names for the algebras under study.

Let $\mathfrak{S}$ be the representation of the symmetric group algebra in End $M(E^m, C)$, as described in Section 12.5. Within End $M(E^m, C)$ we let $\mathfrak{A}_m$ be the algebra of bi-symmetric matrices of $n^m$ rows and columns. By definition, $\mathfrak{A}_m$ is the set of matrices which commute with all of the matrices in $\mathfrak{S}$. Since the inverse of $P_\sigma$ is $P_\sigma^t$, the transpose, $\sigma$ a permutation of $1, \ldots, m$, it follows that $\mathfrak{A}_m$ is closed under transpose and conjugation and hence is an $H^*$ algebra. The same clearly holds of $\mathfrak{S}$.

Lemma 12.6.1 below is a result of Wedderburn which we take from Albert (1939), page 19. See also Weyl (1946), page 93, Theorem (3.5.A).

Lemma 12.6.1. (Wedderburn) Let $\mathfrak{M}$ be a matrix algebra and $\mathfrak{N}$ be a matrix subalgebra of $\mathfrak{M}$ such that the unit of $\mathfrak{N}$ is the identity matrix and $\mathfrak{N}$ is a total matrix algebra. Then $\mathfrak{M} = \mathfrak{N} \otimes \mathfrak{C}$, the tensor product, where $\mathfrak{C}$ is the commutator algebra of $\mathfrak{N}$ in $\mathfrak{M}$.

Proof. $\mathfrak{N}$ contains elements $e_{ij}$, $1 \leq i, j \leq p$, such that $e_{11}, e_{22}, \ldots, e_{pp}$ are orthogonal idempotents, $I = \sum_{i=1}^{p} e_{ii}$, and if $a \in \mathfrak{N}$ then $a = \sum_{i=1}^{p} \sum_{j=1}^{p} \bar{a}_{ij} e_{ij}$, where if $1 \leq i, j \leq p$, $\bar{a}_{ij} \in C$, a complex number.

Given $a$, let

$$a_{ij} = \sum_{k=1}^{p} e_{ki} a e_{jk}. \tag{12.6.1}$$

Then from (12.6.1),

$$a_{ij} e_{rs} = e_{ri} a e_{js} \quad \text{and} \quad e_{rs} a_{ij} = e_{ri} a e_{js}. \tag{12.6.2}$$

Therefore $a_{ij}$ commutes with all the $e_{ij}$ of $\mathfrak{N}$ and the $a_{ij}$ are contained in $\mathfrak{C}$. Using $I = e_{11} + \ldots + e_{pp}$ and (12.6.1),

$$\sum_{i=1}^{p} \sum_{j=1}^{p} a_{ij} e_{ij} = \sum_{i=1}^{p} \sum_{j=1}^{p} \sum_{k=1}^{p} e_{ki} a e_{jk} e_{ij} \tag{12.6.3}$$

$$= \sum_{i=1}^{p} \sum_{j=1}^{p} e_{ii} a e_{jj} = a.$$

Finally, let $a_{ij}$, $1 \leq i, j \leq p$ be $p^2$ elements in $\mathfrak{C}$. Assume

$$a = \sum_{i=1}^{p} \sum_{j=1}^{p} a_{ij} e_{ij} = 0. \tag{12.6.4}$$

Then

$$0 = \sum_{k=1}^{p} e_{ki_0} \Big( \sum_{i=1}^{p} \sum_{j=1}^{p} a_{ij} e_{ij} \Big) e_{j_0 k} \tag{12.6.5}$$

$$= \sum_{k=1}^{p} \sum_{i=1}^{p} \sum_{j=1}^{p} a_{ij} e_{ki_0} e_{ij} e_{j_0 k} = a_{i_0 j_0} \sum_{k=1}^{p} e_{kk} = a_{i_0 j_0}. \#$$

Lemma 12.6.2. Let $\mathfrak{M}$ be the algebra of all endomorphisms of a finite dimensional vector space $\mathfrak{X}$ over C. Let $\mathfrak{N}$ be a subalgebra of $\mathfrak{M}$ which is a $H^*$ algebra and $\mathfrak{C}$ be the commutator algebra of $\mathfrak{N}$ in $\mathfrak{M}$. Then $\mathfrak{N}$ and $\mathfrak{C}$ have the same center which is the linear span of idempotents $e_1,\ldots,e_r$. $\mathfrak{N}$ is the commutator algebra of $\mathfrak{C}$ in $\mathfrak{M}$.

Proof. Since $\mathfrak{N}$ is a finite dimensional $H^*$ algebra $\mathfrak{N}$ is completely reducible. We let $e_1,\ldots,e_r$ be the irreducible idempotents of the center of $\mathfrak{N}$. It is easily verified that $e_i\mathfrak{M}e_i$ is the algebra of endomorphisms of the vector space $e_i\mathfrak{X}$, $1 \leq i \leq r$. Therefore the center of $e_i\mathfrak{M}e_i$ is one dimensional. We use this fact below.

An element $e_ime_i$ commutes with all the elements of $\mathfrak{N}e_i$ if and only if

$$(12.6.6) \qquad (e_ime_i)a = (e_ime_i)(e_ia) = (e_ime_i)(ae_i) = (ae_i)(e_ime_i) = a(e_ime_i).$$

It follows that $(e_ime_i)$ is an element of $\mathfrak{C}$, which implies $(e_ime_i) \in \mathfrak{C}e_i$. Since $e_i$ is in the center of $\mathfrak{C}$, it follows that

$$(12.6.7) \qquad \mathfrak{C}e_i = e_i\mathfrak{C}e_i \subset e_i\mathfrak{M}e_i$$

and that $\mathfrak{C}e_i$ is the commutator algebra of $\mathfrak{N}e_i$ in $e_i\mathfrak{M}e_i$. Therefore by Wedderburn, Lemma 12.6.1, since $\mathfrak{N}e_i$ is a total matrix algebra (see Loomis on $H^*$ algebras) and $e_i$ is the identity of $e_i\mathfrak{M}e_i$,

$$(12.6.8) \qquad e_i\mathfrak{M}e_i = (\mathfrak{N}e_i) \otimes (\mathfrak{C}e_i).$$

The centers of $e_i\mathfrak{M}e_i$ and $\mathfrak{N}e_i$ are one dimensional. It is easy to see that if the center of $\mathfrak{C}e_i$ is more than one dimensional then so is the center of $e_i\mathfrak{M}e_i$. Hence the center of $\mathfrak{C}e_i$ is one dimensional and is generated by $e_i$.

Within the algebra $e_i \mathfrak{M} e_i$ every element is representable as (c.f. Lemma 12.6.1)

(12.6.9) $$\Sigma_p \Sigma_q e_{pq} c_{pq}$$

where the $e_{pq}$ are fixed elements of the total matrix algebra $\mathfrak{N} e_i$. An element of the form (12.6.9) commutes with every element $c \in \mathfrak{C} e_i$ if and only if

(12.6.10) $$c \Sigma_p \Sigma_q e_{pq} c_{pq} = \Sigma_p \Sigma_q (c e_{pq}) c_{pg} = \Sigma_p \Sigma_q (e_{pq} c) c_{pq}$$

$$= \Sigma_p \Sigma_q e_{pq} (c c_{pq}) = \Sigma_p \Sigma_q e_{pq} (c_{pq} c) .$$

Since this representation is unique, for all $p, q$

(12.6.11) $$c c_{pq} = c_{pq} c .$$

In words, the $c_{pq}$ are in the center of $\mathfrak{C} e_i$ and there exist complex numbers $\phi_{pq}$ such that

(12.6.12) $$c_{pq} = \phi_{pq} e_i \quad \text{and} \quad \Sigma_p \Sigma_q e_{pq} c_{pq} = \Sigma_p \Sigma_q \phi_{pq} e_{pq} \in \mathfrak{N} e_i .$$

Hence $\mathfrak{N} e_i$ is the commutator of $\mathfrak{C} e_i$ in $e_i \mathfrak{M} e_i$. Since,

(12.6.13) $$\mathfrak{C} = \bigoplus_{i=1}^{r} \mathfrak{C} e_i$$

it follows that the center of $\mathfrak{C}$ is the linear span of the $e_i$ and equals the center of $\mathfrak{N}$. And, since

(12.6.14) $$\mathfrak{N} = \bigoplus_{i=1}^{r} \mathfrak{N} e_i ,$$

it follows $\mathfrak{N}$ is the commutator algebra of $\mathfrak{C}$ in $\mathfrak{M}$. #

Corollary 12.6.3. The center of $\mathfrak{N}_m$, the algebra of bi-symmetric matrices, is generated by the idempotents $P_{\epsilon_{(m)}}$, as $(m)$ runs over all the partitions of $m$ into $\leq n$ parts.

## 12.7. Homogeneous polynomials II. Two sided unitary invariance.

The matrices $P_{\epsilon_{(m)}}$ are directly related to unitarily invariant polynomials. For, if $X$ is a $n \times n$ matrix and $U$ is $n \times n$ unitary, then, with $A^*$ the conjugate transpose of $A$ and $\overset{m}{\underset{i=1}{\otimes}} X = X \otimes \ldots \otimes X$,

$$(12.7.1) \qquad \operatorname{tr} P_{\epsilon_{(m)}} \overset{m}{\underset{i=1}{\otimes}} (UXU^*) = \operatorname{tr}\left( \overset{m}{\underset{i=1}{\otimes}} U^* P_{\epsilon_{(m)}} \overset{m}{\underset{i=1}{\otimes}} U\right) \overset{m}{\underset{i=1}{\otimes}} X$$

$$= \operatorname{tr} P_{\epsilon_{(m)}} \overset{m}{\underset{i=1}{\otimes}} X .$$

We use here the fact that $P_{\epsilon_{(m)}}$ is in the center of the algebra $\mathfrak{A}_m$. By the results of Section 12.6, the center of $\mathfrak{A}_m$ is spanned by the matrices $P_{\epsilon_{(m)}}$ taken over all partitions $(m)$ of $m$ into $n$ or fewer parts. Being orthogonal idempotents, these matrices are clearly linearly independent. As polynomials, suppose

$$(12.7.2) \qquad \operatorname{tr}\left(\Sigma_{(m)} a_{(m)} P_{\epsilon_{(m)}}\right) \overset{m}{\underset{i=1}{\otimes}} X = 0,$$

identically in $X$. Since the coefficient matrix is bi-symmetric, by Lemma 12.1.11, it follows that

$$(12.7.3) \qquad \operatorname{tr}\left(\Sigma_{(m)} a_{(m)} P_{\epsilon_{(m)}}\right)\overline{\left(\Sigma_{(m)} a_{(m)} P_{\epsilon_{(m)}}\right)} = 0,$$

and hence that the coefficient matrix is zero. As noted this implies $a_{(m)} = 0$ for all partitions $(m)$.

Conversely, suppose a homogeneous polynomial has the invariance property that if $U$ is unitary then

$$(12.7.4) \qquad \operatorname{tr} A \overset{m}{\underset{i=1}{\otimes}} (UXU^*) = \operatorname{tr} A \overset{m}{\underset{i=1}{\otimes}} X .$$

The matrix $A$ is assumed to be a bi-symmetric matrix as explained in Theorem 12.1.12. The relation (12.7.4) clearly implies that if $U$

is unitary then for all $n \times n$ matrices $X$

$$\text{tr}\, A \bigotimes_{i=1}^{m} (UX) - \text{tr}\, A \bigotimes_{i=1}^{m} (XU) = 0. \tag{12.7.5}$$

If we fix $A$ and $X$ and define a polynomial by

$$f(Y) = \text{tr}\, A \bigotimes_{i=1}^{m} (YX) - \text{tr}\, A \bigotimes_{i=1}^{m} (XY)\ , \tag{12.7.6}$$

then (12.7.5) says that $f(U) = 0$ for all unitary matrices $U$. By Weyl (1946) Lemma (7.1.A), page 177, it follows that $f(Y)$ vanishes for all $n \times n$ matrices $Y$ over the complex numbers. Therefore the bi-symmetric coefficient matrix

$$A(\bigotimes_{i=1}^{m} X) - (\bigotimes_{i=1}^{m} X)A = 0, \tag{12.7.7}$$

and this holds for all $X$. Therefore by Lemma 12.1.11, it follows that $A$ is in the center of $\mathfrak{M}_m$. Thus we have proven

<u>Theorem 12.7.1</u>. The polynomial $\text{tr}\, A \bigotimes_{i=1}^{m} X$, with $A \in \mathfrak{M}_m$, has the invariance property (12.7.4) if and only if $A$ is in the center of $\mathfrak{M}_m$.

If the polynomial $\text{tr}\, A \bigotimes_{i=1}^{m} X$ satisfies (12.7.4) then for Hermitian $X$, there exists unitary $U$ such that $UXU^*$ is a diagonal matrix. Therefore the value of $\text{tr}\, A \bigotimes_{i=1}^{m} X$ depends only on the eigenvalues of Hermitian $X$. In the sequel we shall want to know that the polynomials

$$\text{tr}\, P_{\epsilon_{(m)}} \bigotimes_{i=1}^{m} X \tag{12.7.8}$$

are linearly independent when $X$ is restricted to be a real diagonal matrix. This assertion says something about the diagonal elements of $P_{\epsilon_{(m)}}$. Necessary theory is developed in Section 12.8. It will then appear on reading Constantine (1963) that this same theory is crucial in the evaluation of important integrals.

As a final result in the Section we prove the important reproducing property.

Theorem 12.7.2.

$$(12.7.9)\quad \int \operatorname{tr} P_{\varepsilon(m)} \bigotimes_{i=1}^{m} (UXU^{*}Y)\,dU$$

$$= (\operatorname{tr} P_{\varepsilon(m)} \bigotimes_{i=1}^{m} X)(\operatorname{tr} P_{\varepsilon(m)} \bigotimes_{i=1}^{m} Y)/\operatorname{tr} P_{\varepsilon(m)},$$

where the integral is with respect to the Haar measure of unit mass on the unitary group.

Proof. The polynomial

$$(12.7.10)\qquad \operatorname{tr} P_{\varepsilon(m)} \bigotimes_{i=1}^{m} (UXU^{*}Y),$$

as a polynomial in $X$, has coefficient matrix

$$(12.7.11)\qquad (\bigotimes_{i=1}^{m} U^{*}Y) P_{\varepsilon(m)} \bigotimes_{i=1}^{m} U,$$

which, since $P_{\varepsilon(m)}$ is in the center of $\mathfrak{U}_m$, is a matrix in the (closed) ideal $\mathfrak{U}_m P_{\varepsilon(m)}$. Therefore if we view (12.7.9) as an integral of matrices in $\mathfrak{U}_m P_{\varepsilon(m)}$, then the coefficient matrix lies in this ideal. The polynomial (12.7.9) is invariant in the sense of (12.7.4) so the coefficient matrix is, by Theorem 12.7.1, in the center of $\mathfrak{U}_m$, and by Lemma 12.6.2 as applied to $\mathfrak{U}$ and $\mathfrak{G}$, the only elements of the center of $\mathfrak{U}_m$ in $\mathfrak{U}_m P_{\varepsilon(m)}$ are $\phi P_{\varepsilon(m)}$, $\phi$ a complex number. Therefore

$$(12.7.12)\qquad P_{\varepsilon(m)} \int (\bigotimes_{i=1}^{m} U^{*}YU)\,dU = \phi(Y) P_{\varepsilon(m)}$$

and (12.7.9) is given by

$$(12.7.13) \qquad \emptyset(Y)\operatorname{tr} P_{\epsilon_{(m)}} \bigotimes_{i=1}^{m} X .$$

A similar argument shows that (12.7.9) is given by

$$(12.7.14) \qquad \emptyset'(X)\operatorname{tr} P_{\epsilon_{(m)}} \bigotimes_{i=1}^{m} Y ,$$

where

$$(12.7.15) \qquad \emptyset'(X) = \int \left( \bigotimes_{i=1}^{m} UXU^{*} \right) dU .$$

Set $Y$ = identity matrix and find $\emptyset(I_n) = 1$ and

$$(12.7.16) \qquad \emptyset'(X)\operatorname{tr} P_{\epsilon_{(m)}} = \operatorname{tr} P_{\epsilon_{(m)}} \bigotimes_{i=1}^{m} X .$$

Substitution of (12.7.16) into (12.7.14) yields (12.7.9). #

## 12.8. Diagonal matrices.

If the partition $(m)$ is $m_1 \geq m_2 \geq \cdots \geq m_p \geq 1$, then we will say the index set $i_1,\ldots,i_m$ belongs to $(m)$ if and only if there exists a permutation $\pi$ of $1,\ldots,m$ such that $i_{j_1} = i_{j_2}$ if and only if $\pi(j_1)$ and $\pi(j_2)$ are in the same row set of $T(m)$.

In this section we work with the canonical basis elements $e_1,\ldots,e_n$ of $E$ and define

$$(12.8.1) \quad E^m_{(m)} = \{(e_{i_1},\ldots,e_{i_m}) \mid \text{such that } i_1,\ldots,i_m \text{ belongs to } (m)\}.$$

In what follows recall that if $f \in M(E^m, C)$ then $P_\sigma f$ is defined by $(P_\sigma f)(e_{i_1},\ldots,e_{i_m}) = f(e_{i_{\sigma(1)}},\ldots,e_{i_{\sigma(m)}})$, so that $P_\sigma P_\tau = P_{\sigma\tau}$.

It is easy to see that $(e_{i_1},\ldots,e_{i_m})$ belongs to $(m)$ if and only if $(e_{i_{\sigma(1)}},\dots,e_{i_{\sigma(m)}})$ belongs to $(m)$, so the action of $\mathcal{S}$ on $E^m_{(m)}$ is $\mathcal{S}E^m_{(m)} \subset E^m_{(m)}$. Likewise if $G$ is a $n\times n$ permutation matrix (in the basis $e_1,\dots,e_n$) then $(\bigotimes_{i=1}^{m} G)E^m_{(m)} \subset E^m_{(m)}$ since $Ge_{i_1},\dots,Ge_{i_m}$ belongs to $(m)$ if and only if $e_{i_1},\dots,e_{i_m}$ does.

In the sequel we use the same notation $P_\sigma$ as an operator on m-tuples $(e_{i_1},\dots,e_{i_m})$ so that $(P_\sigma f)(e_{i_1},\dots,e_{i_m}) = f(P_\sigma(e_{i_1},\dots,e_{i_m}))$.

Lemma 12.8.1. Let $(m) \geq (m)'$ and $(m) \neq (m)'$. Let $c' = \Sigma_{p'}\Sigma_{q'}\epsilon(q')p'q'$ be the Young's symmetrizer of $(m)'$. Let $i_1,\dots,i_m$ belong to $(m)$. Define $d$ by

(12.8.2) $$\Sigma_{p'}\Sigma_{q'}\epsilon(q')P_{p'q'}(e_{i_1},\dots,e_{i_m}) = \Sigma_\sigma d(\sigma)P_\sigma(e_{i_1},\dots,e_{i_m}).$$

Then $d \equiv 0$.

Proof. By linear independence of the basis elements, it follows that

(12.8.3) $$d(\sigma) = \Sigma\Sigma\epsilon(q') \quad \{p',q' \mid P_{p'q'}(e_{i_1},\dots,e_{i_m}) = P_\sigma(e_{i_1},\dots,e_{i_m})\}.$$

The condition in (12.8.3) is equivalent to the condition

(12.8.4) $$i_{p'q'(j)} = i_{\sigma(j)}, \quad 1 \leq j \leq m .$$

Since $i_1,\dots,i_m$ belongs to $(m)$ there exists a permutation $\pi$ of $1,\dots,m$ depending only on $(m)$ and $i_1,\dots,i_m$, such that if $1 \leq j \leq m$,

(12.8.5) $\pi p'q'(j)$ and $\pi\sigma(j)$ are in the same row set of $T(m)$.

Thus there exists a permutation $p$ of type $p$ such that

(12.8.6) $$p\pi p'q' = \pi\sigma \quad\text{and}\quad \sigma = (\pi^{-1}p\pi)(p'q').$$

Therefore the condition (12.8.3) is

(12.8.7) $$d(\sigma) = \Sigma\epsilon(q') \quad \{p',q' \mid \text{for some } p,\ \sigma = (\pi^{-1}p\pi)(p'q')\}.$$

Then

(12.8.8) $$d(\pi^{-1}p_0\pi\sigma) = \Sigma\epsilon(q') \quad \{p',q' \mid \text{for some } p, \pi^{-1}p_0\pi\sigma = \pi^{-1}p\pi(p',q')\}$$

$$= \Sigma\epsilon(q') \quad \{p',q' \mid \text{for some } p,\ \sigma = \pi^{-1}p_0p\pi(p'q')\}$$

$$= d(\sigma).$$

Also,

(12.8.9) $$d(\sigma q_0') = \Sigma\epsilon(q') \quad \{p',q' \mid \text{for some } p, \sigma q_0' = \pi^{-1}p\pi(p'q')\}$$

$$= \Sigma\epsilon(q_0')\epsilon(q'q_0'^{-1}) \quad \{p',q' \mid \text{for some } p,\ \sigma = \pi^{-1}p\pi p'(q'q_0'^{-1})\}$$

$$= \epsilon(q_0')d(\sigma)\ .$$

By Lemma 12.4.8 and its proof, $d = 0$. #

<u>Corollary 12.8.2</u>. If $(m) \geq (m)'$ and $(m) \neq (m)'$ then $P_{\epsilon_{(m)'}}$ is the zero operator of $E^m_{(m)}$.

<u>Proof</u>. $\phi^2\epsilon_{(m)'} = \Sigma_g gc'g^{-1}$, as shown in the proof of Lemma 12.4.14, first line. As noted above, $P_g E^m_{(m)} \subset E^m_{(m)}$ for all permutations $g$ of $1,\ldots,m$, so that since $P_{c'}E^m_{(m)} = 0$, the result follows. #

<u>Corollary 12.8.3</u>. Let

(12.8.10) $$P_{\epsilon_{(m)'}} u_{i_1}\cdots u_{i_m} = \sum_{j_1,\ldots,j_m} a_{j_1,\ldots,j_m} u_{j_1}\cdots u_{j_m}\ .$$

If $(m) > (m)'$ and $j_1,\ldots,j_m$ belong to $(m)$ then $a_{j_1\cdots j_m} = 0$.

Proof. $a_{j_1\cdots j_m} = (P_{\epsilon_{(m)'}} u_{i_1}\cdots u_{i_m})(e_{j_1},\ldots,e_{j_m})$

$$= u_{i_1}\cdots u_{i_m}(P_{\epsilon_{(m)'}}(e_{j_1},\ldots,e_{j_m})) = 0. \ \#$$

Remark 12.8.4. The canonical basis elements of $M(E^m,C)$ are the m-forms $u_{i_1}\cdots u_{i_m}$, so the effect of Corollary 12.8.3 is to identify certain basis elements that map into zero.

Lemma 12.8.5. The matrices $P_{\epsilon_{(m)}}$ are symmetric.

Proof. Elements of the center are obtained as $\emptyset^2 \epsilon_{(m)} = \Sigma_g gcg^{-1}$ $= \Sigma_p \Sigma_q \epsilon(q) \Sigma_g g(pq)g^{-1}$. In the permutation groups, since $pq$ and $(pq)^{-1}$ have the same decomposition as a product of cycles, they are conjugate. Since $P_{(pq)^{-1}} = P^t_{pq}$, it follows that the matrix $\Sigma_g P_{gpqg^{-1}}$ is a symmetric matrix and the result follows. #

Corollary 12.8.6. If the $(i_1,\ldots,i_m)$, $(j_1,\ldots,j_m)$ entry of the matrix $P_{\epsilon_{(m)}}$ is nonzero then $i_1,\ldots,i_m$ belongs to $(m)'$ and $j_1,\ldots,j_m$ belongs to $(m)''$ such that $(m)' \leq (m)$ and $(m)'' \leq (m)$.

Lemma 12.8.7. The diagonal of $P_{\epsilon_{(m)}}$ is constant for those indices $i_1,\ldots,i_m$ which belong to $(m)'$.

Proof. The maps $i_1,\ldots,i_m \to \tau(i_{\sigma(1)}),\ldots,\tau(i_{\sigma(m)})$ are transitive on the index set that belongs to $(m)'$ and all these maps are realizable by matrix multiplication

$$(12.8.11) \qquad P_{\epsilon_{(m)}} = (\bigotimes_{i=1}^{m} G) P_{\epsilon_{(m)}} (\bigotimes_{i=1}^{m} G^t), \text{ and } P_{\epsilon_{(m)}} = P_\sigma P_{\epsilon_{(m)}} P^t_\sigma ,$$

with $\sigma$ a permutation of $1,\ldots,m$ and $\tau$ a permutation of $1,\ldots,n$. The Lemma then follows. #

Lemma 12.8.8. If $c$ is the Young's symmetrizer for the partition $(m)$ then $P_c$ is not zero on $E^m_{(m)}$.

Proof. Let $i_1,\ldots,i_m$ belong to $(m)$ such that $i_{j_1} = i_{j_2}$ if and only if $j_1$ and $j_2$ are in the same row set of $T(m)$. Then

(12.8.12) $$(e_{i_{pq(1)}},\ldots,e_{i_{pq(m)}}) = (e_{i_1},\ldots,e_{i_m})$$

if and only if $i_{pq(j)}$ and $i_j$ are equal,
if and only if $pq(j)$ and $j$ are in the same row set,
if and only if $q(j)$ and $j$ are in the same row set,
which implies $q(j) = j$ for all $j$.

Therefore $\Sigma_p\Sigma_q\epsilon(q)(e_{i_{pq(1)}},\ldots,e_{i_{pq(m)}}) = (\#$ permutations of type $p)(e_{i_1},\ldots,e_{i_m})$ plus other terms. By linear independence this is not zero. #

Lemma 12.8.9. Let $d$ be in the group algebra of the symmetric group. If $i_1,\ldots,i_m$ belong to $(m)$ and if $P_d u_{i_1}\ldots u_{i_m} = \Sigma a_{j_1\ldots j_m} u_{j_1}\ldots u_{j_m}$, then $a_{j_1\ldots j_m} \neq 0$ implies $j_1,\ldots,j_m$ belongs to $(m)$.

Proof. It suffices to prove the lemma for $d$ which are permutations. Then

$$P_\sigma u_{i_1}\ldots u_{i_m}(e_{k_1},\ldots,e_{k_m}) = u_{i_1}\ldots u_{i_m}(e_{k_{\sigma(1)}},\ldots,e_{k_{\sigma(m)}})$$

$$= \Sigma a_{j_1\ldots j_m} u_{j_1}\ldots u_{j_m}(e_{k_1},\ldots, e_{k_m}).$$

The left side vanishes except for the single case $i_1 = k_{\sigma(1)},\ldots,i_m = k_{\sigma(m)}$. Hence the only nonzero term on the right is $u_{i_{\sigma^{-1}(1)}}\ldots u_{i_{\sigma^{-1}(m)}}$ for which the index set $i_{\sigma^{-1}(1)},\ldots,i_{\sigma^{-1}(m)}$ belongs to $(m)$. #

Lemma 12.8.10. Let $M(E^m, C, (m))$ be the linear span of those basis elements $u_{i_1} \dots u_{i_m}$ such that $i_1, \dots, i_m$ belong to $(m)$. Then if $d$ is in the group algebra of the symmetric group, $P_d M(E^m, C, (m)) \subset M(E^m, C, (m))$. Then $P_{\epsilon_{(m)}}$ is not zero on $M(E^m, C, (m))$.

Proof. By Lemma 12.8.9, the operator $P_d$ is reduced by $M(E^m, C, (m))$. As in the proof of Lemma 12.8.8 we compute that $P_c u_{j_1} \dots u_{j_m} (e_{i_1}, \dots, e_{i_m})$ $= u_{j_1} \dots u_{j_m}$ ((# of permutations of type p)$(e_{i_1}, \dots, e_{i_m})$ + other terms) = (# permutations of type p) if $(i_1, \dots, i_m) = (j_1, \dots, j_m)$, and is $= 0$ if $(j_1, \dots, j_m)$ does not belong to $(m)$. Since $P_{\epsilon_{(m)}} P_c$ $= P_{\epsilon_{(m)} c} = P_c$, it follows that $P_{\epsilon_{(m)}} (P_c u_{i_1} \dots u_{i_m}) \neq 0$, and since $P_c u_{i_1} \dots u_{i_m}$ is an element of $M(E^m, C, (m))$, the result follows. #

Corollary 12.8.11. If $i_1, \dots, i_m$ belongs to $(m)$ then $P_{\epsilon_{(m)}} u_{i_1} \dots u_{i_m} = \phi u_{i_1} \dots u_{i_m}$ + other terms, $\phi \neq 0$. $\phi$ is a rational number.

Proof. $\phi$ is the diagonal entry of $P_{\epsilon_{(m)}}$. By Lemma 12.8.7 $P_{\epsilon_{(m)}}$ has the same diagonal entry for all $u_{i_1} \dots u_{i_m}$ such that $i_1, \dots, i_m$ belong $(m)$, it suffices to prove $\phi \neq 0$. $P_{\epsilon_{(m)}}$ is an orthogonal projection which maps $M(E^m, C, (m))$ into itself, with the matrix given in terms of an orthonormal basis. Since $P_{\epsilon_{(m)}}$ is not the zero operator on $M(E^m, C, (m))$, tr $P_{\epsilon_{(m)}} \neq 0$. Since tr $P_{\epsilon_{(m)}}$ $= \phi$ (dimension $M(E^m, C, (m))$), $\phi$ is rational. #

## 12.9. Polynomials of diagonal matrices.

We let $A \in \mathfrak{M}_m$ and $B \in \mathfrak{M}_m$ be a diagonal matrix, $B = \bigotimes_{i=1}^{m} X$, $X$ a diagonal matrix. The polynomial

$$\text{(12.9.1)} \qquad \operatorname{tr} AB = \operatorname{tr} A \bigotimes_{i=1}^{m} X$$

depends only on the diagonal of $A$ for its value.

Lemma 12.9.1. Let $X$ be a $n \times n$ diagonal matrix and $A \in \mathfrak{N}_m$ be a diagonal matrix. If

$$(12.9.2) \qquad \operatorname{tr} A \overset{m}{\underset{i=1}{\otimes}} X = 0 \quad \text{for all } n \times n \text{ diagonal } X,$$

then $A = 0$.

Proof. (12.9.2) is a homogeneous polynomial of $n$ variables $x_1, \ldots, x_n$, and the coefficients of the polynomial are all zero. Clearly $x_{i_1} \ldots x_{i_m} = x_{j_1} \ldots x_{j_m}$ if and only if there exists a permutation $\sigma$ of $1, \ldots, m$ such that $i_{\sigma(1)} = j_1, \ldots, i_{\sigma(m)} = j_m$. Since $A$ is bi-symmetric $a_{(i_1, \ldots, i_m), (i_1, \ldots, i_m)}$ $= a_{(i_{\sigma(1)}, \ldots, i_{\sigma(m)}), (i_{\sigma(1)}, \ldots, i_{\sigma(m)})}$. Hence the diagonal terms all vanish and $A = 0$.

Theorem 12.9.2. The polynomials $\operatorname{tr} P_{\epsilon_{(m)}} \overset{m}{\underset{i=1}{\otimes}} Y$, $Y$ diagonal, are linearly independent.

Proof. If $\operatorname{tr}(\Sigma_{(m)} a_{(m)} P_{\epsilon_{(m)}}) \overset{m}{\underset{i=1}{\otimes}} Y = 0$ identically on $Y$ then by Lemma 12.9.1, the diagonal of $\Sigma_{(m)} a_{(m)} P_{\epsilon_{(m)}}$ is identically zero. Order the partitions as $(m)_1 < (m)_2 < \ldots < (m)_r$. By Lemma 12.8.2 the matrices $P_{\epsilon_{(m)_1}}, \ldots, P_{\epsilon_{(m)_{r-1}}}$ are zero on $u_{i_1} \ldots u_{i_m}$ such that $i_1, \ldots, i_m$ belong to $(m)_r$. Thus, since the diagonal of $P_{\epsilon_{(m)}}$ is not zero for those $u_{i_1} \ldots u_{i_m}$ such that $i_1, \ldots, i_m$ belongs to $(m)_r$, it follows that $a_{(m)_r} = 0$. Using the obvious backward induction together with the nested character of the matrices $P_{\epsilon_{(m)}}$ as described in Corollary 12.8.6 and Corollary 12.8.11, the result that $a_{(m)_i} = 0$, $1 \leq i \leq r$, follows. #

If $G$ is a $n \times n$ permutation matrix then since $P_{\epsilon_{(m)}}$ is in the center of $\mathfrak{N}_m$,

$$(12.9.3)\quad \operatorname{tr} P_{\epsilon_{(m)}} \bigotimes_{i=1}^{m} (GYG^t) = \operatorname{tr}\Big(\big(\bigotimes_{i=1}^{m} G^t\big) P_{\epsilon_{(m)}} \big(\bigotimes_{i=1}^{m} G\big)\Big) \bigotimes_{i=1}^{m} Y$$

$$= \operatorname{tr} P_{\epsilon_{(m)}} \bigotimes_{i=1}^{m} Y \ .$$

If $Y$ is a diagonal matrix with diagonal entries $y_1,\ldots,y_n$ then the polynomial (12.9.3) can be expressed as

$$(12.9.4)\qquad f(y_1,\ldots,y_n) = \operatorname{tr} P_{\epsilon_{(m)}} \bigotimes_{i=1}^{m} Y \ ,$$

and (12.9.3) says that $f$ is a symmetric function in the variables $y_1,\ldots,y_n$, homogeneous of degree $m$. The space of these polynomials has dimension $r$ = number of partitions of $m$ into less than or equal $n$ parts. This is also the number of idempotents $P_{\epsilon_{(m)_1}},\ldots,P_{\epsilon_{(m)_r}}$ in the center of $\mathfrak{A}_m$. Therefore

Theorem 12.9.3. The space of polynomials

$$(12.9.5)\qquad \operatorname{tr}\big(\Sigma_{(m)} a_{(m)} P_{\epsilon_{(m)}}\big) \bigotimes_{i=1}^{m} Y \ ,$$

$Y$ a $n \times n$ diagonal matrix with diagonal entries $y_1,\ldots,y_n$, is the space of homogeneous symmetric polynomials of degree $m$ in $y_1,\ldots,y_n$.

## 12.10. Zonal polynomials of real matrices.

The theory of the preceding sections has assumed complex numbers as coefficients. In the process we have obtained a theory in Section 12.7 of polynomials $f$ with the unitary invariance

$$(12.10.1)\qquad f(UXU^*) = f(X), \ U \text{ unitary.}$$

In this section we wish to study polynomials which satisfy the orthogonal invariance

(12.10.2) $$f(UXU^t) = f(X),\ U \in \underline{O}(n),\ X \text{ a real } n \times n \text{ matrix.}$$

The object is to obtain a representation of $f$ as a linear combination of zonal polynomials. The polynomials will enjoy a reproducing property similar to (12.7.9) but in which integration is over the group $\underline{O}(n)$ of $n \times n$ orthogonal matrices. The polynomials however are not those of Section 12.7 and a more elaborate construction is required.

Related to the construction given are polynomials $f$ with a two-sided invariance property

(12.10.3) $$f(U_1 X U_2) = f(X),\ U_1,\ U_2 \in \underline{O}(n),\ X \in GL(n).$$

The examples of (12.10.3) arise from integrals

(12.10.4) $$\int_{\underline{O}(n)} f(\text{tr } UX)\,dU$$

which do not vanish if $f$ is a positive function. A function satisfying (12.10.3) clearly satisfies

(12.10.5) $$f(X) = f((X^tX)^{-1/2}X^t)X) = f((X^tX)^{1/2}) = f(D)$$

with $D$ a $n \times n$ diagonal matrix whose entries are the eigenvalues of $(X^tX)^{1/2}$. If $f$ is a homogeneous polynomial of degree $m$ then

(12.10.6) $$f(X) = f(-X) = (-1)^m f(X).$$

Thus if $m$ is odd the polynomial vanishes. In the sequel we write $2m$ for the degree. By the theory of orthogonal invariants in Weyl (1946), $f((X^tX)^{1/2})$ is a polynomial $f_1(S)$ of degree $m$ in the entries of $S = X^tX$. Thus

Lemma 12.10.1. If $f$ is a homogeneous polynomial of degree $2m$ in the entries of $X$ such that (12.10.3) holds then there is a polynomial $f_1$ of $n$ variables such that

$$(12.10.7) \qquad f(X) = f_1(\operatorname{tr} X^tX, \operatorname{tr}(X^tX)^2, \ldots, \operatorname{tr}(X^tX)^n)$$

and such that every term is homogeneous of degree 2m.

Proof. $f(X) = f_2(S)$ and $f_2$ satisfies (12.10.2), so that $f_2(S) = f_3(D^2)$, with $D$ as in (12.10.5). Therefore the value of $f(X)$ is a symmetric function of the eigenvalues of $X^tX$, hence the form (12.10.7). #

It is clear that if $f$ satisfies (12.10.2) then the polynomial $f(X^tX)$ satisfies (12.10.3). Lemma 12.10.1 states the converse, and implies the dimension of the space of polynomials in question is again $r$, the number of partitions of $m$ into $\leq n$ parts. See Theorem 12.2.5 and (12.10.7).

A homogeneous polynomial of degree $2m$ satisfying (12.10.3) must have the form

$$(12.10.8) \qquad \operatorname{tr} A \bigotimes_{i=1}^{2m} X = \operatorname{tr} A \bigotimes_{i=1}^{2m} U_1 \bigotimes_{i=1}^{2m} X \bigotimes_{i=1}^{2m} U_2 .$$

Thus we are led to define an operator

$$(12.10.9) \qquad E = \int_{\underline{O}(n)} \left(\bigotimes_{i=1}^{2m} U\right) dU .$$

The coefficient matrix $A$ must satisfy, $A$ is bi-symmetric, i.e., $A \in \mathfrak{N}_{2m}$, and

$$(12.10.10) \qquad A = EAE .$$

The main result of this Section is that the polynomials satisfying (12.10.3) are given uniquely by bi-symmetric matrices $A$ satisfying (12.10.10). One of the main results of James (1961b) is that the set of matrices $EAE$ is a commutative algebra of dimension $r$ containing $r$ mutually orthogonal idempotents. These idempotents provide the definition of zonal polynomials.

Lemma 12.10.2.

(12.10.11) $E^2 = E$, $E^t = E$ and $\overline{E} = E$. If $U \in \underline{O}(n)$ then

$$\left(\bigotimes_{i=1}^{2m} U\right)E = E = E\left(\bigotimes_{i=1}^{2m} U\right).$$

Proof. The entries of $E$ are real numbers so that $\overline{E} = E$ follows automatically. Since the Haar measure of unit mass on $\underline{O}(n)$ is invariant,

$$E\left(\bigotimes_{i=1}^{2m} V\right) = \int_{\underline{O}(n)} \left(\bigotimes_{i=1}^{2m} U\right)\left(\bigotimes_{i=1}^{2m} V\right)dU = \int_{\underline{O}(n)} \left(\bigotimes_{i=1}^{2m} UV\right)dU = E. \tag{12.10.12}$$

Therefore

$$E = \int_{\underline{O}(n)} E\,dV = \int_{\underline{O}(n)} E\left(\bigotimes_{i=1}^{2m} V\right)dV = EE = E^2. \tag{12.10.13}$$

Last, the mapping $U \to U^t$ is a measure preserving map of Haar measure on $\underline{O}(n)$ so that

$$\int_{\underline{O}(n)} f(U)dU = \int_{\underline{O}(n)} f(U^t)dU, \tag{12.10.14}$$

$f$ an arbitrary integrable function. In particular $E^t = E$. #

We now formalize the opening discussion in a Lemma.

Lemma 12.10.3. The two sided invariant polynomials which are homogeneous of degree $2m$ are the polynomials

$$\operatorname{tr}(EAE)\left(\bigotimes_{i=1}^{2m} X\right),\ A \in \mathfrak{M}_{2m}. \tag{12.10.15}$$

The coefficient matrices $EAE$ clearly form an algebra $E\,\mathfrak{M}_{2m}E$. Since $(EAE)^t = E^tA^tE^t = EA^tE \in E\,\mathfrak{M}_{2m}E$, and $\overline{EAE} = E\overline{A}E \in E\,\mathfrak{M}_{2m}E$, the algebra is a $H^*$ algebra. As noted above, if $X$ is $n \times n$ nonsingular and $f$ satisfies (12.10.3) then $f(X) = f_1(X^tX)$. If $X$ is singular take $\epsilon \neq 0$ so that $X + \epsilon I_n$ is nonsingular. Then

$f(X+\epsilon I_n) = f_1((X+\epsilon I_n)^t(X+\epsilon I_n))$. By continuity as $\epsilon \to 0$, $f(X) = f_1(X^tX)$ follows for all $n \times n$ matrices $X$. In particular if $(X^tX)^{1/2}$ is a positive semidefinite square root then $f((X^tX)^{1/2}) = f_1((X^tX)^{1/2}(X^tX)^{1/2}) = f_1(X^tX) = f(X)$. The matrix $A \in \mathfrak{M}_{2m}$ can be written as

(12.10.16) $$A = \Sigma_i a_i(\overset{2m}{\underset{j=1}{\otimes}} X_i) ,$$

so that

(12.10.17) $$EAE = \Sigma_i a_i(E(\overset{2m}{\underset{j=1}{\otimes}} X_i)E) = \Sigma_i a_i(E(\overset{2m}{\underset{j=1}{\otimes}} (X_i^tX_i)^{1/2})E)$$
$$= \Sigma_i a_i(E(\overset{2m}{\underset{j=1}{\otimes}} D_i)E) = E(\Sigma_i a_i \overset{2m}{\underset{j=1}{\otimes}} D_i)E .$$

In this expression the matrices $D_i$ are diagonal matrices. Therefore to every $A \in \mathfrak{M}_{2m}$ there is a diagonal matrix $D \in \mathfrak{M}_{2m}$ such that $EAE = EDE$. All the entries of $D$ with index $i_1,\ldots,i_m$ belonging to $(m)$ have the same value. See Section 12.8. Therefore $D$ is a sum $D = \Sigma_{(2m)}D_{(2m)}I_{(2m)}$ in which $I_{(2m)}$ is the identity on $M(E^{2m},C,(2m))$ defined in Lemma 12.8.10, and $I_{(2m)}$ is zero otherwise.

<u>Theorem 12.10.4</u>. The algebra $E\,\mathfrak{M}_{2m}E$ is commutative. The irreducible idempotents are those elements $P_{\epsilon_{(2m)}}E$ which are not zero, and $E\,\mathfrak{M}_{2m}E$ is the linear span of these idempotents. The number of idempotents is the number of partitions of $m$ into $\leq n$ parts. The algebra $E\,\mathfrak{M}_{2m}E$ is also the linear span of the matrices $EI_{(2m)}E$, some of which are zero.

<u>Proof</u>. As seen above the elements of $E\,\mathfrak{M}_{2m}E$ are expressible as $EDE$ with $D$ a diagonal matrix. Therefore

(12.10.18) $$(EDE)^t = E^tD^tE^t = EDE .$$

If in an algebra $x = x^t$ for all $x$, then

(12.10.19) $$xy = (xy)^t = y^t x^t = yx ,$$

and the algebra is commutative.

As noted above the algebra $E\mathfrak{N}_{2m}E$ is a $H^*$ algebra and is therefore a direct sum of its minimal ideals each of which is a total matrix algebra. Thus the minimal ideals are all one-dimensional, and $E\mathfrak{N}_{2m}E$ is the linear span of its irreducible idempotents. We know the dimension of $E\mathfrak{N}_{2m}E$ is to be the dimension of the space of homogeneous polynomials of degree $m$ in $n$ variables, call this $r$, so that $E\mathfrak{N}_{2m}E$ is the linear span of $r$ idempotents.

We now identify the irreducible idempotents. In the process we shall find that the representation $a \to P_{\epsilon a}E$ of the symmetric group algebra is irreducible. Let $\mathfrak{a}$ be the space of $n^m \times n^m$ matrices acting as linear transformations of $M(E^{2m},C)$ in the canonical basis of this space. First we find the endomorphisms of the subspace $E\,M(E^{2m},C)$. A transformation $T$ of $E\,M(E^{2m},C)$ may be extended to a transformation $T$ of $M(E^{2m},C)$ by defining $T(f) = 0$ on the orthogonal complement $(I_{n^{2m}} - E)M(E^{2m},C)$. Then $T$ extended has a matrix $A$ in the canonical basis, and since

(12.10.20) $$T(f) = T(Ef) + T((I-E)f) = T(Ef) = E(T(Ef)),$$

it follows that $A = EAE$.

Within the matrix set $E\mathfrak{a}E$ we seek the commutator algebra of the matrix set $\{EP_aE,\ a$ in the symmetric group algebra$\}$. If $g$ is a permutation of $1,\ldots,2m$ and $(EAE)$ is in the commutator algebra then

(12.10.21) $$(EP_gE)(EAE) = (EAE)(EP_gE) .$$

The matrix $E$ is bi-symmetric so that by definition of the bi-symmetric matrices

(12.10.22) $P_aE = EP_a$, $a \in$ the symmetric group algebra. It follows that

(12.10.23) $P_g(EAE) = (EAE)P_g$, $g$ a permutation of $1,\ldots,2m$.

Therefore $EAE$ is a bi-symmetric matrix and

(12.10.24) $$EAE = E(EAE)E.$$

Conversely, if $A$ is a bi-symmetric matrix and $g$ a permutation of $1,\ldots,2m$ then

(12.10.25) $$(EP_gE)(EAE) = (EAE)(EP_gE).$$

Therefore the matrix set $EAE$ with $A$ bi-symmetric is the commutator algebra of

(12.10.26) $$\{EP_aE \mid a \in \text{symmetric group algebra}\}$$
$$= \{P_aE \mid a \in \text{sym. group algebra}\}$$

The centers of the algebras are the same, by Lemma 12.6.2. Clearly $\{a \mid P_aE = 0\}$ is an ideal of the symmetric group algebra. Thus either $P_{\epsilon_{(2m)}}E = 0$ or the map $a\epsilon_{(2m)} \to P_aP_{\epsilon_{(2m)}}E$ is one to one into. Therefore the center of the matrix set (12.10.26) is the linear span of the idempotents $P_{\epsilon_{(2m)}}E$ that are not zero. These are the irreducible idempotents of the center.

Since the commutator algebra is a commutative algebra, the representation $g \to P_gP_{\epsilon_{(2m)}}E$ as an endomorphism on the vector space $E\,M(E^{2m},C)$ is an irreducible representation of the group of permutations of $1,\ldots,2m$.

(12.10.27) $\operatorname{tr} P_{\epsilon_{(2m)}}E = 0$ or $\operatorname{tr} P_{\epsilon_{(2m)}}E =$ dimension of the representation
$$= \chi_{(2m)}(1).$$

See Lemma 12.11.4 and Problem 13.3.11. #

The question remains, which $P_{\epsilon_{(2m)}} E = 0$? This question is resolved as the last result of this Section by showing that if (2m) has all even summands then $P_{\epsilon_{(2m)}} E \neq 0$. A dimensionality argument will then show that if (2m) has an odd summand then $P_{\epsilon_{(2m)}} E = 0$. This leads to the natural definition 2(m) as the partition $2m_1 \geq 2m_2 \geq \ldots \geq 2m_k$ where (m) is the partition $m_1 \geq m_2 \geq \ldots \geq m_k$.

Theorem 12.10.5.

$$(12.10.28) \qquad \int_{\underline{O}(n)} \operatorname{tr} P_{\epsilon_{(2m)}} E \bigotimes_{i=1}^{2m} (UXU^tY)\,dU$$

$$= (\operatorname{tr} P_{\epsilon_{(2m)}} E \bigotimes_{i=1}^{2m} X)(\operatorname{tr} P_{\epsilon_{(2m)}} E \bigotimes_{i=1}^{2m} Y)/\operatorname{tr} P_{\epsilon_{(2m)}} E.$$

Proof. We assume $P_{\epsilon_{(2m)}} E \neq 0$ and write $f(X) = \operatorname{tr} P_{\epsilon_{(2m)}} E \bigotimes_{i=1}^{2m} X$. Then

$$(12.10.29) \qquad \int_{\underline{O}(n)} f(UXU^tY)\,dU$$

has coefficient matrix in $\mathfrak{N}_{2m} P_{\epsilon_{(2m)}} E$, and, as a polynomial in X, is two sided invariant in the sense of (12.10.3). Consequently, since the only coefficient matrices of two sided invariant polynomials in $\mathfrak{N}_{2m} P_{\epsilon_{(2m)}} E$ are the matrices $\emptyset P_{\epsilon_{(2m)}} E$, $\emptyset$ a complex number, it follows that the integral in (12.10.29) has value

$$(12.10.30) \qquad \emptyset_1(Y) f(X) = \emptyset_2(X) f(Y) .$$

Take X = identity to obtain

$$(12.10.31) \qquad \emptyset_1(Y)(\operatorname{tr} P_{\epsilon_{(2m)}} E) = \emptyset_2(I) f(Y) = f(Y). \ \#$$

Lemma 12.10.6. If the partition $(2m)$ involves only even summands then $P_{\epsilon(2m)} E \neq 0$. If this partition involves at least one odd summand then $P_{\epsilon(2m)} E = 0$.

Proof. We let $\delta(i,j) = 1$ if $i = j$ and $\delta(i,j) = 0$ if $i \neq j$. We examine the two-sided invariant polynomial

$$(12.10.32) \quad (\operatorname{tr} X^t X)^m = \sum_{i_1 \dots i_{2m}} \dots \sum_{k_1 \dots k_{2m}} a_{i_1 \dots i_{2m} k_1 \dots k_{2m}} x_{i_1 k_1} \dots x_{i_{2m} k_{2m}} ,$$

with coefficients

$$a_{i_1 \dots i_{2m} k_1 \dots k_{2m}}$$

$$= \delta(i_1, i_2)\delta(i_3, i_4) \dots \delta(k_1, k_2) \dots \delta(k_{2m-1}, k_{2m}) .$$

The coefficient matrix $A$ in (12.10.32) is symmetric but not bi-symmetric. However it is easy to see that $A$ is positive semi-definite. Consequently if $B$ is a $n^{2m} \times n^{2m}$ matrix of real entries it follows that $AB = 0$ if and only if $B^t AB = 0$. We let $\tilde{A} = (m!)^{-1} \Sigma_g P_g A P_g^t$ so that the bi-symmetric matrix $\tilde{A}$ is the coefficient matrix of the same polynomial as (12.10.32). Further it is easy to see that $(m!)\tilde{A} \geq A$ and $(m!) P_{\epsilon(2m)} \tilde{A} P_{\epsilon(2m)} \geq P_{\epsilon(2m)} A P_{\epsilon(2m)}$. Since $P_{\epsilon(2m)} = P^t_{\epsilon(2m)}$, it follows that if $c = \Sigma_p \Sigma_q \epsilon(q) pq$ is the Young's symmetrizer for $(2m)$ then $P_c P_{\epsilon(2m)} = P_c$ and $P_c A \neq 0$ implies $P_{\epsilon(2m)} A \neq 0$ implies $P_{\epsilon(2m)} A P_{\epsilon(2m)} \neq 0$, implies $P_{\epsilon(2m)} \tilde{A} P_{\epsilon(2m)} \neq 0$. Since $\tilde{A} \in \mathfrak{N}_{2m}$ and $P_{\epsilon(2m)}$ is in the center of $\mathfrak{N}_{2m}$, the above implies $P_{\epsilon(2m)} \tilde{A} P_{\epsilon(2m)} = \tilde{A} P_{\epsilon(2m)} \neq 0$.

To show $P_c A \neq 0$ is equivalent to showing $A P_c^t = (P_c A)^t \neq 0$. Further $P_c^t = \Sigma_p \Sigma_q \epsilon(q) (P_p P_q)^t = \Sigma_p \Sigma_q \epsilon(q) P_{q^{-1}} P_{p^{-1}}$. Then

$$(12.10.33) \qquad P_c^t(u_{i_1} \cdots u_{i_{2m}}) = \Sigma_p \Sigma_q \epsilon(q) u_{i_{pq(1)}} \cdots u_{i_{pq(2m)}} .$$

When $A$ is applied to a term of (12.10.33) then the entries of the row of $A$ are zero except in the case

$$(12.10.34) \qquad 1 = \delta(i_{pq(1)}, i_{pq(2)}) \cdots \delta(i_{pq(2m-1)}, i_{pq(2m)}) .$$

Choose the index set $i_1, \ldots, i_{2m}$ to satisfy (see (12.4.2))

(12.10.35) if $1 \leq j \leq 2m$ then $i_j$ = the number of the row of $T(2m)$ containing $j$.

Then (12.10.34) requires $pq(2j+1)$ and $pq(2j+2)$ to be in the same row of $T(2m)$, hence $q(2j+1)$ and $q(2j+2)$ are in the same row. Since $2j+1$ and $2j+2$ are necessarily in the same row of $T(2m)$, and since $q$ maps each column of $T(2m)$ into itself, it follows that

$$(12.10.36) \qquad q(2j+1) + 1 = q(2j+2), \; 0 \leq j \leq m-1.$$

This is because a column of $T(2m)$ contains only even integers or only odd integers. Therefore it follows that $q$ is an even permutation and $\epsilon(q) = 1$. In particular all coefficients of $AP_c^t u_{i_1} \cdots u_{i_{2m}}$ are nonnegative and $AP_c^t u_{i_1} \cdots u_{i_{2m}} \neq 0$.

Thus $\tilde{A}P_{\epsilon_{(2m)}} \neq 0$. Since $\tilde{A}$ is the bi-symmetric coefficient matrix of a two-sided invariant polynomial, it follows from Theorem 12.10.4 that $\tilde{A} = \Sigma_{(2m)'} a_{(2m)'} P_{\epsilon_{(2m)'}} E$. This clearly implies $P_{\epsilon_{(2m)}} E \neq 0$. The space $E \mathfrak{N}_{2m} E$ has dimension $r$, the number of partitions of $m$ into $\leq n$ parts. This is the number of partitions $(2m)$ into even summands. Since the nonzero matrices $P_{\epsilon_{(2m)}} E$ are linearly independent, it follows that $P_{\epsilon_{(2m)}} E = 0$ if $(2m)$ involves an odd summand. #

## 12.11. Alternative definitions of zonal polynomials. Group Characters.

James (1960) defined zonal polynomials in terms of group representations as follows. A homogeneous polynomial of degree $m$ in the entries of an $n \times n$ symmetric matrix $S$, say $f(S)$, allows substitutions

$$(12.11.1) \qquad A \to f(A^t SA)$$

of the real general linear group. Since $f(A^t SA)$ is again a homogeneous polynomial of degree $m$, this substitution is a representation of $GL(n)$. The space of such polynomials, $V_m$, decomposes into a direct sum $V_m = \bigoplus_{i=1}^{r} V_{mi}$ of invariant subspaces, $r$ = the number of partitions of $m$ into $\leq n$ parts, and each $V_{mi}$ contains a one-dimensional subspace generated by an orthogonally invariant polynomial $\phi_i$, that is, $\phi_i(U^t SU) = \phi_i(S)$ for all $U \in \underline{O}(n)$. The polynomial $\phi_i$, normalized in the manner described below, is the zonal polynomial for the representation of $GL(n)$ in $V_{mi}$.

A polynomial $f(S)$ of degree $m$ determines a polynomial $g(X)$ of degree $2m$ by

$$(12.11.2) \qquad g(X) = f(X^t X).$$

Clearly if $U \in \underline{O}(n)$ then $g(UX) = g(X)$. If

$$(12.11.3) \qquad g(X) = \operatorname{tr} A \bigotimes_{i=1}^{2m} X ,$$

with $A \in \operatorname{Re} \mathcal{U}_{2m}$, the real bi-symmetric matrices, then clearly

$$(12.11.4) \qquad g(X) = g(UX) = \operatorname{tr}(A \bigotimes_{i=1}^{2m} U)(\bigotimes_{i=1}^{2m} X),$$

so on integration over $\underline{O}(n)$ by Haar measure of unit mass, it follows that

(12.11.5) $$AE = A.$$

The converse is obvious, modulo Weyl (1946), since by the theory of orthogonal invariants, if $A \in \mathrm{Re}\,\mathfrak{U}_{2m}E$ there exists a polynomial $f$ such that

(12.11.6) $$f(X^tX) = \mathrm{tr}\, A \bigotimes_{i=1}^{2m} X .$$

Hence $V_m$ is isomorphic to $\mathrm{Re}\,\mathfrak{U}_{2m}E$ and the invariant subspaces of $V_m$ are clearly given by the ideals

(12.11.7) $$\mathrm{Re}\,\mathfrak{U}_{2m}P_{\epsilon_{(2m)}}E$$

such that $P_{\epsilon_{(2m)}}E \neq 0$. Theorem 12.10.4 clearly implies the invariant subspace corresponding to the ideal $\mathrm{Re}\,\mathfrak{U}_{2m}P_{\epsilon_{(2m)}}E$ has a one-dimensional subspace generated by $P_{\epsilon_{(2m)}}E$ which is the coefficient matrix of the uniquely determined two-sided invariant polynomial. Therefore the following theorem holds.

Theorem 12.11.1. If $f$ is a zonal polynomial of degree $m$ in the entries of the real positive definite matrices $S$, then there exists $\emptyset \in C$ and a partition $(2m)$ of even summands such that if $X$ is a $n \times n$ matrix,

(12.11.8) $$f(X^tX) = \emptyset\, \mathrm{tr}(P_{\epsilon_{(2m)}}E)(\bigotimes_{i=1}^{2m} X).$$

Conversely, every such polynomial, except for normalization, is a zonal polynomial.

In tables of zonal polynomials given by James (1960, 1961a, 1964, 1968) and by James and Parkhurst (1974) the normalization used by James is to make the coefficient of the term $(\mathrm{tr}\, S)^m$ equal one.

Definition 12.11.2. Let the partition $(m)$ be $m_1 \geq \cdots \geq m_p$. The zonal polynomial $Z_{(m)}$ of real positive semidefinite matrices $X^tX$

is the polynomial (12.11.8) with $\emptyset$ choosen to make the coefficient of $(\operatorname{tr} X^tX)^m$ equal one, where $2(m) = (2m)$ is the partition $2m_1 \geq 2m_2 \geq \cdots \geq 2m_p$.

The fact that the term $(\operatorname{tr} X^tX)^m$ always has a nonzero coefficient follows from the proof of Lemma 12.10.6. The polynomials are unknown in explicit closed form, but have been tabled for degrees $m \leq 12$. See James and Parkhurst (1974). In James (1960, 1961a, 1964) the polynomials are expressed as polynomials of the symmetric functions $\operatorname{tr} X^tX$, $\operatorname{tr} (X^tX)^2, \ldots, \operatorname{tr} (X^tX)^m$. In James (1964) the zonal polynomials were also given as polynomial functions of the elementary symmetric functions. In James (1968) a recurrence relation is derived from the Laplace-Beltrami operator. Use of the recurrence relation leads naturally to expression of the zonal polynomials in terms of the symmetric functions

$$\Sigma_\sigma x_{\sigma(1)} \cdots x_{\sigma(k)}, \quad 1 \leq k \leq n, \tag{12.11.9}$$

and James (1968) tables the zonal polynomials for degrees $\leq 5$ as functions of the symmetric functions (12.11.9).

It was realized by Constantine (1963) and James (1964) that more elegantly expressed formulas result if a different normalization is used.

Definition 12.11.3. The zonal polynomial $C_{(m)}$ is the polynomial $\emptyset_{(m)} Z_{(m)}$, with $\emptyset_{(m)} \in C$, such that

$$\Sigma_{(m)} C_{(m)}(S) = (\operatorname{tr} S)^m, \tag{12.11.10}$$

$S$ a real positive semidefinite matrix. By Theorem 12.10.4 the expansion (12.11.10) is unique.

The numbers $\emptyset_{(m)}$ used to pass from one definition to the other are not easy to compute but were computed by Constantine (1963) and are restated by James (1964). We state here without proof the following Lemma.

Lemma 12.11.4. The numbers $\phi_{(m)}$ of Definition 12.11.3 have the value

$$(12.11.11) \qquad \chi_{2(m)}(1)2^m(m!)/(2m!) ,$$

and $\chi_{2(m)}(1)$ is the dimension of the representation $2(m)$ of the symmetric group on $2m$ symbols, given by

$$(12.11.12) \qquad \chi_{2(m)}(1) = (2m)! \prod_{i<j}^{p} (2m_i-2m_j-i+j)/\prod_{i=1}^{p} (2m_i+p-i)! ,$$

the partition being $2m_1 \geq 2m_2 \geq \cdots \geq 2m_p \geq 2$.

In regard (12.11.12), c.f. Weyl (1946) page 213, Theorem (7.7.B).

Aside from the two methods of definition mentioned a third definition arises from the fact that zonal polynomials are spherical functions in the sense of Helgason (1962). We finish this section with a discussion of group representations and spherical functions.

If $\mathfrak{A}$ is an algebra over the complex numbers $C$ a mapping of a group $\mathfrak{G}$,

$$(12.11.13) \qquad A \to \pi(A) \in \mathfrak{A}$$

is a group representation provided $\mathfrak{A}$ has a multiplicative identity, $\pi(\text{identity}) = \text{identity}$ and $\pi(AB) = \pi(A)\pi(B)$. $\pi(A)$ determines an endomorphism $T_A$ of the vector space $\mathfrak{A}$ by $T_A(B) = \pi(A)B$ satisfying $T_{AB} = T_A T_B$. We consider here the case that $\mathfrak{A}$ is completely reducible and $e$ is an idempotent of the center of $\mathfrak{A}$ such that $\mathfrak{A}e$ is a minimal ideal. Then $\mathfrak{A}e$ is a full matrix algebra spanned by elements $e_{ij}$ satisfying

$$(12.11.14) \qquad e_{ik}e_{kj} = e_{ij},\ e = e_{11} + \cdots + e_{pp},\ 1 \leq i,\ j \leq p .$$

Thus we write

(12.11.15) $$\pi(A)e = \sum_{i=1}^{p} \sum_{j=1}^{p} \bar{a}_{ij} e_{ij}$$

and note that

(12.11.16) $$\pi'(A) = \pi(A)e$$

is a representation. Acting on $\mathfrak{N} e_{kk}$ the matrix $\pi'(A)$ may be considered to be an endomorphism whose matrix in the basis $e_{1k}, \ldots, e_{kk}$ is

(12.11.17) $$\pi'(A)e_{qk} = \sum_{i=1}^{p} \sum_{j=1}^{p} \bar{a}_{ij} e_{ij} e_{qk} = \sum_{i=1}^{p} \bar{a}_{iq} e_{ik} ,$$

that is, the matrix is $(\bar{a}_{ij})^t$. Therefore these representations of $\mathfrak{G}$ in End $\mathfrak{N} e_{kk}$, $1 \leq k \leq p$, are conjugate and each have the same matrix (12.11.17).

The <u>group character</u> is the trace of the matrix of $\pi(A)$ as a linear transformation, so that the character is

(12.11.18) $$\chi(A) = p(\bar{a}_{11} + \ldots + \bar{a}_{pp}) .$$

In the case $\mathfrak{N}_m$ of bi-symmetric $n^m \times n^m$ matrices and the symmetric group, p is the number of distinct conjugates of a Young's diagram corresponding to a particular (m), which we denote

(12.11.19) $$\chi_{(m)}(1) = p.$$

• We now turn to a specific interpretation.

<u>Lemma 12.11.5</u>. To each zonal polynomial $C_{(m)}$ define

(12.11.20) $$D_{(m)}(X) = C_{(m)}(X^t X)/C_{(m)}(I_n) ,$$

then

(12.11.21) $$\int_{\underline{O}(n)} D_{(m)}(XHY)\,dH = D_{(m)}(X)D_{(m)}(Y) .$$

Proof. That is Theorem 12.10.5 and Definitions 12.11.2, 12.11.3. #

Functions satisfying (12.11.21) are spherical functions in the sense of Helgason (1962).

As noted in Theorem 12.11.1, there is a complex number $\phi_{2(m)}$ such that

$$D_{(m)}(X) = \phi_{2(m)} \operatorname{tr} P_{\varepsilon_{2(m)}} E \bigotimes_{i=1}^{2m} X , \tag{12.11.22}$$

where, if $(m)$ is $m_1 \geq m_2 \geq \cdots \geq m_n$ then $2(m)$ is $2m_1 \geq 2m_2 \geq \cdots \geq 2m_n$. Therefore

$$\begin{aligned} D_{(m)}(XA) &= \phi_{2(m)} \operatorname{tr} P_{\varepsilon_{2(m)}} E\Big(\bigotimes_{i=1}^{2m} (XA)\Big) \\ &= \phi_{2(m)} \operatorname{tr} \Big(\bigotimes_{i=1}^{2m} A\Big) P_{\varepsilon_{2(m)}} E\Big(\bigotimes_{i=1}^{2m} X\Big) . \end{aligned} \tag{12.11.23}$$

Restricting ourselves to real matrices, from (12.11.7), the invariant subspace $V_{2(m)}$ of homogeneous polynomials of degree $m$ in $X^tX$ is in one to one correspondence with $\phi_{2(m)} \mathfrak{N}_{2m} P_{\varepsilon_{2(m)}} E$ which is isomorphic to the space spanned by the right translates $D_{(m)}(\cdot A)$ of $D_{(m)}$.

Theorem 12.11.6. The representation $A \to \pi(A)$ of $GL(n)$ in $V_{2(m)}$ defined by

$$(\pi(A)f)(X) = f(XA) \tag{12.11.24}$$

is an irreducible representation and

$$D_{(m)}(A) = \int_{\underline{O}(n)} \operatorname{tr} \pi(HA)\, dH . \tag{12.11.25}$$

Proof. (Following Helgason.) As just shown, the space $V_{2(m)}$ is finite dimensional. Therefore the operator

$$\overline{E} = \int_{\underline{O}(n)} \pi(H)\, dH \tag{12.11.26}$$

is well defined by the definition

$$(\overline{E}f)(X) = \int_{\underline{O}(n)} (\pi(H)f)(X)dH = \int_{\underline{O}(n)} f(XH)dH . \tag{12.11.27}$$

If $f$ is a translate of $D_{(m)}$, say $f(X) = D_{(m)}(XA)$, and if $H \in \underline{O}(n)$, then

$$f(XH) = D_{(m)}(XHA) , \tag{12.11.28}$$

and by (12.11.21),

$$(\overline{E}f)(X) = D_{(m)}(X)D_{(m)}(A). \tag{12.11.29}$$

By linearity, $\overline{E}$ maps $V_{(2m)}$ onto the one-dimensional space spanned by $D_{(m)}$ and if $X \in GL(n)$ then

$$(\overline{E}D_{(m)})(X) = D_{(m)}(X). \tag{12.11.30}$$

The fact that the right translates of $D_{(m)}$ span follows from (12.11.23) and the assertion (12.11.30) follows from (12.11.29). Thus $\overline{E} \neq 0$ and clearly $\overline{E}^2 = \overline{E}$, and $\overline{E}$ is a rank one transformation.

To show $\pi$ is irreducible, suppose $V_{2(m)} = \bigoplus_{i=1}^{p} V_i'$, with each $V_i'$ an invariant subspace under $\pi$. Then each $V_i'$ reduces $\overline{E}$, as follows from (12.11.26), so that since $\overline{E} \neq 0$, for some $i$, $\overline{E}V_i' \neq 0$. Then $D_{(m)} \in V_i'$ and since $\pi(A)D_{(m)}$ is the right translate of $D_{(m)}$, it follows that $V_i'$ contains all the right translates of $D_{(m)}$, and hence their linear span which is $V_{2(m)}$.

The linear transformation $\overline{E}\pi(A)$ maps $V_{2(m)}$ onto the one-dimensional space generated by $D_{(m)}$, so that in any basis including $D_{(m)}$, since

$$(\overline{E}\pi(A))D_{(m)} = D_{(m)}(A)D_{(m)} , \tag{12.11.31}$$

the matrix of the transformation has trace $D_{(m)}(A)$. Thus, since the trace is invariant under the choice of basis,

$$(12.11.32) \qquad D_{(m)}(A) = \operatorname{tr} \bar{E}\pi(A) = \operatorname{tr} \int_{\underline{O}(n)} \pi(HA)dH = \int_{\underline{O}(n)} \operatorname{tr} \pi(HA)dH,$$

which verifies (12.11.25). #

Remark 12.11.7. The space $\mathfrak{M}_{2m}P_{\epsilon_{2(m)}}E$ defined by invariance on the left, and corresponds to the space of right translates of $D_{(m)}$. The space of left translates results in $\int_{\underline{O}(n)} \operatorname{tr} \pi(AH)dH = D_{(m)}(A)$.

Remark 12.11.8. The endomorphism $\bar{E}$ has rank one but the matrix $P_{\epsilon_{2(m)}}$ in general does not have rank 1. Instead, as noted in (12.11.17), there will be several equivalent (conjugate) representations which result from consideration of non-bisymmetric coefficient matrices.

Remark 12.11.9. Formula (34) of James (1964) follows from (12.11.32). For the complex zonal polynomial with respect to the unitary group discussed in Section 12.7, the zonal polynomial cannot result from integration of a group character. For, if $\operatorname{tr} \pi(A)$ is a polynomial group character then $f(A) = \int \operatorname{tr} \pi(AH)dH$ over the unitary group is a constant function. See Weyl (1946), page 177, Lemma (7.1.A).

Remark 12.11.10. Formulas (35), (36), (37), and (38) of James (1964) give expressions for group characters. The source of these formulas is not stated by James. The important formula (35) can be deduced from Littlewood (1950), page 188, together with formula (6.3;1) on page 87 of Littlewood. Formula (44) of James, op. cit., is then a direct translation of Littlewood (6.2;15), which in our notation would read

$$(12.11.33) \qquad (\operatorname{tr} X_1^{m_1})(\operatorname{tr} X_2^{m_2})\dots(\operatorname{tr} X_p^{m_p}) = \sum_{(m)'} \chi_{(m)'}((m))\chi_{(m)'}(X),$$

where $\chi_{(m)'}(X)$ is the character of the representation of weight $(m)'$, and $\chi_{(m)'}((m))$ is the character of the permutation $(m)$ in the representation $(m)'$ of the symmetric group. For us, $X$ is $n \times n$ and $p \leq m$. The case of special statistical interest is

$$(12.11.34) \qquad (\operatorname{tr} X)^m = \sum_{(m)} \chi_{(m)}(1^m)\chi_{(m)}(X).$$

Remark 12.11.11. Formula (41) of James (1964) states that if any of $m_1,\ldots,m_p$ are odd in the partition $(m): m_1 \geq m_2 \geq \cdots \geq m_p$ of $m$, then $\int_{\underline{O}(n)} \chi_{(m)}(XH)\,dH = 0$. In these notes we do not obtain quite so strong a result. For us the representations

$$(12.11.35) \qquad A \to \operatorname{tr} P_{\epsilon_{2(m)}} E \bigotimes_{i=1}^{2m} (XA) = T_A(\operatorname{tr} P_{\epsilon_{2(m)}} E \bigotimes_{i=1}^{2m} X)$$

has the property that $\int_{\underline{O}(n)} T_{AH}\,dH = \int_{\underline{O}(n)} T_{HA}\,dH = 0$ if the partition $(2m)$ has an odd summand. See Lemma 12.10.6. This together with Theorem 12.11.6 or James (1964) formula (34) is sufficient to derive James (1964) formula (43).

These comments yield the special case

$$(12.11.36) \qquad \int_{\underline{O}(n)} (\operatorname{tr} XH)^{2m}\,dH = \sum_{2(m)} \chi_{2(m)}(1^{2m}) \int_{\underline{O}(n)} \chi_{2(m)}(XH)\,dH$$

$$= \sum_{2(m)} \chi_{2(m)}(1^{2m}) \frac{C_{(m)}(X^tX)}{C_{(m)}(I_n)}.$$

Here we let $2(m)$ be the partition $2m_1 \geq 2m_2 \geq \cdots \geq 2m_p$ of $2m$, with $(m)$ given by $m_1 \geq m_2 \geq \cdots \geq m_p$. See Problem 13.3.2.

## 12.12. Third construction of zonal polynomials. The converse theorem.

The first method of construction of the zonal polynomials was due to James (1961b) and is sketched in Section 13.5 with the details left to the reader to fill in. An entirely new second method of construction appeared in James (1968) in which the polynomials were obtained as solutions of a differential equation. A third, and again entirely different method, has been presented in Saw (1975), and is the outgrowth of trying to use generating functions to define zonal polynomials in an elementary teachable manner. The paper by Saw starts with a seemingly arbitrary definition from which the essential properties of zonal polynomials can be proven. After the mathematical development of Chapter 12 is made, looking back, it is clear such an arbitrary definition must be correct. In this Section we use hindsight to explore the possibility of Saw's approach.

Given a partition $(m) = m_1 \geq m_2 \geq \cdots \geq m_n$ of $m$ we let the monomial symmetric function $M_{(m)}(S)$, use the eigenvalues of S, be

$$(12.12.1) \qquad \Sigma\, s_{i_1}^{m_1} s_{i_2}^{m_2} \cdots s_{i_m}^{m_n}$$

with the summation being taken over the distinct monomial terms. Let the ordered partitions be $(m)_1 \geq (m)_2 \geq \cdots \geq (m)_r$ and define $r \times 1$ vectors

$$(12.12.2) \qquad C^t = (C_{(m)_1}, \ldots, C_{(m)_r}), \text{ see } (12.11.10);$$

$$M^t = (M_{(m)_1}, \ldots, M_{(m)_r});$$

$$J^t = (J_{(m)_1}, \ldots, J_{(m)_r}) .$$

The functions $J_{(m)}$, sometimes called s-functions, are defined in (13.3.14). As is a consequence of Section 12.8 and as is discussed further in the Problems of Section 13.1, there is a lower triangular

$r \times r$ matrix $T_1$ with rows and columns indexed by the partitions of $m$ such that

$$C^t = M^t T_1, \tag{12.12.3}$$

where the row vectors are as defined above.

If $A$ is a $r \times r$ diagonal matrix then

$$C^t A C = M^t (T_1 A T_1^t) M, \tag{12.12.4}$$

and from the matrix of the quadratic form on the right side of (12.12.4) the matrix $T_1$ is recoverable except for the proportionalities introduced by $A$. That is, given an equation

$$C^t A C = M^t B M \tag{12.12.5}$$

with $A$ a diagonal positive definite matrix and $B$ a $r \times r$ positive definite matrix, one has uniquely that there exists $T \in \underline{T}(r)$ such that

$$TT^t = B; \text{ and } C^t = M^t T A^{-1/2}. \tag{12.12.6}$$

<u>Proof</u>. By 12.12.3, $C^t = M^t T_1$ and by (12.12.4) we have

$$C^t A C = M^t (T_1 A^{1/2})(T_1 A^{1/2})^t M = M^t B M. \tag{12.12.7}$$

$M$ is a function of the eigenvalues $s_1, \ldots, s_n$ of $S$ (this variable is suppressed in these expressions) and both sides of (12.12.7) are polynomials in $s_1, \ldots, s_n$. The identity holds for all positive values of $s_1, \ldots, s_n$ and hence for all complex numbers $s_1, \ldots, s_n$. By Section 12.2 the values of $M(S)$ cover $C \times C \times \ldots \times C$ so that it follows the quadratic forms in (12.12.7) have the same matrix. That is,

$$B = (T_1 A^{1/2})(T_1 A^{1/2})^t . \tag{12.12.8}$$

Since B is positive definite the decomposition (12.12.8) into a product of a lower triangular matrix and its transpose is unique. Therefore $T = T_1 A^{1/2}$, which proves (12.12.6). #

In order to compute the zonal polynomials as components of the vector C in principle one needs only determine a suitable equation of the form (12.12.4). The matrix product $T_1 A T_1^t$ must be known but A need not be known in order to determine C up to proportionality. In part this is what Saw, op. cit., has done, using the generating function

(12.12.9) $$E \exp (\mathrm{tr}\ S\underline{X}T\underline{X}^t).$$

We will consider the case of square matrices but the arguments can be done for S and T square, and $\underline{X}$ a rectangular matrix. Orthogonal invariance is introduced by assuming $\underline{X}$ to be a matrix of $n^2$ mutually independent normal 0, 1 random variables, and S, T symmetric with eigenvalues $s_1 \geq \cdots \geq s_n$, $t_1 \geq \cdots \geq t_n$ respectively, so that

(12.12.10) $$E(\mathrm{tr}\ S\underline{X}T\underline{X}^t)^m = E\left(\sum_{i=1}^{n} \sum_{j=1}^{n} s_i t_j ((\underline{X})_{ij})^2\right)^m.$$

Then (12.12.9) is the Laplace transform of the joint density function of $n^2$ independent $\chi_1^2$ random variables and

(12.12.11) $$E \exp(\mathrm{tr}\ \theta\ S\underline{X}T\underline{X}^t) = \prod_{i=1}^{n} \prod_{j=1}^{n} (1 - 2\theta s_i t_j)^{-1/2}.$$

This Laplace transform is considered in Problems 13.3.5 and 13.3.6 where it is shown that

(12.12.12) $$E \exp(\mathrm{tr}\ \theta S\underline{X}T\underline{X}^t) = \sum_{m=0}^{\infty} (\theta^m/(m!)) \Sigma_{(m)} \lambda((m)) J_{(m)}(S) J_{(m)}(T).$$

The coefficients $\lambda((m))$ are defined in (13.3.15). On the other hand

$$(12.12.13) \qquad E \exp(\mathrm{tr}\, \theta S\underline{X}T\underline{X}^t) = \sum_{m=0}^{\infty} (\theta^m/(m!))E(\mathrm{tr}\, S\underline{X}T\underline{X}^t)^m,$$

so that

$$(12.12.14) \qquad \Sigma_{(m)} \lambda((m)) J_{(m)}(S) J_{(m)}(T) = E\, \Sigma_{(m)} C_{(m)}(S) C_{(m)}(\underline{X}T\underline{X}^t)$$

$$= E\, \Sigma_{(m)} C_{(m)}(S) C_{(m)}(T(\underline{X}^t\underline{X})).$$

The random variable $\underline{X}^t\underline{X}$ has a central Wishart distribution with $n$ degrees of freedom and the value of the expectation $EC_{(m)}(T(\underline{X}^t\underline{X}))$ is known. See Problem 13.2.6. That is,

$$(12.12.15) \qquad \Sigma_{(m)} \lambda((m)) J_{(m)}(S) J_{(m)}(T)$$

$$= \Sigma_{(m)} 2^m \frac{\Gamma_n(\frac{1}{2}n, (m))}{\Gamma_n(\frac{1}{2}n)} C_{(m)}(S) C_{(m)}(T).$$

Aside from (12.12.3) one has also that there exists $T_2 \in \underline{T}(r)$ such that

$$(12.12.16) \qquad J^t = M^t T_2^t .$$

The fact that one uses an upper triangular matrix $T_2^t$ is easily shown. Hence if $\Lambda$ is the diagonal matrix with diagonal entries $\lambda((m))$,

$$(12.12.17) \qquad J^t \Lambda J = M^t(T_2^t \Lambda T_2)M = (12.12.15).$$

Thus, if $T_3 \in \underline{T}(r)$ and $T_3 T_3^t = (T_2^t \Lambda T_2)$, one has that

$$(12.12.18) \qquad M^t T_3 \text{ is proportional to } C^t.$$

The numbers $\lambda((m))$ are computable from the formula and tables of $T_2$, $2 \leq m \leq 12$, are available in David, Kendall and Barton (1966).

The converse theory due to Saw (1975) proceeds as follows. Start with the known equation

$$(12.12.19)\qquad \Sigma_{(m)}\lambda((m))J_{(m)}(S)J_{(m)}(T) = E\,(\mathrm{tr}\ \underline{S}\underline{X}T\underline{X}^t)^m.$$

Since the $r$ functions which are the components of $M^t$ are a basis of the space of homogeneous symmetric polynomials of degree $m$,

$$(12.12.20)\qquad E(\mathrm{tr}\ \underline{S}\underline{X}T\underline{X}^t)^m = \Sigma_i\Sigma_j a_{ij}M_{(m)_i}(S)M_{(m)_j}(T).$$

That is, the left side of (12.12.20) is a symmetric function in the eigenvalues of $S$, which, when expressed in the basis $M$ is a sum whose coefficients are symmetric functions in the eigenvalues of $T$. Linear independence of the basis functions is sufficient to show the coefficient polynomials are each homogeneous of degree $m$. It follows from (12.12.19), since both sides of the expression are polynomials in the eigenvalues, that the quadratic form (12.12.20) is positive definite. Thus uniquely one may write $(a_{ij}) = TT^t$ and define up to proportionality that

$$(12.12.21)\qquad C'^t = M^tT.$$

By elementary calculations Saw then shows the component functions of $C'$ have the properties of zonal polynomials including the reproducing property. One knows from hindsight that this must be so. The proofs starting from (12.12.20) lead to an elementary theory that is easily presentable. See Section 13.6.

We do not pursue the converse theory and instead continue with the hindsight point of view. Let $k_1 \geq k_2 \geq \ldots \geq k_p \geq 1$ be integers such that $k_1 + 2k_2 + \ldots + pk_p = m$. A permutation $g$ is said to be of class $k = (k_1,\ldots,k_p)$ if $g$ is the product of $k_1 + \ldots + k_p$ cycles such that $k_i$ of the cycles have length $i$, $1 \leq i \leq p$. There are $\lambda'((m))$ permutations of class $k$, where

$$(12.12.22)\qquad (m!)/\left(\prod_{i=1}^{p} i^{k_p}(k_p!)\right) = \lambda'((m)),$$

where $(m)$ is the partition determined by the numbers $k_1,\ldots,k_p$. To each partition $(m)$ we let $I_{(m)}$ be the idempotent defined as $(m_1!\ldots m_n!)^{-1}\Sigma p$ over all permutations of type $p$ for the diagram $T((m))$. This idempotent determines a left ideal of the group algebra so each group element acts as an endomorphism $\pi(g)(aI_{(m)}) = gaI_{(m)}$. It is known that the group character $\psi$ of this group representation is

$$(12.12.23)\qquad (m_1!\ldots m_n!)^{-1}\ \#\ \text{permutations } h \text{ such that } hgh^{-1} \text{ is a } p$$
$$= \psi((m),k).$$

The value of this character depends only on the class $k$ of $g$. From Weyl (1946) Chapter VII, Section 7, formula (7.6), which is similar to James (1964) formula (44), see also (12.11.34), one obtains

$$(12.12.24)\qquad J_{(m)}(S) = (\operatorname{tr} S)^{k_1}(\operatorname{tr} S^2)^{k_2}\ldots(\operatorname{tr} S^n)^{k_n}$$
$$= \Sigma_{(m)'}\psi((m)',k)M_{(m)'}(S).$$

From (12.12.22) we obtain

$$(12.12.25)\qquad J^t(S)\Lambda' J(T) = \Sigma_{(m)}\Sigma_{(m)'}\Sigma_{(m)''}\psi((m),k)\psi((m)',k)\lambda'((m)'')$$
$$\times M_{(m)}(S)M_{(m)'}(T)\ .$$

In this expression $\lambda'((m)'')$ is the number of permutations in the class $k$, so that $\Sigma_{(m)''}\lambda'((m)'')$ amounts to a sum over all permutations $g$ with equal weighting. Thus the matrix of the right side of (12.12.25) is

$$(12.12.26)\qquad \Sigma_g\psi((m),g)\psi((m)',g)\ ,$$

where $\psi((m),g)$ is the character of the representation of $g$ in the left ideal generated by $I_{(m)}$. The matrix (12.12.26) is immediately computable using the orthogonality relations for group characters. An example is given below.

If $T \in \underline{T}(r)$ and

$$T^t T = (\Sigma_g \psi((m),g)\psi((m)',g)) \tag{12.12.27}$$

then $M^t T^t$ is proportional to $J^t$ and in fact

$$M^t T^t (\Lambda')^{-1/2} = J^t. \tag{12.12.28}$$

We compare the numbers $\lambda'((m))$ with the numbers $\lambda((m))$, see (13.3.15) and (12.12.22), so that if $(m)$ has $k_1$ 1's,...$k_p$ p's then

$$\lambda((m)) = 2^{m-\Sigma k_i}\lambda'((m)). \tag{12.12.29}$$

If $F$ is the diagonal matrix with diagonal entries $2^{m-\Sigma k_i}$ then

$$\Lambda = F\Lambda'; \quad J^t \Lambda^{1/2} = M^t T^t (\Lambda')^{-1/2}(\Lambda')^{1/2}F^{1/2}; \tag{12.12.30}$$

$$J^t \Lambda J = M^t (T^t F T) M.$$

Therefore the lower triangular decomposition of $T^t FT$ defines the zonal polynomials.

For computational purposes we may write $\psi((m),\cdot) = \Sigma_{(m)''} n((m),(m)'')e_{(m)''}$ where the $e_{(m)''}$ are mutually orthogonal idempotents. Then by the orthogonality relations for group characters

$$\Sigma_g \psi((m),g)\psi((m)',g) = \Sigma_{(m)''} n((m),(m)'')n((m)',(m)''), \tag{12.12.31}$$

which is an inner product of the two characters.

For example, in the symmetric group on 3 letters, in terms of the three primative characters $\chi_1$, $\chi_2$, $\chi_3$, one obtains for characters

(12.12.32) $\quad (3) \quad \chi_1(g);$

$(21) \quad \chi_1(g) + \chi_2(g);$

$(1^3) \quad \chi_1(g) + 2\chi_2(g) + \chi_3(g).$

These generate the matrix

$$\begin{pmatrix} 1 & 0 & 0 \\ 1 & 1 & 0 \\ 1 & 2 & 1 \end{pmatrix}\begin{pmatrix} 1 & 1 & 1 \\ 0 & 1 & 2 \\ 0 & 0 & 1 \end{pmatrix} = \begin{pmatrix} 1 & 1 & 1 \\ 1 & 2 & 3 \\ 1 & 3 & 6 \end{pmatrix}. \tag{12.12.33}$$

We find that

$$J^t = M^t \begin{pmatrix} 1 & 1 & 1 \\ 0 & 1 & 3 \\ 0 & 0 & 6 \end{pmatrix} \text{ and } \Lambda' = \begin{pmatrix} 2 & 0 & 0 \\ 0 & 3 & 0 \\ 0 & 0 & 1 \end{pmatrix}, \tag{12.12.34}$$

so that

$$J^t\Lambda' J = 6M^t \begin{pmatrix} 1 & 1 & 1 \\ 1 & 2 & 3 \\ 1 & 3 & 6 \end{pmatrix} M. \tag{12.12.35}$$

To obtain the zonal polynomials one needs $J^t\Lambda J$ where

$$\Lambda = \begin{pmatrix} 8 & 0 & 0 \\ 0 & 6 & 0 \\ 0 & 0 & 1 \end{pmatrix}; \quad J^t\Lambda J = M^t \begin{pmatrix} 15 & 9 & 6 \\ 9 & 15 & 18 \\ 6 & 18 & 36 \end{pmatrix} M. \tag{12.12.36}$$

From this one obtains the lower triangular matrix

$$\begin{pmatrix} 1 & 0 & 0 \\ 3/5 & 4/5 & 0 \\ 2/5 & 6/5 & (2/5)\sqrt{5} \end{pmatrix}. \tag{12.12.37}$$

Except for the proportionality factors this gives for the zonal polynomials

(12.12.38) $\quad M_3 + (3/5)M_{21} + (2/5)M_{1^3} = (1/15)(15M_3 + 9M_{21} + 6M_{1^3});$

$4M_{21} + 6M_{1^3};$

$M_{1^3}.$

These agree with the published polynomials.

## Chapter 13. Problems for users of zonal polynomials.

### 13.0. Introduction.

In this Section we present a series of results that are useful to users of zonal polynomials. In as much as many of these results are relatively easy applications of the preceeding theory many of these results are presented as problems. Some are original to the author. Others are based on various articles by James (1960, 1961a, 1961b, 1964, 1968) and Constantine (1963). In the use of generating functions there is some overlap with a recent paper by Saw (1975).

The problems have been grouped, roughly as follows. A Section of problems related to the mathematics of zonal polynomials is first. Next follows a Section on identities. This is followed by some problems on numerical coefficients in series of zonal polynomials, followed by constructions.

The literature now has many examples of multivariate density functions that have been expressed as infinite sums of zonal polynomials. It was felt that the reader would be better off reading the examples in the existing literature and for illustration purposes only one example, the non-central Wishart distribution, has been included in this chapter.

### 13.1. Theory.

Problem 13.1.1. Let $f$ be a two-sided invariant homogeneous polynomial of degree $2m$ in the variables $X$. Suppose that if $X, Y \in GL(n)$ then

$$\int_{\underline{O}(n)} f(XHY)\,dH = f(X)f(Y). \tag{13.1.1}$$

Show that $f$ is a (unnormalized) zonal polynomial in $X$, i.e., the induced polynomial of $X^tX$ is a zonal polynomial.

Hint: This can be done by writing $f(X) = \Sigma_{(m)} a_{(m)} \operatorname{tr} P_{\epsilon_{(m)}} E(\bigotimes_{i=1}^{2m} X)$. #

Problem 13.1.2. If C is a zonal polynomial of degree m then

$$(13.1.2) \qquad (\det S)^k C(S) \quad \text{and} \quad (\det S)^{-k} C(S)$$

are zonal polynomials of degrees $m+nk$, and $m-nk$ respectively, where $S \in \underline{S}(n)$, provided $m-nk \geq 0$, and $(\det S)^{-1}C(S)$ is a polynomial.

Problem 13.1.3. Let $I_{n-1}$ be the $(n-1) \times (n-1)$ identity matrix and $I'_{n-1}$ be the $n \times n$ matrix $\begin{pmatrix} I_{n-1} & 0 \\ 0 & 0 \end{pmatrix}$. Let $Y \in GL(n)$ and $X' = I'_{n-1} Y I'_{n-1}$. The polynomial

$$(13.1.3) \qquad \operatorname{tr} P_{\epsilon_{2(m)}} E \bigotimes_{i=1}^{2m} X'$$

is a (unnormalized) zonal polynomial of X if and only if it is not the zero polynomial, where

$$X' = \begin{pmatrix} X & 0 \\ 0 & 0 \end{pmatrix}.$$

Hint: $\bigotimes_{i=1}^{2m} I'_{n-1} \mathfrak{N}_{2m} \bigotimes_{i=1}^{2m} I'_{n-1}$ is a subalgebra of $\mathfrak{N}_{2m}$ which is a $H^*$ algebra. The idempotents in the center determined by the symmetric group algebra are the idempotents $P_{\epsilon_{2(m)}} \bigotimes_{i=1}^{2m} I'_{n-1}$, as readily follows from the matrix definitions. From Section 12.10 write the matrix form of the coefficient matrix of a zonal polynomial in X. #

Remark 13.1.4. Zonal polynomials expressed as polynomials of the elementary symmetric functions of the eigenvalues have an expression that is independent of the size of the matrix. The equations for the coefficients in James (1968) again exhibit the independence from the matrix size. #

Remark 13.1.5. Among possible choices of symmetric functions to work with, $\operatorname{tr} S$, $\operatorname{tr} S^2, \ldots, \operatorname{tr} S^n$ have the advantage that the matrix size does not explicitly enter. Thus if $S$ is $n \times n$ and $m > n$, the polynomials of degree $m$ in $\operatorname{tr} S, \ldots, \operatorname{tr} S^n$ still span the space of homogeneous symmetric functions of degree $m$.

Problem 13.1.6. Let $f$ be a polynomial of $n$ variables such that

$$C(S) = f(\operatorname{tr} S, \ldots, \operatorname{tr} S^n) \tag{13.1.4}$$

is a homogeneous polynomial of degree $m$ in the eigenvalues of $S$. Suppose the coefficients of $f$ are real but the entries of $S$ may be complex numbers. Show

$$C(\overline{S}^t) = \overline{C(S)} \quad \text{and} \quad C(ST) = C(TS). \tag{13.1.5}$$

Further, show the coefficient matrix of the polynomial $C$ is in the center of $\mathfrak{N}_m$.

Hint: The matrices $ST$ and $TS$, even if singular, have the same characteristic equation, hence have the same eigenvalues, multiplicities included. Therefore if $L$ is a nonsingular matrix and $C(S) = \operatorname{tr} A \bigotimes_{i=1}^{m} S$, then

$$\left(\bigotimes_{i=1}^{m} L\right) A = A \left(\bigotimes_{i=1}^{m} L\right). \tag{13.1.6}$$

The matrix $L$ may have complex number entries. #

Problem 13.1.7. By the theory of orthogonal invariants for real matrices,

$$\operatorname{tr} P_{\varepsilon_{2(m)}} E \bigotimes_{i=1}^{2m} X = \operatorname{tr} A \bigotimes_{i=1}^{m} (X^t X) \tag{13.1.7}$$

for some bi-symmetric matrix $A$. Does Problem 13.1.6 imply that $A$ is in the center of $\mathfrak{N}_m$?

Problem 13.1.8. In Section 12.8 the subspace $E^m_{(m)}$ were defined as follows. Let $(m)$ be the partition $m_1 \geq m_2 \geq \cdots \geq m_n \geq 0$ and let the set of indices $i_1, \ldots, i_m$ belong to $(m)$. Then the basis of

$E^m_{(m)}$ is the set of all m-tuples

(13.1.8) $$(e_{\tau(i_{\sigma(1)})},\ldots,e_{\tau(i_{\sigma(m)})})$$

where $\tau$ is a permutation of $1,\ldots,n$ and $\sigma$ is a permutation of $1,\ldots,m$. It was shown that $E^m_{(m)}$ is an invariant subspace under all $P_a$, $a \in$ the symmetric group algebra, together with being invariant under $\otimes_{i=1}^m U$, $U \in \underline{O}(n)$ a permutation matrix.

Problem 13.1.9. Let $M^m_{(m)}$ be the space of m-forms generated by the canonical basis elements

(13.1.9) $$u_{\tau(i_{\sigma(1)})}\cdots u_{\tau(i_{\sigma(m)})}$$

corresponding to the basis elements (13.1.8). If a bi-symmetric matrix $A$ commutes with the matrices $\otimes_{i=1}^m U$, $U \in \underline{O}(n)$, $U$ a permutation matrix, then the diagonal elements of the part of $A$ acting on $M^m_{(m)}$ are all equal.

Problem 13.1.10. For each partition $(m)$ of $m$ let $i_1^{(m)},\ldots,i_m^{(m)}$ be an index set belonging to $(m)$. Suppose $A$ is in the center of $\mathfrak{A}_m$ and that $X$ is a $n \times n$ diagonal matrix with diagonal entries $x_1,\ldots,x_n$. Then show

(13.1.10) $$\operatorname{tr} A \otimes_{i=1}^m X = \Sigma_{(m)} a_{(m)} \Sigma_\tau \Sigma_\sigma x_{\tau(i^{(m)}_{\sigma(1)})} \cdots x_{\tau(i^{(m)}_{\sigma(m)})}.$$

Problem 13.1.11. If $A$ is in the center of $\mathfrak{A}_{2m}$ and $X \in GL(n)$ let $d_1,\ldots,d_n$ be the eigenvalues of $(X^tX)^{1/2}$. Then

(13.1.11) $$\operatorname{tr} AE \otimes_{i=1}^{2m} X = \Sigma_{(m)} a_{(m)} \Sigma_\tau \Sigma_\sigma d^2_{\tau(i^{(m)}_{\sigma(1)})} \cdots d^2_{\tau(i^{(m)}_{\sigma(m)})}.$$

Problem 13.1.12. If $U \in \underline{O}(n)$ show

$$(13.1.12)\qquad (\bigotimes_{i=1}^{2m} U)P_{\epsilon_{2(m)}}E = P_{\epsilon_{2(m)}}E(\bigotimes_{i=1}^{2m} U);$$

$$P_\tau P_{\epsilon_{2(m)}}E = P_{\epsilon_{2(m)}}EP_\tau \ ;$$

$P_{\epsilon_{2(m)}}E$ is a symmetric matrix and is an idempotent.

(See Section 12.10.)

Problem 13.1.13. Continue Problems 13.1.10 to 13.1.12. If $X$ is a diagonal $n \times n$ matrix with diagonal entries $x_1^{1/2} \geq 0, \ldots, x_n^{1/2} \geq 0$, then

$$(13.1.13)\qquad \operatorname{tr} P_{\epsilon_{2(m)}}E(\bigotimes_{i=1}^{2m} X) = \sum_{(m)' \leq (m)} a_{(m)'}\Sigma_\tau\Sigma_\sigma x_{\tau(i_{\sigma(1)}^{(m)'})}\cdots x_{\tau(i_{\sigma(m)}^{(m)'})}.$$

The coefficients $a_{(m)'}$ are nonnegative real numbers since the diagonal elements of $P_{\epsilon_{2(m)'}}E$ are nonnegative. (Constantine (1963), James (1968)).

Hint: See Section 12.8 on the nested character of the matrices for the $P_{\epsilon_{(m)}}$. #

Problem 13.1.14. To each partition $(m)$ of $m$ into less than or equal $n$ parts define a monomial symmetric function $M(X,(m))$ as follows. Let $i_1,\ldots,i_m$ belong to $(m)$. (See Section 12.8.) Then

$$(13.1.14)\qquad M(X,(m)) = \Sigma_\tau\Sigma_\sigma x_{\tau(i_{\sigma(1)})}\cdots x_{\tau(i_{\sigma(m)})},$$

$x_1,\ldots,x_n$ the eigenvalues of $X$. Let $M(X)$ be the column vector whose $(m)$-th entry is $M(X,(m))$. Show there exists a upper triangular matrix $A$ with rows and columns indexed by the partitions $(m)$ such that each entry of the vector

$$(13.1.15)\qquad A\,M(X) = C(X)$$

is a (unnormalized) zonal polynomial of $X^tX$. That is, the (m)-th entry of $C(X^tX)$ is $a_{(m)}C_{(m)}(X^tX)$.

<u>Problem 13.1.15</u>. By (13.1.13) we may write

$$\text{(13.1.16)} \qquad \operatorname{tr} P_{\epsilon_{2(m)}} E\left(\overset{2m}{\underset{i=1}{\otimes}} X\right) = \sum_{(m)' \leq (m)} a_{(m)} M(X^tX, (m)').$$

Corresponding to a partition $(m)'$ being $m_1' \geq m_2' \geq \ldots \geq m_n'$ the monomial symmetric function is a sum of terms

$$\text{(13.1.17)} \qquad x_1^{m_1'} x_2^{m_2'} \cdots x_n^{m_n'} = (x_1 \cdots x_n)^{m_n'} (x_1 \cdots x_{n-1})^{m_{n-1}' - m_n'} \cdots (x_1)^{m_1' - m_2'}$$

taken over all permutations of the bottom indices.

<u>Problem 13.1.16</u>. Let the elementary symmetric functions be defined by

$$\text{(13.1.18)} \qquad \prod_{i=1}^{n} (x - x_i) = x^n - u_1 x^{n-1} + \ldots + (-1)^n u_n .$$

Let $(p)$ be given by $p_1 \geq p_2 \geq \ldots \geq p_n$ and $(m)$ be given by $m_1 \geq m_2 \geq \ldots \geq m_n$. Show that if $x_1^{p_1} \cdots x_n^{p_n}$ is a term of the polynomial

$$\text{(13.1.19)} \qquad u_1^{m_1 - m_2} u_2^{m_2 - m_3} \cdots u_{n-1}^{m_{n-1} - m_n} u_n^{m_n}$$

then $(p) \leq (m)$.

<u>Problem 13.1.17</u>. (Constantine (1963)). The zonal polynomial $C_{(m)}$, except for normalization, is given by (13.1.13). Evaluate $C_{(m)}(TS)$ as a function of $T$ given that $S \in \underline{S}(n)$ and that $T$ is a $n \times n$ diagonal matrix. Show that

$$\text{(13.1.20)} \qquad C_{(m)}(TS) = \sum_{(m)' \leq (m)} \Sigma_\tau \Sigma_\sigma t_{\tau(i_{\sigma(1)})}^{(m)'} \cdots t_{\tau(i_{\sigma(m)})}^{(m)'} f(S, \tau, \sigma, (m)').$$

Constantine, op. cit., needed the coefficient of the term

$t_1^{m_1} \dots t_n^{m_n}$. Show this coefficient is

$$(13.1.21) \qquad a_{(m)} t_1^{m_1} \dots t_n^{m_n} |S_1|^{m_1 - m_2} |S_2|^{m_2 - m_3} \dots |S_n|^{m_n} ,$$

where the determinant $|S_r|$ is the determinant of the principal $r \times r$ minor of $S$.

Hint: The elementary symmetric function $u_r$ of $\det(TS - \lambda I_n) = 0$ is given by the sum of the $r \times r$ principal minors of $TS$, that is, the sum of the determinants of these minors. #

Problem 13.1.18. If a zonal polynomial is expressed as a sum of monomial symmetric functions then the coefficients are rational if rational normalization is used.

Hint: The diagonal entries of the projections $P_{\epsilon_{2(m)}} E$ are rational. See (13.1.13).

## 13.2. Numerical identities.

Problem 13.2.1. Show that

$$(13.2.1) \qquad \int_{\underline{S}(n)} \exp(-\mathrm{tr}\, S)(\det S)^{t-(n+1)/2} C_{(m)}(ST) \prod_{j \leq i} ds_{ij}$$

$$= \Gamma_n(t,(m)) C_{(m)}(T).$$

In this problem, only show the existence of the constant $\Gamma_n(t,(m))$, whose value is determined in the next problem.

Hint: In (3.3.16) it is shown that $\underline{S}(n)$ is the factor space $GL(n)/\underline{O}(n)$ and a measure is induced by Haar measure on $GL(n)$ which is invariant under the transformations $S \to ASA^t$ of $\underline{S}(n)$. Except for changes in normalization the differential form for the invariant measure is given in (3.3.16), which accounts for the factor $(\det S)^{-1/2(n+1)}$ in (13.2.1). Since $\mathrm{tr}\, USU^t = \mathrm{tr}\, S$ and

$\det(USU^t) = \det S$, the invariance replaces $C_{(m)}(ST)$ by $C_{(m)}(USU^tT)$. Integrate over $U$. #

Problem 13.2.2. (Constantine (1963)). Show that

$$\Gamma_n(t,(m)) = \pi^{n(n-1)/4} \prod_{i=1}^{n} \Gamma(t + m_i - (i-1)/2) \quad . \tag{13.2.2}$$

Hint: Both sides of (13.2.1) are polynomials of the diagonal elements $t_1,\ldots,t_n$ of $T$. To determine the value of $\Gamma_n(t,(m))$ it is sufficient to find the coefficient of $t_1^{m_1}\ldots t_n^{m_n}$ on both sides of (13.2.1). Using (13.1.21) the required equation is

$$\int_{\underline{S}(n)} \exp(-\operatorname{tr} S)(\det S)^{t-(n+1)/2} |S_1|^{m_1-m_2}\ldots|S_n|^{m_n} \prod_{j\leq i} ds_{ij} \tag{13.2.3}$$

$$= \Gamma_n(t,(m)).$$

Write $S = RR^t$ with $R \in \underline{T}(n)$ and let the elements of $R$ be $r_{ij}$, $j \leq i$. The Jacobian of the transformation is

$$2^n \prod_{i=1}^{n} (r_{ii})^{n-i+1} , \tag{13.2.4}$$

and the integral (13.2.3) in the new variables is

$$\Gamma_n(t,(m)) = \int\ldots\int \exp(-\operatorname{tr} RR^t) \prod_{i=1}^{n} (r_{ii}^2)^{t+m_i-(i+1)/2} \tag{13.2.5}$$

$$\times \prod_{i=1}^{n} d(r_{ii}^2) \prod_{j<i} dr_{ij}.$$

Evaluate this integral to obtain (13.2.2). #

Definition 13.2.3.

$$\Gamma_n(u) = \left(\prod_{i=1}^{n} \Gamma(u - \tfrac{1}{2}(i-1))\right)\pi^{\frac{1}{4}n(n-1)} \quad . \tag{13.2.6}$$

Problem 13.2.4. (Constantine (1963), Theorem 1.)

$$(13.2.7)\qquad \int_{\underline{S}(n)} \exp(-\mathrm{tr}\, RS)(\det S)^{t-(n+1)/2}\, C_{(m)}(ST) \prod_{j\le i} ds_{ij}$$

$$= \Gamma_n(t,(m))(\det R)^{-t} C_{(m)}(TR^{-1}).$$

Remark 13.2.5. (See Problem 4.2.9.) If $X$ is $n\times h$ and consists of independent rows each distributed as normal $(0,\Sigma)$, then the Wishart density function for $n$ degrees of freedom is

$$(13.2.8)\qquad \frac{(\det S)^{(n-(h+1))/2}\exp(-\mathrm{tr}\,(2\Sigma)^{-1}S)}{(\det 2\Sigma)^{n/2}\,\Gamma_h(n/2)}\ .$$

Problem 13.2.6. If $\underline{S}$ has as density function the Wishart density (13.2.8) then

$$(13.2.9)\qquad E\, C_{(m)}(\underline{S}) = \frac{2^m\,\Gamma_h(n/2\ ,(m))}{\Gamma_h(\frac{1}{2}n)}\, C_{(m)}(\Sigma).$$

13.3. Coefficients of series.

Remark 13.3.1. Constantine (1963) defined the zonal polynomials $C_{(m)}$ by requiring the normalization to satisfy

$$(13.3.1)\qquad \text{if}\quad S\in \underline{S}(n)\quad \text{then}\quad (\mathrm{tr}\, S)^m = \Sigma_{(m)} C_{(m)}(S).$$

Since the polynomials $C_{(m)}$ are linearly independent (c.f. Problem 13.1.12) this identity uniquely determines the values of $C_{(m)}(I_n)$. Thus this normalization does depend on the size $n$ of $S$. See Problem 13.1.3.

We now consider the function $\int_{\underline{O}(n)} (\mathrm{tr}\, HX)^{2m} dH$, with $X \in GL(n)$, and write

$$(13.3.2)\qquad \int_{\underline{O}(n)} (\mathrm{tr}\, HX)^{2m} dH = \Sigma_{(m)} b_{(m)} C_{(m)}(X^tX).$$

Note that this is one term of the Laplace transform

(13.3.3) $$\int_{\underline{O}(n)} \exp(HX)\,dH ,$$

and this is the integral that arose in Chapter 5 in the discussion of the noncentral Wishart density function.

Problem 13.3.2. Determine the $b_{(m)}$ by letting $\underline{X}$ have independently and identically distributed rows each normal $(0,\Sigma)$. Obtain the answer

(13.3.4) $$b_{(m)} = ((2m)!)/(m!))2^{-2m}\Gamma_n(n/2\ )/\Gamma_n(n/2\ ,(m)).$$

Hint: If $H \in \underline{O}(n)$ then $\underline{X}$ and $H\underline{X}$ have the same distribution. Thus

(13.3.5) $$\exp(\mathrm{tr}\ \tfrac{1}{2}s^2\Sigma) = E\int_{\underline{O}(n)} \exp(\mathrm{tr}\ sH\underline{X})\,dH.$$

Expand the left side in an infinite series and the integrand of the right side is an infinite series and match the coefficients of the powers of $s$, first, and then the functions $C_{(m)}(\Sigma)$, second. Obtain

(13.3.6) $$\Sigma_{(m)}C_{(m)}(\ \tfrac{1}{2}\Sigma) = (m!)/(2m!)\Sigma_{(m)}\frac{b_{(m)}2^m\Gamma_n(n/2\ ,(m))}{\Gamma_n(n/2\ )}\ C_{(m)}(\Sigma),$$

and from (13.3.6) obtain (13.3.4). #

Definition 13.3.3. (James (1964)).

(13.3.7) $$(a)_m = a(a+1)\ldots(a+m-1);$$

$$(n/2\ )_{(m)} = \Gamma_n(n/2\ ,(m))/\Gamma_n(n/2\ )\ .$$

Problem 13.3.4. The coefficients $b_{(m)}$ in (13.3.4) are

(13.3.8) $$b_{(m)} = (1/2)_m/(n/2\ )_{(m)}\ .$$

Hint: $2^{-2m}(2m!) = (m-1/2)(m-(3/2))\dots(1/2)(m!) = (\frac{1}{2})_m(m!)$ . #

Problem 13.3.5. If $X$ is a $n \times n$ matrix consisting of $n^2$ mutually independent normal (0,1) random entries and if $S$ and $T$ are $n \times n$ diagonal matrices (possibly with complex numbers on the diagonals) then

$$(13.3.9) \qquad E \exp(\operatorname{tr} \theta SXTX^t) = \prod_{i=1}^{n} \prod_{j=1}^{n} (1 - 2\theta s_i t_j)^{-1/2},$$

where $s_1, \dots, s_n$, $t_1, \dots, t_n$ are the diagonal entries of $S$, $T$ respectively. Show (13.3.9) and that

$$(13.3.10) \qquad \ln E \exp(\operatorname{tr} \theta SXTX^t) = (1/2) \sum_{n=1}^{\infty} \frac{(2\theta)^n}{n} (\operatorname{tr} S^n)(\operatorname{tr} T^n).$$

Use (13.3.10) to show that

$$(13.3.11) \qquad E(\operatorname{tr} SX\overline{S}X^t)^m \geq 0 ,$$

where $\overline{S}$ is the conjugate of $S$.

Problem 13.3.6 (Saw (1975)). Compute the exponential of (13.3.10), that is, compute

$$(13.3.12) \qquad \sum_{m=0}^{\infty} (1/m!)\left( \frac{1}{2} \sum_{n=1}^{\infty} \frac{(2\theta)^n}{n} (\operatorname{tr} S^n)(\operatorname{tr} T^n)\right)^m$$

and show this is the series

$$(13.3.13) \qquad \sum_{m=0}^{\infty} \frac{\theta^m}{m!} \Sigma_{(m)} \lambda((m)) J_{(m)}(S) J_{(m)}(T) .$$

Interpret this formula as follows. Let $\pi_j$ be the number of times $j$ occurs in the partition $(m)$, $1 \leq j \leq m$. Then

$$(13.3.14) \qquad J_{(m)}(S) = \prod_{j=1}^{m} (\operatorname{tr} S^j)^{\pi_j} .$$

The coefficient $\lambda((m))$ may be computed, using the multinomial theorem, to be

(13.3.15) $$\lambda((m)) = m!\,2^{m-\Sigma_{j=1}^{m}\pi_j} \Big/ \prod_{j=1}^{m} (j^{\pi_j}\pi_j!) .$$

Problem 13.3.7. Formulas (13.3.10) and (13.3.13) hold if $S, T \in \underline{S}(n)$ or if $S$, $T$ are diagonal matrices with possibly complex numbers on the diagonal.

Problem 13.3.8. (Saw (1975)). Let $S$, $T \in \underline{S}(n)$ and $\underline{X}$ be as in Problem 13.3.5. Show that

(13.3.16) $$E(\mathrm{tr}\ SXTX^t)^m = \Sigma_{(m)} 2^m \frac{\Gamma_n(n/2\ ,(m))}{\Gamma_n(n/2\ )} \frac{C_{(m)}(S)C_{(m)}(T)}{C_{(m)}(I_n)} .$$

Hence, from (13.3.13),

(13.3.17) $$\Sigma_{(m)}\lambda((m))J_{(m)}(S)J_{(m)}(T)$$

$$= \Sigma_{(m)} 2^m \frac{\Gamma_n(n/2\ ,(m))C_{(m)}(S)C_{(m)}(T)}{\Gamma_n(n/2\ )C_{(m)}(I_n)} .$$

Problem 13.3.9. Derive orthogonality relations as follows. Write

(13.3.18) $$C_{(m)} = \Sigma_{(m)'}k((m),(m)')J_{(m)'}$$

and let $K$ be the coefficient matrix in (13.3.18) so that the rows and columns of $K$ are indexed by the partitions of $m$, ordered in the partial ordering of partitions. Substitute (13.3.18) into (13.3.17) and use the linear independence of the functions $J_{(m)}$ to obtain the system of equations

(13.3.19) $$\lambda((m)') = \Sigma_{(m)}k((m),(m)')^2 2^m \frac{\Gamma_n(n/2\ ,(m))}{\Gamma_n(n/2\ )C_{(m)}(I_n)} ;$$

$$0 = \Sigma_{(m)}k((m),(m)')k((m),(m)'')2^m \frac{\Gamma_n(n/2\ ,(m))}{\Gamma_n(n/2\ )C_{(m)}(I_n)} ,$$

$$(m)' \neq (m)'' .$$

The coefficient matrix $K$ is dependent on the number of rows in $S$, $T$ only if $m > n$ and (13.3.19) holds for all matrix sizes $n$.

Problem 13.3.10. (Constantine (1963)). Verify the following evaluation:

$$(13.3.20) \qquad (\Gamma_n(a))^{-1} \int_{S>0} \exp(\mathrm{tr} - S)\exp(\mathrm{tr}\, SZ)(\det(S))^{a - \frac{1}{2}(n+1)} dS$$

$$= (\det(I_n - Z))^{-a}, \quad a > 1/2(n+1).$$

Expand $\exp(SZ)$ as an infinite sum of zonal polynomials, assuming $Z$ is symmetric. Obtain the identity

$$(13.3.21) \qquad (\det(I_n - Z))^{-a} = \sum_{m=0}^{\infty} (m!\Gamma_n(a))^{-1}\Sigma_{(m)}\Gamma_n(a,(m))C_{(m)}(Z).$$

Problem 13.3.11. In the notation of Section 12.10, show that

$$(13.3.22) \qquad \Sigma_{(m)} \mathrm{tr}\, P_{\epsilon_{2(m)}} E \bigotimes_{i=1}^{2m} X = \int_{\underline{O}(n)} (\mathrm{tr}\, HX)^{2m} dH.$$

Use Problem 13.3.2 to obtain

$$(13.3.23) \qquad \Sigma_{(m)} \mathrm{tr}\, P_{\epsilon_{2(m)}} E \bigotimes_{i=1}^{2m} X = \Sigma_{(m)} b_{(m)} C_{(m)}(X^t X).$$

Therefore,

$$(13.3.24) \qquad \mathrm{tr}\, P_{\epsilon_{2(m)}} E = b_{(m)} C_{(m)}(I_n).$$

Use (13.3.4) and the value (c.f. James (1964), formula (21))

$$(13.3.25) \quad C_{(m)}(I_n) = 2^{2m} m!(n/2)_{(m)} \prod_{i<j}^{p} (2m_i - 2m_j - i + j) / \prod_{i=1}^{p} (2m_i + p - i)!,$$

where $p$ is the number of nonzero summands in the partition $(m)$, to obtain

$$(13.3.26)\quad \operatorname{tr} P_{\epsilon_{2(m)}} E = (2m)! \prod_{i<j} (2m_i-2m_j-i+j)/\prod_{i=1}^{p} (2m_i+p-i)! \,.$$

Compare this value with James (1964), formula (19).

Problem 13.3.12. If $\underline{X}$ is $n \times h$ with independently distributed rows each normally distributed with covariance matrix $\Sigma$, and if $E\,\underline{X} = M$, then the joint density function is

$$(13.3.27)\quad f(X) = (2\pi)^{-nh/2} (\det \Sigma)^{-n/2} \exp -1/2 \operatorname{tr} \Sigma^{-1} X^t X$$

$$\times \exp -1/2 \operatorname{tr} \Sigma^{-1} M^t M \quad \exp \operatorname{tr} \Sigma^{-1} X^t M \,.$$

By the results of Chapter 5, if a random variable $\underline{Y}$ has density function

$$(13.3.28)\quad \int_{O(n)} f(H^t Y)\, dH \,,$$

then $\underline{X}^t\underline{X}$ and $\underline{Y}^t\underline{Y}$ have the same distribution, namely the noncentral Wishart density function. The evaluation of (13.3.28) requires evaluation of the integral

$$(13.3.29)\quad \int_{\underline{O}(n)} \exp(\Sigma^{-1} X^t H M)\, dH \,,$$

which may be expressed as an infinite sum of zonal polynomials. Using the idea of Chapter 4, the noncentral Wishart density function is then (13.3.27) multiplied by a normalization. Hence, write down the density function of $\underline{X}^t\underline{X}$.

Problem 13.3.13. Refer to (12.10.27), (12.11.36), (13.3.2) and (13.3.4). Infer that

$$(13.3.30)\quad \chi_{2(m)}(1^{2m}) = b_{(m)} C_{(m)}(I_n) = \operatorname{tr} P_{\epsilon_{2(m)}} E \,.$$

See (13.3.24).

## 13.4. On group representations.

Part of Theorem 12.11.6 may be given a more algebraic proof along the lines of the proof to Theorem 12.10.4. As shown in Section 12.11 the coefficient matrices of the homogeneous polynomials $f(X^tX)$ of degree $2m$ are $AE$, $A \in \mathfrak{N}_{2m}$. Thus the representation is $X \to \pi(X) \in \mathrm{End}\,(\mathfrak{N}_{2m}E)$ defined by

$$(13.4.1) \qquad \pi(X)(AE) = (\bigotimes_{i=1}^{2m} X)(AE).$$

By linearity the transformations $\pi(X)$ extend to transformations $T_A$, $A \in \mathfrak{N}_{2m}$, and the $T_A$ form an $H^*$ algebra, $T_A(XE) = A(XE)$, $A, X \in \mathfrak{N}_{2m}$.

Problem 13.4.1. If $S \in \mathrm{End}(\mathfrak{N}_{2m}E)$ and $ST_A = T_AS$ for all $A \in \mathfrak{N}_{2m}$ then there exists $B \in \mathfrak{N}_{2m}$ such that if $A \in \mathfrak{N}_{2m}$ then

$$(13.4.2) \qquad S(AE) = (AE)(EBE) = AEBE.$$

This problem simply expresses the well known result that left multiplication commutes with right multiplication when considered as endomorphisms. Write $S_{EBE}$ for the transformation defined in (13.4.2).

Problem 13.4.2. The map $EBE \to S_{EBE}$ is a one to one algebra homomorphism.

Problem 13.4.3. The algebras $\{T_A \mid A \in \mathfrak{N}_{2m}\}$ and $\{S_{EBE} \mid B \in \mathfrak{N}_{2m}\}$ have the same center.

Hint: See Lemma 12.6.2. #

That is, if $A \in \mathfrak{N}_{2m}$ and $T_A$ is in the center then there exists $B \in \mathfrak{N}_{2m}$ such that

$$(13.4.3) \qquad A(XE) = (XE)(EBE) = XEBE \quad \text{for all } X \in \mathfrak{N}_{2m}.$$

Problem 13.4.4. Determine the $A \in \mathfrak{N}_{2m}$ satisfying (13.4.3).

Hint: Either $T_A = 0$ for all $A \in \mathfrak{N}_{2m} P_{\epsilon_{2(m)}}$ or the map $A \to T_A$ is one to one on this set. If not zero, then $T_A T_B = T_B T_A$ if and only if $AB = BA$. The solutions to (13.4.3) must satisfy, therefore, $AB = BA$ for all $B \in \mathfrak{N}_{2m}$. Therefore

$$(13.4.4) \qquad A = \sum_{\{2(m) \mid P_{\epsilon_{2(m)}} E \neq 0\}} a_{2(m)} P_{\epsilon_{2(m)}} .$$

Problem 13.4.5. The representation $X \to \pi(X) \in \text{End}\,(\mathfrak{N}_{2m} E)$ is either zero or is irreducible.

Hint: The commutator algebra is Abelian. #

## 13.5. First construction of zonal polynomials.

This Section presents in outline a method by which, in principle, zonal polynomials may be constructed. The method is based on James (1961b) and is essentially the method by which James first constructed the polynomials. However we do not make the detailed analysis of the equivalences of various group representations that is to be found in James (1961b). The sketch of this Section is a problem in that the reader may try to fill in some or all the missing arguments.

As in Section 12.10, if $A$ is the bi-symmetric coefficient matrix of the trace function, that is,

$$(13.5.1) \qquad \text{tr}\, A \bigotimes_{i=1}^{2m} X = (\text{tr}\, XX^t)^m, \text{ then}$$

$$(13.5.2) \qquad \phi^{-1} P_c A = \phi^{-1} b_{(m)}^{-1} P_c E$$

where $c$ is the Young's symmetrizer for the diagram $T(2(m))$ and the number $\phi^{-1} = \chi_{2(m)}(1)/(2m)!$ as shown in Lemma 12.4.11.

From Section 12.10 the $(i_1,\ldots,i_{2m})$, $(j_1,\ldots,j_{2m})$-entry of $A$ is

(13.5.3) $$((2m)!)^{-1}\Sigma_g \delta(i_{g(1)},i_{g(2)})\ldots\delta(j_{g(2m-1)},j_{g(2m)}),$$

the summation being over all permutations $g$ of $1, 2,\ldots,2m$. For brevity we write

(13.5.4) $$i = (i_1,\ldots,i_{2m}) \text{ and } j = (j_1,\ldots,j_{2m}); \text{ and,}$$

$$g(i) = (i_{g(1)},\ldots,i_{g(2m)}), \text{ similarly for } g(j).$$

For a basis element $u_{i_1}u_{i_2}\ldots u_{i_m}$ we write for short $u_i$ with the transformed basis element $P_g u_i = u_{g^{-1}(i)}$. Then $P_c A$ applied to the basis element $u_i$ yields

(13.5.5) $$\Sigma_j (A)_{i_1\ldots i_{2m}j_1\ldots j_{2m}} P_c(u_j) = \Sigma_j (A)_{ij} P_c(u_j)$$

$$= \Sigma_j \Sigma_{p,q} \epsilon(q) A_{ij} u_{(pq)^{-1}(j)}$$

$$= \Sigma_j \Sigma_{p,q} \epsilon(q) A_{i(pq)((pq)^{-1}(j))} u_{(pq)^{-1}(j)}$$

$$= \Sigma_j \Sigma_{p,q} \epsilon(q) A_{i(pq)(j)} u_j \,.$$

In James' construction one notes that the matrix entry $(A)_{i\ pq(j)}$ is

(13.5.6) $$((2m)!)^{-1}\Sigma_g \delta(i_{g(1)},i_{g(2)})\ldots\delta(j_{g(pq(1))},j_{g(pq(2))})\ldots$$

One then writes down the pairs

(13.5.7) $$(1,2),(3,4),\ldots,(2m-1,2m) \text{ and}$$
$$(pq(1),pq(2)),\ldots,(pq(2m-1),pq(2m)).$$

From these 2m pairs one constructs cycles as follows. We explain by example. In James' example $2m = 6$ and the pairs are (1,3), (2,4), (5,6) and (1,2), (3,6), (4,5). Then $1 \to 3 \to 6 \to 5 \to 4 \to 2 \to 1$, a cycle of length 6. Call the set of 2m pairs in (13.5.7) #pq, and let $2m_1', \ldots, 2m_n' = 2(m)'$ be the partition determined by the cycle lengths. Then, with $S = XX^t$ the term

$$b_{(m)} \emptyset^{-1} \epsilon(q) (\mathrm{tr}\, S^{m_1'})(\mathrm{tr}\, S^{m_2'}) \ldots (\mathrm{tr}\, S^{m_n'}) \tag{13.5.8}$$

occurs in the polynomial $P_{\epsilon_{2(m)}} E(\overset{2m}{\underset{i=1}{\otimes}} X)$ and this polynomial is the sum of the terms (13.5.8) taken over all permutations pq such that p is of type-p and q is of type-q for the diagram T(2(m)).

In this construction of the zonal polynomial the factor $\emptyset^{-1}$ is common to all terms and can be ignored so long as proportionality is the only consideration. In the polynomial the term $(\mathrm{tr}\, S)^m$ results from those permutations pq which map the pairs (1,2), (3,4), ..., (2m-1, 2m) among themselves. There are

$$2^m (m_1)! \ldots (m_n)! (1!)^{p_1} (2!)^{p_2} \ldots (n!)^{p_n} \tag{13.5.9}$$

such permutations pq where the $p_1, \ldots, p_n$ are the number of times $1, \ldots, n$ occur in the diagram T((m)) as a column length.

For example, if $m = 3$ and the diagram T(2(m)) has the single row 1,2,...,6, then $2^3(3!) = 48$ and the symmetric group of 6 letters factored by this subgroup has 15 cosets corresponding to the 15 isomorphic cycles produced. These 15 classes then break down into

(13.5.10)

1 class produces three cycles of length 2;

6 classes produce two cycles, one of length 2, one of length 4;

8 classes produce a single cycle of length 6.

In this Young's diagram there is only one permutation of type $q$ so $\epsilon(q) = +1$. Thus the desired polynomial, up to proportionality, is

$$(13.5.11) \qquad (\text{tr } S)^3 + 6(\text{tr } S)(\text{tr } S^2) + 8(\text{tr } S^3).$$

The polynomial in (13.5.11) is the zonal polynomial $Z_{(3)}(S)$, where (3) is the partition of 3 into 3, 0, 0, ... .

Problem 13.5.1. In the zonal polynomials $Z_{(m)}$, when expressed as a weighted sum of terms $(\text{tr } S^{m_1'})(\text{tr } S^{m_2'})\ldots(\text{tr } S^{m_n'})$, the coefficients are integers. Show this.

Problem 13.5.2. Look in a table of symmetric functions and determine the number of partitions of $m = 12$, hence the number of zonal polynomials of degree 12 in the variables of $S$.

## 13.6. A teaching version.

The sequel sets out in a series of problems a sequence of steps whereby zonal polynomials may be defined and some of the basic properties proven. The computations required are relatively easy thus leading to a presentation that is teachable, i.e., not requiring undue mathematical training. The sequel is based on Saw (1975) but does not follow his paper exactly. Notations are defined in Section 12.12. It is important to note that the following argument works because certain definitions can be made independently of the size $n$ of the matrices involved. Thus an immediate consequence of this method of definition is that the zonal polynomials have a definition not depending on the size of the matrices. See Problem 13.1.4.

Problem 13.6.1. If $0 = \sum_{i=1}^{r} a_i M_{(m)_i}(S)$ for all positive definite matrices $S$ then $a_i = 0$, $1 \leq i \leq r$.

Problem 13.6.2. Let $T \in \underline{T}(r)$ and $C^t = M^t T$. Let $C^t = (C_{(m)_1}, \ldots, C_{(m)_r})$. If $a$ is a $r \times 1$ vector such that $C^t(S)a = 0$ for all positive definite matrices $S$ then $a = 0$.

Problem 13.6.3. There exists a $r \times r$ matrix $(a_{ij})$ such that if $n \geq 1$ and if $S$, $T$, $\underline{X}$ are $n \times n$ matrices, if the entries of $\underline{X}$ are independent normal $0, 1$ random variables, and if $S$, $T$ are symmetric, then

$$(13.6.1) \quad E(\operatorname{tr} S\underline{X}T\underline{X}^t)^m = \sum_{i=1}^{r} \sum_{j=1}^{r} a_{ij} M_{(m)_i}(S) M_{(m)_j}(T).$$

Show further that since $\underline{X}$ and $\underline{X}^t$ have the same distribution,

$$(13.6.2) \qquad E(\operatorname{tr} S\underline{X}T\underline{X}^t)^m = E(\operatorname{tr} T\underline{X}S\underline{X}^t)^m.$$

Therefore, by Problem 13.6.1, the matrix $(a_{ij})$ is symmetric.

Problem 13.6.4. The matrix $(a_{ij})$ of (13.6.1) is positive definite.

Hint: By (13.3.13),

$$(13.6.3) \qquad E(\operatorname{tr} S\underline{X}T\underline{X}^t)^m = \Sigma_{(m)} \lambda((m)) J_{(m)}(S) J_{(m)}(T).$$

Further by (12.12.16),

$$(13.6.4) \qquad \Sigma_{(m)} \lambda((m)) J_{(m)}(S) J_{(m)}(T) = M^t(S)(T_2^t \Lambda T_2) M(T).$$

The matrix $T_2^t \Lambda T_2$ is positive definite. Use Problem 13.6.1. #

Definition 13.6.5. Let $T_3 \in \underline{T}(r)$ and $(a_{ij}) = T_3 T_3^t$. Then

$$(13.6.5) \qquad C^t =_{\text{def}} M^t T_3 A^{-1/2},$$

where the diagonal matrix $A \in \underline{D}(r)$ is to be specified.

Corollary 13.6.6. Except possibly for the choice of $A$, the definition (13.6.5) is independent of the matrix size $n$, and as functions

(13.6.6) $$C^t AC = M^t T_3 T_3^t M.$$

Evaluated, if $n \geq 1$ and $S$ and $T$ are symmetric matrices in $\underline{S}(n)$ then

(13.6.7) $$C^t(S)AC(T) = E(\text{tr } S\underline{X}T\underline{X}^t)^m .$$

Problem 13.6.7. If $T$ is a diagonal $n \times n$ and if $\underline{W}$ is a $n \times n$ matrix which has a Wishart distribution then $E\, C_{(m)_i}(T\underline{W})$ is a symmetric homogeneous polynomial in the diagonal entries of $T$. There exists $D^t \in \underline{T}(r)$ such that $D = (d_{ij})$ and

(13.6.8) $$\sum_{j=i}^{r} d_{ij} C_{(m)_j}(T) = E\, C_{(m)_i}(T\underline{W}) .$$

Hint: This result is similar to a Lemma in Constantine (1963) and has a similar proof. #

Problem 13.6.8. There exists a matrix $B = (b_{ij})$ such that

(13.6.9) $$\int_{\underline{O}(n)} (\text{tr } SHTH^t)dH = \sum_{i=1}^{r} \sum_{j=1}^{r} b_{ij} C_{(m)_i}(S) C_{(m)_j}(T) .$$

The matrix $B$ is symmetric. The $b_{ij}$ depend on the matrix size $n$ of $S$, and $T$.

Problem 13.6.9. If $S, T \in \underline{S}(n)$ and $\underline{X}$ is $n \times n$ with mutually independent normal $0, 1$ entries then

(13.6.10) $$\begin{aligned} E(\text{tr } S\underline{X}T\underline{X}^t)^m &= E\Big( \int_{\underline{O}(n)} (\text{tr } SH(\underline{X}T\underline{X}^t)H^t) dH \\ &= \sum_{i=1}^{r} \sum_{j=1}^{r} b_{ij} C_{(m)_i}(S) E C_{(m)_j}(T\underline{X}^t\underline{X}) \\ &= \sum_{i=1}^{r} \sum_{j=1}^{r} \sum_{k=j}^{r} b_{ij} d_{jk} C_{(m)_i}(S) C_{(m)_j}(T) . \end{aligned}$$

That is, $A = BD$. Show the matrices $B$ and $D$ are diagonal matrices.

Problem 13.6.10. If $\underline{W}$ is a $n \times n$ symmetric matrix which has a Wishart distribution with covariance matrix $\Sigma$ then

$$(13.6.11) \qquad d_{ii} C_{(m)_i}(T\Sigma) = E\, C_{(m)_i}(T\underline{W}) .$$

Remark 13.6.11. If $\underline{W}$ is a $n \times n$ symmetric matrix that has a Wishart distribution with covariance matrix $\Sigma$ and $h$ degrees of freedom then

$$(13.6.12) \qquad E\, C_{(m)}(\underline{W}) = (2^m \Gamma_n(h/2\ ,(m))/\Gamma_n(h/2\ ))C_{(m)}(\Sigma).$$

See (13.2.9). This must be proven by an argument similar to that of Constantine (1963).

Problem 13.6.12. Given two partitions $(m)_i \neq (m)_j$ of $m$ there exists an $h$ such that $\Gamma_n(h/2\ ,(m)_i)$ and $\Gamma_n(h/2\ ,(m)_j)$ are distinct.

Hint: Form the ratio. The ratio may be irrational due to $\Gamma(1/2)$ and hence not equal 1, or else the ratio is the ratio of two integers. Choose $n$ in the later case so one of the numerator or denominator has a prime factor not in the other.

Problem 13.6.13. The equation (13.6.6) is universal. That is, $T_3$ is chosen independent of the matrix size and the normalization $A$ may be chosen so as not to depend on $n$, for example $A = I_r$.

Problem 13.6.14. Show that

$$(13.6.13) \qquad \int_{\underline{O}(n)} C_{(m)_i}(SHTH^t)\,dH = \sum_{j=1}^{r} \sum_{k=1}^{r} e_{jk} C_{(m)_j}(S) C_{(m)_k}(T),$$

where the constants $e_{ij}$ do depend on the size of the matrices $S,T$.

Problem 13.6.15. Show that

(13.6.14) $$\int_{\underline{O}(n)} C_{(m)_i}(SHTH^t)\,dH = C_{(m)_i}(S)C_{(m)_i}(T)/C_{(m)_i}(I_n).$$

Hint: By definition of the functions $C_{(m)}$, $C_{(m)}(AB) = C_{(m)}(BA)$. Therefore, if $\underline{T}$ has a Wishart distribution with covariance $\Sigma$ then

(13.6.15) $$d_{ii}\int_{\underline{O}(n)} C_{(m)_i}(H^t SH\Sigma)\,dH = \int_{\underline{O}(n)} E\, C_{(m)_i}(H^t SH\underline{T})\,dH$$

$$= E\int_{\underline{O}(n)} C_{(m)_i}(SH\underline{T}H^t)\,dH = \sum_{j=1}^{r}\sum_{k=1}^{r} e_{jk}C_{(m)_j}(S)E\, C_{(m)_k}(\underline{T})$$

$$= \sum_{j=1}^{r}\sum_{k=1}^{r} e_{jk}d_{kk}C_{(m)_j}(S)C_{(m)_k}(\Sigma)\ .$$

Thus

(13.6.16) $$d_{ii}e_{jk} = e_{jk}d_{kk}\ .$$

The coefficients $e_{ij}$ are dependent on $n$ while the $d_{kk}$ are given by (13.6.12). Use Problem 13.6.12, and let the number of degrees of freedom tend over all sufficiently large positive integers. #

# Chapter 14. References.


Aitkin, Murray A. (1969). Some tests for correlation matrices. Biometrika 56 443-446.

(Use of "delta" method in computation of asymptotic distributions.)

Albert, A.A. (1939). Structure of Algebra. American Mathematical Society Collequium Publications xxiv.

Anderson, George A. (1965). An asymptotic expansion for the distribution of the latent roots of the estimated covariance matrix. Ann. Math. Statist. 36 1153-1173.

Anderson, George A. (1970). An asymptotic expansion for the noncentral Wishart distribution. Ann. Math. Statist. 41 1700-1707.

(Uses zonal polynomials. As the roots go to $\infty$.)

Anderson, T. W. (1946). The noncentral Wishart distribution and certain problems of multivariate statistics. Ann. Math. Statist. 17 409-431.

Anderson, T. W. (1955). The integral of a symmetric unimodel function over a symmetric convex set and some probability inequalities. Proc. Amer. Math. Soc. 6 170-176.

(Useful for showing multivariate tests are unbiased.)

Anderson, T. W. (1958). An Introduction to Multivariate Statistical Analysis. John Wiley & Sons, New York.

Anderson, T. W. (1963). Asymptotic theory for principal component analysis. Ann. Math. Statist. 34 122-148.

Anderson, T. W. and Das Gupta, S. (1964). A monotonicity property of the power functions of some tests of the equality of two covariance matrices. Ann. Math. Statist. 35 1059-1063.

Anderson, T. W., Gupta, S. S., and Styon, G. P. H. (1972). A Bibliography of Multivariate Statistical Analysis. John Wiley & Sons, Inc. New York.

Bartlett, M. S. (1933). On the theory of statistical regression. Proc. Royal Soc. of Edinburgh 53 260-283.

(Bartlett decomposition.)

Bechhofer, R. E., Kiefer, J., and Sobel, M. (1968). Sequential Identification and Ranking Procedures. The University of Chicago Press.

Bhattacharya, R. N. (1971). Rates of weak convergence and asymptotic expansions for classical central limit theorems. Ann. Math. Statist. 42 241-259.

(This paper contains a list of useful references.)

Bingham, Christopher. (1972). An asymptotic expansion for the distribution of the eigenvalues of a 3 by 3 Wishart matrix. Ann. Math Statist. 43 1498-1506.

Birnbaum, A. (1955). Characterization of complete classes of tests of some multivariate hypotheses, with applications to likelihood ratio tests. Ann. Math. Statist. 26 21-36.

Box, G. E. P. (1949). A general distribution theory for a class of likelihood criteria. Biometrika 36 317-346.

(Obtains asymptotic series by use of Fourier transforms.)

Braaksma, L. J. (1964). Asymptotic expansions and analytic continuations for a class of Barnes-integrals. Compositio Mathematica 15 239-341.

(Defines H-functions.)

Brillinger, D. R. (1963). Necessary and sufficient conditions for a statistical problem to be invariant under a Lie group. Ann. Math. Statist. 34 492-500.

Chambers, J. M. (1967). On methods of asymptotic approximation for multivariate distributions. Biometrika 54 367-383.

(Edgeworth expansions, perturbation approximations.)

Cohen, A. and Strawderman, W. E. (1971). Unbiasedness of tests for homogeneity of variances. Ann. Math. Statist. 42 355-360.

Constantine, A. G., and James, A. T. (1958). On the general canonical correlation distribution. Ann. Math. Statist. 29 1146-1166.

Constantine, A. G. (1960). Multivariate Distributions. CSIRO Report.

Constantine, A. G. (1963). Noncentral distribution problems in multivariate analysis. Ann. Math. Statist. 34 1270-1285.

Constantine, A. G. (1966). The distribution of Hotelling's generalized $T_0^2$. Ann. Math. Statist. 37 215-225.

Consul, P. C. (1969). The exact distributions of likelihood criteria for different hypotheses. Pages 171-181 of Multivariate Analysis, Edited by P. R. Krishnaiah, Academic Press.

(Meijer's functions, Mellin inversion.)

Das Gupta, S. (1969). Properties of power functions of some tests concerning dispersion of matrices of multivariate normal distributions. Ann. Math. Statist. 40 697-701.

(Unbiasedness proven.)

Davis, A. W., and Hill, G. W. (1968). Generalized asymptotic expansions of Cornish-Fisher type. Ann. Math. Statist. 39 1264-1273.

Davis, A. W. (1972). On the marginal distributions of the latent roots of the multivariate beta matrix. Ann. Math. Statist. 43 1664-1670.

Deemer, W. L., and Olkin, I. (1951). The Jacobians of certain matrix transformations useful in multivariate analysis. Biometrika 38 345-367.

Dempster, A. P. (1969). Elements of Continuous Multivariate Analysis. Addison Wesley.

De Waal, D. J. (1972). On the expected values of the elementary asymmetric functions of a noncentral Wishart matrix. Ann. Math. Statist. 43 344-347.

Dieudonné, J. (1972). Treatise on Analysis, Vol. III. Academic Press, New York.

Doob, J. L. (1958). Lecture notes on probability theory. Personal notes of R. Farrell, unpublished.

Eaton, M. L. (1972). Multivariate Statistical Analysis. Institute of Mathematical Statistics University of Copenhagen, August 1972.

Erdelyi, A., Magnus, W. and Oberhettinger, F. (1953). Higher Transcendental Functions. McGraw Hill, New York.

Erdelyi, A. (1956). Asymptotic Expansions. Dover Publications, Inc.

Farrell, R. H. (1968). Towards a theory of generalized Bayes tests. Ann. Math. Statist. 39 1-22.

Feller, W. (1966). An Introduction to Probability Theory and its Applications, Vol. II. John Wiley & Sons, Inc.

Frankel, L. R., and Hotelling, H. (1938). The transformation of statistics to simplify their distributions. Ann. Math. Statist. 9 87-96.

Gelfond, A. O. (1971). Residues and their applications. Moscow: MIR Publishers.

Giri, N., Kiefer, J., and Stein, C. (1963). Minimax character of Hotelling's $T^2$ test in the simplest case. Ann. Math. Statist. 34 1524-1535.

Giri, N., and Kiefer, J. (1964). Local and asymptotic minimax properties of multivariate tests. Ann. Math. Statist. 35 21-35.

Giri, N. (1968). Locally and asymptotically minimax tests of a multivariate problem. Ann. Math. Statist. 39 171-178.

Giri, N. (1968). On tests of the equality of two covariance matrices. Ann. Math. Statist. 39 275-277.

Gleser, L. J. (1966). A note on the sphericity test. Ann. Math. Statist. 37 464-467.
(Proves unbiasedness.)

Gleser, L. J. (1968). On testing a set of correlation coefficients for equality: Some asymptotic results. Biometrika 55 513-517.

Good, I. J. (1969). Conditions for a quadratic form to have a Chi-squared distribution. Biometrika 56 215-216.

Graybill, F. A., and Milliken, G. (1969). Quadratic forms and idempotent matrices with random elements. Ann. Math. Statist. 40 1430-1438.

Gupta, S. S. (1963). Bibliography on the multivariate normal integral and related topics. Ann. Math. Statist. 34 829-838.

Halmos, P. R. (1950). Measure Theory. New York: D. van Nostrand Company, Inc.

Helgason, S. (1962). Differential Geometry and Symmetric Spaces. Academic Press, New York.

Herz, C. S. (1955). Bessel functions of a matrix argument. Annals of Mathematics 61 474-523.

Hodges, J. L. (1955). On the noncentral beta distribution. Ann. Math. Statist. 26 648-653.

Hsu, P. L. (1938). Notes on Hotelling's generalized T. Ann. Math. Statist. 9 231-243.

Itô, K. (1956). Asymptotic formula for the distribution of Hotelling's generalized $T_0^2$ statistic. Ann. Math. Statist. 27 1091-1101.

Itô, K. (1960). Asymptotic formula for the distribution of Hotelling's generalized $T_0^2$ statistic II. Ann. Math. Statist. 31 1148-1153.

James, A. T. (1954). Normal multivariate analysis and the orthogonal group. Ann. Math. Statist. 25 40-75.

James, A. T. (1955a). The noncentral Wishart distribution. Proc. Royal Society Series A 229 364-366.

James, A. T. (1955b). A generating function for averages over the orthogonal group. Proc. Royal Society Series A 229 367-375.

James, A. T. (1957). The relationship algebra of an experimental design. Ann. Math. Statist. 28 993-1002.

James, A. T. (1960). The distribution of the latent roots of the covariance matrix. Ann. Math. Statist. 31 151-158.

James, A. T. (1961a). The distribution of noncentral means with known covariance. Ann. Math. Statist. 32 874-882.

James, A. T. (1961b). Zonal polynomials of the real positive definite symmetric matrices. Annals of Mathematics 74 456-479.

James, A. T. (1964). Distributions of matrix variates and latent roots derived from normal samples. Ann. Math. Statist. 35 475-501.

James, A. T. (1966). Inference on latent roots by calculation of hypergeometric functions of matrix argument. Multivariate Analysis, Edited by P. R. Krishnaiah. Academic Press.

James, A. T. (1968). Calculation of zonal polynomial coefficients by use of the Laplace-Beltrami operator. Ann. Math. Statist. 39 1711-1718.

James, A. T. (1969). Tests of equality of latent roots of the covariance matrix. Multivariate Analysis, Edited by P. R. Krishnaiah. Academic Press.

James, A. T. and Wilkinson, G. N. (1971). Factorization of the residual operator and canonical decomposition of non orthogonal factors in the analysis of variance. Biometrika 58 279-294.

James, A. T., and Parkhurst, A. M. (1974). Zonal polynomials of order 1 through 12. Selected Tables in Mathematical Statistics, Vol. 2. American Mathematical Society, Providence.

John, S. (1971). Some optimal multivariate tests. Biometrika 58 123-127.

John, S. (1972). The distribution of a statistic used for testing sphericity of a normal distribution. Biometrika 59 169-173.

Kabe, D. G. (1965). Generalization of Sverdrup's lemma and its applications to multivariate distribution theory. Ann. Math. Statist. 36 671-676.

Karlin, S. (1960). Notes on Multivariate Analysis. Preprint form, distributed to statistics students at Stanford University, Spring 1960.

Kettenring, J. R. (1971). Canonical analysis of several sets of variables. Biometrika 58 433-451.

Khatri, C. G. (1967). Some distribution problems connected with the characteristic roots of $S_1S_2^{-1}$. Ann. Math. Statist. 38 944-948.

Khatri, C. G., and Pillai, K. C. S. (1968). On the noncentral distributions of two test criteria in multivariate analysis of variance. Ann. Math. Statist. 39 215-226.

Kiefer, J. (1957). Invariance, minimax sequential estimation, and continuous time processes. Ann. Math. Statist. 28 573-601.

Kiefer, J. (1958). On the nonrandomized optimality and randomized nonoptimality of symmetrical designs. Ann. Math. Statist. 29 675-699.

Kiefer, J. and Schwartz, R. (1965). Admissible Bayes character of $T^2$-, $R^2$-, and other fully invariant tests for classical multivariate normal problems. Ann. Math. Statist. 36 747-770.

Kiefer, J. (1966). Multivariate optimality results. Multivariate Analysis, Edited by P. R. Krishnaiah. Academic Press.

Koehn, U. (1970). Global cross sections and the densities of maximal invariants. Ann. Math. Statist. 41 2045-2056.

Korin, B. P. (1968). On the distribution of a statistic used for testing a covariance matrix. Biometrika 55 171-178.

(Obtains asymptotic series and checks accuracy of previous results.)

Kruskal, W. (1954). The monotonicity of the ratio of two noncentral t density functions. Ann. Math. Statist. 25 162-165.

Kruskal, W. K. (1961). The coordinate-free approach to Gauss-Markov estimation and its application to missing and extra observations. Fourth Berkeley Symposium on Mathematical Statistics and Probability, Vol. 1, 435-451.

Kshirsagar, A. M. (1961). The non-central multivariate beta distribution. Ann. Math. Statist. 32 104-111.

(Special functions and random variable methods.)

Kullback, S. (1934). An application of characteristic functions to the distribution problem of statistics. Ann. Math. Statist. 5 263-307.

(Obtains the distribution of a product of 3 or more chi-squares.)

Kullback, S. (1959). Information Theory and Statistics. John Wiley & Sons, Inc. New York.

Lee, Y.-s. (1971). Asymptotic formulae for the distribution of a multivariate test statistic: Power comparisons of certain multivariate tests. Biometrika 58 647-651.

Lehmann, E. L. (1959). Testing Statistical Hypotheses. John Wiley & Sons, Inc. New York.

Littlewood, D. E. (1940). The Theory of Group Characters and Matrix Representations of Groups. Oxford, The Clarendon Press.

Littlewood, D. E. (1950). The Theory of Group Characters and Matrix Representations of Groups. Oxford at the Clarendom Press.

Loomis, L. H. (1953). An Introduction to Abstract Harmonic Analysis. D. van Nostrand Company, Inc. New York.

Mathai, A. M. and Rathie, P. N. (1971). The exact distributions of Wilks' criterion. Ann. Math. Statist. 42 1010-1019.

(An example of the inversion of Mellin transforms.)

Mathai, A. M. and Saxena, R. K. (1969). Distribution of a product and the structural set up of densities. Ann. Math. Statist. 40 1439-1448.

(Use of Mellin transforms and H-functions to obtain the distribution of a product of noncentral chi-squares.)

Mauldon, J. G. (1955). Pivotal quantities for Wishart's and related distributions. Jour. Royal Statistical Society Series B. 17 79-85.

(Also obtained the lower triangular decomposition of the covariance matrix.)

Miller, K. S. (1964). Multidimensional Gaussian Distributions. John Wiley & Sons, Inc. New York.

Moran, P. A. N. (1970). On asymptotically optimal tests of composite hypotheses. Biometrika 57 47-55.

Mudholkar, G. S. (1965). A class of tests with monotone power functions for two problems in multivariate statistical analysis. Ann. Math. Statist. 36 1794-1831.

Mudholkar, G. S. (1966a). On confidence bounds associated with multivariate analysis of variance and non-independence between two sets of variates. Ann. Math. Statist. 37 1736-1746.

(Matrix Inequalities.)

Mudholkar, G. S. (1966b). The integral of an invariant unimodal function over an invariant convex set - an inequality and applications. Proc. Amer. Math. Soc. 17 1327-1333.

(Unbiasedness.)

Mudholkar, G. S. and Rao, P. S. R. S. (1967). Some sharp multivariate Tchebycheff inequalities. Ann. Math. Statist. 38 393-401.

Muirhead, R. (1970a). Systems of partial differential equations for hypergeometric functions of a matrix argument. Ann. Math. Statist. 41 991-1001.

Muirhead, R. (1970b). Asymptotic distributions of some multivariate tests. Ann. Math. Statist. 41 1002-1010.

Muirhead, R. (1972a). On the test of independence between two sets of variates. Ann. Math. Statist. 43 1491-1497.

Muirhead, R. (1972b). The asymptotic non-central distribution of Hotelling's generalized $T_0^2$. Ann. Math. Statist. 43 1671-1677.

Nachbin, L. (1965). The Haar Integral. New York: D. van Nostrand.

Naimark, M. A. (1970). Normed Rings. Groningen: Walters-Noordhoff.

Okamoto, M. (1963). An asymptotic expansion for the distribution of the linear discriminant function. Ann. Math. Statist. 34 1286-1301.

Olkin, I. (1951). On distribution problems in multivariate analysis. Institute of Statistics Mimeograph Series, No. 43, pp.1-126.

Olkin, I. (1952). Note on the Jacobians of certain matrix transformations useful in multivariate analysis. Biometrika 40 43-46.

Olkin, I. and Roy, S. N. (1954). On multivariate distribution theory. Ann. Math. Statist. 25 329-339.

Olkin, I. (1959). A class of integral identities with matrix argument. Duke Math. J. 26 207-213.

Olkin, I. and Rubin, H. (1964). Multivariate beta distributions and independence properties of the Wishart distributions. Ann. Math. Statist. 35 261-263.

Olkin, I. and Siotani, M. (1964). Asymptotic distribution of functions of a correlation matrix. Technical Report No. 6, Stanford University.

Olkin, I. (1966). Special topics in matrix theory and inequalities. Mimeographed notes recorded by M. L. Eaton, Department of Statistics, Stanford University.

Ord, J. K. (1968). Approximations to distribution functions which are hypergeometric series. Biometrika 55 243-248.

Pearson, E. S. (1968). Studies in the history of probability and statistics xx. Some early correspondence between W. S. Gosset, R. A. Fisher, and Karl Pearson, with notes and comments. Biometrika 55 445-467.

Penrose, R. A. (1955). A generalized inverse for matrices. Proc. Camb. Phil. Soc. 51 406-413.

Perlman, M. D. (1969). One-sided testing problems in multivariate analysis. Ann. Math. Statist. 40 549-567.

Pillai, K. C. S. and Jouris, G. M. (1969). On the moments of elementary symmetric functions of the roots of two matrices. Ann. Inst. Statist. Math. 21 309-320.

(Table of constants for expressing functions of $I_n - S$ as series of polynomials in S.)

Pillai, K. C. S. and Sugiyama, T. (1969). Non-central distributions of the largest latent roots of three matrices in multivariate analysis. Ann. Inst. Statist. Math. 21 321-327.

(More constants.)

Pillai, K. C. S. (1966). Noncentral multivariate beta distribution and the moments of traces of some matrices. Multivariate Analysis, Edited by P. R. Krishnaiah, Academic Press, New York.

Pitman, E. J. G. (1939). Tests of hypotheses concerning location and scale parameters. Biometrika 31 200-215.

(Unbiasedness of tests.)

Press, S. J. (1966). Linear combinations of non-central chi-square variates. Ann. Math. Statist. 37 480-487.

Salaevskii, Y. (1971). Essay in Investigations in Classical Problems of Probability Theory and Mathematical Statistics by V. M. Kalinin and O. V. Shalaevskii. Translated from Russian, New York, Consultants Bureau 1971 (V. A. Steklov Mathematical Institute, Leningrad Seminars in Mathematics Vol. 13.)

Saw, John G. (1975). Zonal polynomials: An alternative approach. Submitted to Ann. Statist.

Scheffé, H. (1959). The Analysis of Variance. John Wiley & Sons, Inc. New York

Schwartz, R. E. (1966a). Properties of invariant multivariate tests. Cornell University, Ph. D. Thesis.

Schwartz, R. E. (1966b). Fully invariant proper Bayes tests. Multivariate Analysis, Edited by P. R. Krishnaiah, Academic Press.

Schwartz, R. E. (1967a). Admissible tests in multivariate analysis of variance. Ann. Math. Statist. 38 698-710.

Schwartz, R. E. (1967b). Locally minimax tests. Ann. Math. Statist. 38 340-360.

Schwartz, R. E. (1969). Invariant proper Bayes tests for exponential families. Ann. Math. Statist. 40 270-283.

Shah, B. K. (1970). Distribution theory of positive definite quadratic forms with matrix argument. Ann. Math. Statist. 41 692-697.

Shanbhag, D. N. (1970). On the distribution of a quadratic form. Biometrika 57 212-223.

Srivastava, M. S. (1968). On the distribution of a multiple correlation matrix; non-central multivariate beta distributions. Ann. Math. Statist. 39 227-232.

Stein, C. (1956a). Inadmissibility of the usual estimator for the mean of a multivariate normal distribution. Proceedings of the Third Berkeley Symposium on Mathematical Statistics and Probability, Vol. 1.

Stein, C. (1956b). The admissibility of Hotelling's $T^2$-test. Ann. Math. Statist. 27 616-623.

Stein, C. (1956c). Some problems of multivariate analysis I. Technical Report No. 6, Stanford University, Department of Statistics.

(Reference by U. Koehn for the first source of the maximal invariant theory.)

Stein, C. (1959). Notes on Multivariate Analysis. Preprint form distributed to statistics students at Stanford University around 1959.

Stein, C. M. (1966). Multivariate Analysis. Mimeograph notes recorded by M. L. Eaton, Stanford University, Department of Statistics.

Sugiura, N. and Nagao, H. (1968). Unbiasedness of some test criteria for the equality of one or two covariance matrices. Ann. Math. Statist. 39 1686-1692.

Sugiura, N. (1969). Asymptotic expansions of the distributions of the likelihood ratio criteria for covariance matrix. Ann. Math. Statist. 40 2051-2063.

Wasow, W. (1956). On the asymptotic transformations of certain distributions into the normal distribution. Proceedings of the 6th Symposium Applied Mathematics of Amer. Math. Soc. (Numerical Analysis) Vol. vi 251-259, New York: McGraw Hill.

Weyl, H. (1946). The Classical Groups. Princeton University Press, Princeton, N.J.

Widder, D. V. (1941). The Laplace Transform. Princeton University Press, Princeton, N.J.

Wiener, N. (1933). The Fourier Integral and Certain of its Applications. New York: Dover Publications.

Wijsman, R. A. (1957). Random orthogonal transformations and their use in some classical distribution problems in multivariate analysis. Ann. Math. Statist. 28 415-423.

Wijsman, R. A. (1958). Lectures on Multivariate Analysis. Notes recorded by R. Farrell.

Wijsman, R. A. (1966). Cross-sections of orbits and their application to densities of maximal invariants. Proceedings of the Fifth Berkeley Symposium on Mathematical Statistics and Probability, Vol. 1, 389-400.

Wilks, S. S. (1962). Mathematical Statistics. John Wiley & Sons, Inc.

Wishart, J. (1928). The generalized product moment distribution in samples from a normal multivariate population. Biometrika 20A 32-52.

Vol. 399: Functional Analysis and its Applications. Proceedings 1973. Edited by H. G. Garnir, K. R. Unni and J. H. Williamson. II, 584 pages. 1974.

Vol. 400: A Crash Course on Kleinian Groups. Proceedings 1974. Edited by L. Bers and I. Kra. VII, 130 pages. 1974.

Vol. 401: M. F. Atiyah, Elliptic Operators and Compact Groups. V, 93 pages. 1974.

Vol. 402: M. Waldschmidt, Nombres Transcendants. VIII, 277 pages. 1974.

Vol. 403: Combinatorial Mathematics. Proceedings 1972. Edited by D. A. Holton. VIII, 148 pages. 1974.

Vol. 404: Théorie du Potentiel et Analyse Harmonique. Edité par J. Faraut. V, 245 pages. 1974.

Vol. 405: K. J. Devlin and H. Johnsbråten, The Souslin Problem. VIII, 132 pages. 1974.

Vol. 406: Graphs and Combinatorics. Proceedings 1973. Edited by R. A. Bari and F. Harary. VIII, 355 pages. 1974.

Vol. 407: P. Berthelot, Cohomologie Cristalline des Schémas de Caracteristique $p > o$. II, 604 pages. 1974.

Vol. 408: J. Wermer, Potential Theory. VIII, 146 pages. 1974.

Vol. 409: Fonctions de Plusieurs Variables Complexes, Séminaire François Norguet 1970–1973. XIII, 612 pages. 1974.

Vol. 410: Séminaire Pierre Lelong (Analyse) Année 1972–1973. VI, 181 pages. 1974.

Vol. 411: Hypergraph Seminar. Ohio State University, 1972. Edited by C. Berge and D. Ray-Chaudhuri. IX, 287 pages. 1974.

Vol. 412: Classification of Algebraic Varieties and Compact Complex Manifolds. Proceedings 1974. Edited by H. Popp. V, 333 pages. 1974.

Vol. 413: M. Bruneau, Variation Totale d'une Fonction. XIV, 332 pages. 1974.

Vol. 414: T. Kambayashi, M. Miyanishi and M. Takeuchi, Unipotent Algebraic Groups. VI, 165 pages. 1974.

Vol. 415: Ordinary and Partial Differential Equations. Proceedings 1974. XVII, 447 pages. 1974.

Vol. 416: M. E. Taylor, Pseudo Differential Operators. IV, 155 pages. 1974.

Vol. 417: H. H. Keller, Differential Calculus in Locally Convex Spaces. XVI, 131 pages. 1974.

Vol. 418: Localization in Group Theory and Homotopy Theory and Related Topics. Battelle Seattle 1974 Seminar. Edited by P. J. Hilton. VI, 172 pages 1974.

Vol. 419: Topics in Analysis. Proceedings 1970. Edited by O. E. Lehto, I. S. Louhivaara, and R. H. Nevanlinna. XIII, 392 pages. 1974.

Vol. 420: Category Seminar. Proceedings 1972/73. Edited by G. M. Kelly. VI, 375 pages. 1974.

Vol. 421: V. Poénaru, Groupes Discrets. VI, 216 pages. 1974.

Vol. 422: J.-M. Lemaire, Algèbres Connexes et Homologie des Espaces de Lacets. XIV, 133 pages. 1974.

Vol. 423: S. S. Abhyankar and A. M. Sathaye, Geometric Theory of Algebraic Space Curves. XIV, 302 pages. 1974.

Vol. 424: L. Weiss and J. Wolfowitz, Maximum Probability Estimators and Related Topics. V, 106 pages. 1974.

Vol. 425: P. R. Chernoff and J. E. Marsden, Properties of Infinite Dimensional Hamiltonian Systems. IV, 160 pages. 1974.

Vol. 426: M. L. Silverstein, Symmetric Markov Processes. X, 287 pages. 1974.

Vol. 427: H. Omori, Infinite Dimensional Lie Transformation Groups. XII, 149 pages. 1974.

Vol. 428: Algebraic and Geometrical Methods in Topology, Proceedings 1973. Edited by L. F. McAuley. XI, 280 pages. 1974.

Vol. 429: L. Cohn, Analytic Theory of the Harish-Chandra C-Function. III, 154 pages. 1974.

Vol. 430: Constructive and Computational Methods for Differential and Integral Equations. Proceedings 1974. Edited by D. L. Colton and R. P. Gilbert. VII, 476 pages. 1974.

Vol. 431: Séminaire Bourbaki – vol. 1973/74. Exposés 436–452. IV, 347 pages. 1975.

Vol. 432: R. P. Pflug, Holomorphiegebiete, pseudokonvexe Gebiete und das Levi-Problem. VI, 210 Seiten. 1975.

Vol. 433: W. G. Faris, Self-Adjoint Operators. VII, 115 pages. 1975.

Vol. 434: P. Brenner, V. Thomée, and L. B. Wahlbin, Besov Spaces and Applications to Difference Methods for Initial Value Problems. II, 154 pages. 1975.

Vol. 435: C. F. Dunkl and D. E. Ramirez, Representations of Commutative Semitopological Semigroups. VI, 181 pages. 1975.

Vol. 436: L. Auslander and R. Tolimieri, Abelian Harmonic Analysis, Theta Functions and Function Algebras on a Nilmanifold. V, 99 pages. 1975.

Vol. 437: D. W. Masser, Elliptic Functions and Transcendence. XIV, 143 pages. 1975.

Vol. 438: Geometric Topology. Proceedings 1974. Edited by L. C. Glaser and T. B. Rushing. X, 459 pages. 1975.

Vol. 439: K. Ueno, Classification Theory of Algebraic Varieties and Compact Complex Spaces. XIX, 278 pages. 1975

Vol. 440: R. K. Getoor, Markov Processes: Ray Processes and Right Processes. V, 118 pages. 1975.

Vol. 441: N. Jacobson, PI-Algebras. An Introduction. V, 115 pages. 1975.

Vol. 442: C. H. Wilcox, Scattering Theory for the d'Alembert Equation in Exterior Domains. III, 184 pages. 1975.

Vol. 443: M. Lazard, Commutative Formal Groups. II, 236 pages. 1975.

Vol. 444: F. van Oystaeyen, Prime Spectra in Non-Commutative Algebra. V, 128 pages. 1975.

Vol. 445: Model Theory and Topoi. Edited by F. W. Lawvere, C. Maurer, and G. C. Wraith. III, 354 pages. 1975.

Vol. 446: Partial Differential Equations and Related Topics. Proceedings 1974. Edited by J. A. Goldstein. IV, 389 pages. 1975.

Vol. 447: S. Toledo, Tableau Systems for First Order Number Theory and Certain Higher Order Theories. III, 339 pages. 1975.

Vol. 448: Spectral Theory and Differential Equations. Proceedings 1974. Edited by W. N. Everitt. XII, 321 pages. 1975.

Vol. 449: Hyperfunctions and Theoretical Physics. Proceedings 1973. Edited by F. Pham. IV, 218 pages. 1975.

Vol. 450: Algebra and Logic. Proceedings 1974. Edited by J. N. Crossley. VIII, 307 pages. 1975.

Vol. 451: Probabilistic Methods in Differential Equations. Proceedings 1974. Edited by M. A. Pinsky. VII, 162 pages. 1975.

Vol. 452: Combinatorial Mathematics III. Proceedings 1974. Edited by Anne Penfold Street and W. D. Wallis. IX, 233 pages. 1975.

Vol. 453: Logic Colloquium. Symposium on Logic Held at Boston, 1972–73. Edited by R. Parikh. IV, 251 pages. 1975.

Vol. 454: J. Hirschfeld and W. H. Wheeler, Forcing, Arithmetic, Division Rings. VII, 266 pages. 1975.

Vol. 455: H. Kraft, Kommutative algebraische Gruppen und Ringe. III, 163 Seiten. 1975.

Vol. 456: R. M. Fossum, P. A. Griffith, and I. Reiten, Trivial Extensions of Abelian Categories. Homological Algebra of Trivial Extensions of Abelian Categories with Applications to Ring Theory. XI, 122 pages. 1975.

Vol. 457: Fractional Calculus and Its Applications. Proceedings 1974. Edited by B. Ross. VI, 381 pages. 1975.

Vol. 458: P. Walters, Ergodic Theory – Introductory Lectures. VI, 198 pages. 1975.

Vol. 459: Fourier Integral Operators and Partial Differential Equations. Proceedings 1974. Edited by J. Chazarain. VI, 372 pages. 1975.

Vol. 460: O. Loos, Jordan Pairs. XVI, 218 pages. 1975.

Vol. 461: Computational Mechanics. Proceedings 1974. Edited by J. T. Oden. VII, 328 pages. 1975.

Vol. 462: P. Gérardin, Construction de Séries Discrètes p-adiques. »Sur les séries discrètes non ramifiées des groupes réductifs déployés p-adiques«. III, 180 pages. 1975.

Vol. 463: H.-H. Kuo, Gaussian Measures in Banach Spaces. VI, 224 pages. 1975.

Vol. 464: C. Rockland, Hypoellipticity and Eigenvalue Asymptotics. III, 171 pages. 1975.

Vol. 465: Séminaire de Probabilités IX. Proceedings 1973/74. Edité par P. A. Meyer. IV, 589 pages. 1975.

Vol. 466: Non-Commutative Harmonic Analysis. Proceedings 1974. Edited by J. Carmona, J. Dixmier and M. Vergne. VI, 231 pages. 1975.

Vol. 467: M. R. Essén, The Cos $\pi\lambda$ Theorem. With a paper by Christer Borell. VII, 112 pages. 1975.

Vol. 468: Dynamical Systems – Warwick 1974. Proceedings 1973/74. Edited by A. Manning. X, 405 pages. 1975.

Vol. 469: E. Binz, Continuous Convergence on C(X). IX, 140 pages. 1975.

Vol. 470: R. Bowen, Equilibrium States and the Ergodic Theory of Anosov Diffeomorphisms. III, 108 pages. 1975.

Vol. 471: R. S. Hamilton, Harmonic Maps of Manifolds with Boundary. III, 168 pages. 1975.

Vol. 472: Probability-Winter School. Proceedings 1975. Edited by Z. Ciesielski, K. Urbanik, and W. A. Woyczyński. VI, 283 pages. 1975.

Vol. 473: D. Burghelea, R. Lashof, and M. Rothenberg, Groups of Automorphisms of Manifolds. (with an appendix by E. Pedersen) VII, 156 pages. 1975.

Vol. 474: Séminaire Pierre Lelong (Analyse) Année 1973/74. Edité par P. Lelong. VI, 182 pages. 1975.

Vol. 475: Répartition Modulo 1. Actes du Colloque de Marseille-Luminy, 4 au 7 Juin 1974. Edité par G. Rauzy. V, 258 pages. 1975. 1975.

Vol. 476: Modular Functions of One Variable IV. Proceedings 1972. Edited by B. J. Birch and W. Kuyk. V, 151 pages. 1975.

Vol. 477: Optimization and Optimal Control. Proceedings 1974. Edited by R. Bulirsch, W. Oettli, and J. Stoer. VII, 294 pages. 1975.

Vol. 478: G. Schober, Univalent Functions – Selected Topics. V, 200 pages. 1975.

Vol. 479: S. D. Fisher and J. W. Jerome, Minimum Norm Extremals in Function Spaces. With Applications to Classical and Modern Analysis. VIII, 209 pages. 1975.

Vol. 480: X. M. Fernique, J. P. Conze et J. Gani, Ecole d'Eté de Probabilités de Saint-Flour IV-1974. Edité par P.-L. Hennequin. XI, 293 pages. 1975.

Vol. 481: M. de Guzmán, Differentiation of Integrals in $R^n$. XII, 226 pages. 1975.

Vol. 482: Fonctions de Plusieurs Variables Complexes II. Séminaire François Norguet 1974–1975. IX, 367 pages. 1975.

Vol. 483: R. D. M. Accola, Riemann Surfaces, Theta Functions, and Abelian Automorphisms Groups. III, 105 pages. 1975.

Vol. 484: Differential Topology and Geometry. Proceedings 1974. Edited by G. P. Joubert, R. P. Moussu, and R. H. Roussarie. IX, 287 pages. 1975.

Vol. 485: J. Diestel, Geometry of Banach Spaces – Selected Topics. XI, 282 pages. 1975.

Vol. 486: S. Stratila and D. Voiculescu, Representations of AF-Algebras and of the Group U ($\infty$). IX, 169 pages. 1975.

Vol. 487: H. M. Reimann und T. Rychener, Funktionen beschränkter mittlerer Oszillation. VI, 141 Seiten. 1975.

Vol. 488: Representations of Algebras, Ottawa 1974. Proceedings 1974. Edited by V. Dlab and P. Gabriel. XII, 378 pages. 1975.

Vol. 489: J. Bair and R. Fourneau, Etude Géométrique des Espaces Vectoriels. Une Introduction. VII, 185 pages. 1975.

Vol. 490: The Geometry of Metric and Linear Spaces. Proceedings 1974. Edited by L. M. Kelly. X, 244 pages. 1975.

Vol. 491: K. A. Broughan, Invariants for Real-Generated Uniform Topological and Algebraic Categories. X, 197 pages. 1975.

Vol. 492: Infinitary Logic: In Memoriam Carol Karp. Edited by D. W. Kueker. VI, 206 pages. 1975.

Vol. 493: F. W. Kamber and P. Tondeur, Foliated Bundles and Characteristic Classes. XIII, 208 pages. 1975.

Vol. 494: A Cornea and G. Licea. Order and Potential Resolvent Families of Kernels. IV, 154 pages. 1975.

Vol. 495: A. Kerber, Representations of Permutation Groups II. V, 175 pages. 1975.

Vol. 496: L. H. Hodgkin and V. P. Snaith, Topics in K-Theory. Two Independent Contributions. III, 294 pages. 1975.

Vol. 497: Analyse Harmonique sur les Groupes de Lie. Proceedings 1973–75. Edité par P. Eymard et al. VI, 710 pages. 1975.

Vol. 498: Model Theory and Algebra. A Memorial Tribute to Abraham Robinson. Edited by D. H. Saracino and V. B. Weispfenning. X, 463 pages. 1975.

Vol. 499: Logic Conference, Kiel 1974. Proceedings. Edited by G. H. Müller, A. Oberschelp, and K. Potthoff. V, 651 pages 1975.

Vol. 500: Proof Theory Symposion, Kiel 1974. Proceedings. Edited by J. Diller and G. H. Müller. VIII, 383 pages. 1975.

Vol. 501: Spline Functions, Karlsruhe 1975. Proceedings. Edited by K. Böhmer, G. Meinardus, and W. Schempp. VI, 421 pages. 1976.

Vol. 502: János Galambos, Representations of Real Numbers by Infinite Series. VI, 146 pages. 1976.

Vol. 503: Applications of Methods of Functional Analysis to Problems in Mechanics. Proceedings 1975. Edited by P. Germain and B. Nayroles. XIX, 531 pages. 1976.

Vol. 504: S. Lang and H. F. Trotter, Frobenius Distributions in $GL_2$-Extensions. III, 274 pages. 1976.

Vol. 505: Advances in Complex Function Theory. Proceedings 1973/74. Edited by W. E. Kirwan and L. Zalcman. VIII, 203 pages. 1976.

Vol. 506: Numerical Analysis, Dundee 1975. Proceedings. Edited by G. A. Watson. X, 201 pages. 1976.

Vol. 507: M. C. Reed, Abstract Non-Linear Wave Equations. VI, 128 pages. 1976.

Vol. 508: E. Seneta, Regularly Varying Functions. V, 112 pages. 1976.

Vol. 509: D. E. Blair, Contact Manifolds in Riemannian Geometry. VI, 146 pages. 1976.

Vol. 510: V. Poènaru, Singularités $C^\infty$ en Présence de Symétrie. V, 174 pages. 1976.

Vol. 511: Séminaire de Probabilités X. Proceedings 1974/75. Edité par P. A. Meyer. VI, 593 pages. 1976.

Vol. 512: Spaces of Analytic Functions, Kristiansand, Norway 1975. Proceedings. Edited by O. B. Bekken, B. K. Øksendal, and A. Stray. VIII, 204 pages. 1976.

Vol. 513: R. B. Warfield, Jr. Nilpotent Groups. VIII, 115 pages. 1976.

Vol. 514: Séminaire Bourbaki vol. 1974/75. Exposés 453 – 470. IV, 276 pages. 1976.

Vol. 515: Bäcklund Transformations. Nashville, Tennessee 1974. Proceedings. Edited by R. M. Miura. VIII, 295 pages. 1976.